Handbook of Batch Process Design

VISIT OUR FINE CHEMISTRY SITE ON THE WEB

http://www.finechemistry.com

e-mail orders: direct.orders@itps.co.uk

Handbook of Batch Process Design

Edited by

P.N. SHARRATT
Department of Chemical Engineering
UMIST
Manchester, UK

BLACKIE ACADEMIC & PROFESSIONAL
An Imprint of Chapman & Hall
London · Weinheim · New York · Tokyo · Melbourne · Madras

Published by Blackie Academic and Professional,
an imprint of Chapman & Hall, 2–6 Boundary Row, London SE1 8HN, UK

Chapman & Hall, 2–6 Boundary Row, London SE1 8HN, UK

Chapman & Hall GmbH, Pappelallee 3, 69469 Weinheim, Germany

Chapman & Hall USA, 115 Fifth Avenue, New York, NY 10003, USA

Chapman & Hall Japan, ITP-Japan, Kyowa Building, 3F, 2-2-1 Hirakawacho, Chiyoda-ku, Tokyo 102, Japan

DA Book (Aust.) Pty Ltd, 648 Whitehorse Road, Mitcham 3132, Victoria, Australia

Chapman & Hall India, R. Seshadri, 32 Second Main Road, CIT East, Madras 600 035, India

First edition 1997
© 1997 Chapman & Hall

Typeset in 10/12pt Times by Academic & Technical Typesetting, Bristol
Printed in Great Britain by St Edmundsbury Press, Bury St Edmunds, Suffolk

ISBN 0 7514 0369 5

Apart from any fair dealing for the purposes of research or private study, or criticism or review, as permitted under the UK Copyright Designs and Patents Act, 1988, this publication may not be reproduced, stored, or transmitted, in any form or by any means, without the prior permission in writing of the publishers, or in the case of reprographic reproduction only in accordance with the terms of the licences issued by the Copyright Licensing Agency in the UK, or in accordance with the terms of licences issued by the appropriate Reproduction Rights Organization outside the UK. Enquiries concerning reproduction outside the terms stated here should be sent to the publishers at the London address printed on this page.

The publisher makes no representation, express or implied, with regard to the accuracy of the information contained in this book and cannot accept any legal responsibility or liability for any errors or omissions that may be made.

A catalogue record for this book is available from the British Library

Library of Congress Catalog Card Number: 97-71787

∞ Printed on acid-free text paper, manufactured in accordance with ANSI/NISO Z39.48-1992 (Permanence of Paper).

Contents

List of contributors		xi
Preface		xiii
Acknowledgements		xiv

1 Chemicals manufacture by batch processes — 1
P.N. SHARRATT

1.1	Introduction	1
1.2	Industrial background	2
	1.2.1 Definitions	2
	1.2.2 Product life cycles and regulatory influences	5
1.3	Reasons for the use of batch processing	7
1.4	Batch process design for fine and speciality chemical production	8
	1.4.1 Process chemistry	9
	1.4.2 Process design	13
	1.4.3 Unit operations	17
	1.4.4 Process operation and control	18
	1.4.5 Health, safety and environmental issues during operation	20
1.5	Plant design	21
1.6	Summary	22
	References	22

2 Scheduling and simulation of batch processes — 24
G.V. REKLAITIS, J. PEKNY and G.S. JOGLEKAR

2.1	Introduction	24
2.2	Batch process features	24
2.3	The scheduling problem	29
	2.3.1 General features	29
	2.3.2 Generic solution approaches	31
	2.3.3 Uniform discretization scheduling approach	35
	2.3.4 Requirements for implementation	38
2.4	Batch process simulation	40
	2.4.1 The role of simulation	40
	2.4.2 Simulation model features	41
	2.4.3 Elements of combined simulation	41
	2.4.4 Limitations of discrete simulation languages	48
	2.4.5 Features of a process-oriented combined simulator	50
2.5	Design implications	57
	References	58

3 Solvents in chemicals production   61
M. SHEEHAN

3.1	Introduction	61
3.2	Solvent properties	62
	3.2.1 Chemical properties	62
	3.2.2 Physical properties	63
3.3	Solubility	71
	3.3.1 Non-specific intermolecular forces	72
	3.3.2 Specific intermolecular forces	74
	3.3.3 Energy of solvation	75
	3.3.4 Selective solvation	76
	3.3.5 Solvent mixtures	76
	3.3.6 Supercritical solvents	77
3.4	Solvent effects on reactions	77
	3.4.1 Rules of thumb – Hughes–Ingold rules	78
	3.4.2 The transition state approach	79
	3.4.3 Solvation dynamics	81
	3.4.4 Reaction examples	82
	3.4.5 Heterogeneous reactions	87
3.5	Solvent recovery	90
	3.5.1 Origin of solvent wastes	90
	3.5.2 Designing for recovery	91
	3.5.3 Separation of solvents from gaseous wastes	94
	3.5.4 Separation of solvents as liquid	95
3.6	Solvent destruction	101
	3.6.1 Non-biological treatment	101
	3.6.2 Biological treatment	102
3.7	Conclusion	104
3.8	Nomenclature	105
References		105

4 Agitation   107
K.J. CARPENTER

4.1	Agitator selection	107
	4.1.1 Agitator duties	107
	4.1.2 Agitator types	108
4.2	Calculation of agitator power, discharge flow and mixing time	114
	4.2.1 Typical power levels	114
	4.2.2 Calculation of power	114
	4.2.3 Discharge flow	115
	4.2.4 Mixing time	116
4.3	Power and circulation in non-Newtonian fluids	117
	4.3.1 Calculation of power	117
	4.3.2 Circulation	118
4.4	Design to disperse solid particles	119
	4.4.1 Disperse sinking particles	119
	4.4.2 Draw down floating particles	120
4.5	Design for two or more liquid phases	121
	4.5.1 Miscibility	121
	4.5.2 Phase continuity and phase inversion	123
	4.5.3 Phase distribution – the just dispersed condition	125
	4.5.4 Drop sizes	127
	4.5.5 Mass transfer	129

	4.6	Design for dispersing gas	130
		4.6.1 Sparged gas	130
		4.6.2 Sparged with surface incorporation	131
		4.6.3 Specialized gas-inducing impellers	131
		4.6.4 Mass transfer	131
	4.7	Design for heat transfer	132
		4.7.1 Heat transfer surfaces	132
		4.7.2 Service side heat transfer coefficient	133
		4.7.3 Process side heat transfer coefficient	135
		4.7.4 Wall resistance	136
	4.8	Nomenclature	136
	References	137

5 Mixing and the selectivity of fast chemical reactions — 139
J.R. BOURNE

5.1	The problem	139
5.2	Mixing mechanisms and modelling	141
	5.2.1 Semi-batch reactor: micromixing	141
	5.2.2 Extensions	144
5.3	Applications	145
	5.3.1 Model reactions	145
	5.3.2 Characterization of mixers	147
	5.3.3 Scale-up principles	148
5.4	Extensions	149
5.5	Concluding remarks	149
5.6	Nomenclature	150
References	150

6 Batch filtration of solid–liquid suspensions — 153
A. RUSHTON

6.1	Introduction	153
6.2	Filtration process fundamentals	155
	6.2.1 Flow of fluids in filtration	155
	6.2.2 Quantitative relationship for cake filtration	157
	6.2.3 Laboratory tests and filter media in cake filtration	159
	6.2.4 Application of basic relationships to centrifugal filters	165
	6.2.5 Filter cake washing	166
	6.2.6 Filter cake dewatering	168
	6.2.7 Clarification filtration processes	169
	6.2.8 Laboratory tests and filter media in clarification processes	171
	6.2.9 Membrane filtration principles	174
6.3	Batch operated filtration machinery	177
	6.3.1 Pressure–vacuum filters	177
	6.3.2 Centrifugal filters	185
	6.3.3 Membrane filters	186
6.4	Nomenclature	189
References	190

7 Design and engineering of a batch plant — 193
M.J. MAYES

7.1	Introduction	193
7.2	Project definition	193
7.3	Project strategy	194

	7.4	Project organization	195
	7.5	'Fast track' projects	198
		7.5.1 Techniques	198
		7.5.2 Problems	201
	7.6	Regulations and other controls	202
		7.6.1 Hazard and operability (HAZOP) studies	202
		7.6.2 Environmental impact studies	204
		7.6.3 Fire and explosion hazards	204
		7.6.4 Construction safety	205
	7.7	Design techniques	205
		7.7.1 Case study – design for ease of construction	207
	7.8	Layout considerations	208
		7.8.1 Case study – layout	210
	7.9	Plant relocation/reuse of existing equipment	212
		7.9.1 Case study – plant relocation	217
	7.10	Modular plant	217

8 Control
P.E. SAWYER
219

8.1	Introduction	219
8.2	Control of continuous processes	220
8.3	Control of batch processes	220
	8.3.1 A simple example	220
	8.3.2 Multiproduct, multistream and multipurpose operations	223
8.4	Batch control systems – structure and functions	225
	8.4.1 Models and terminology – the SP88 standard	225
	8.4.2 Models for batch processing	226
8.5	Computer control	235
	8.5.1 Introduction	235
	8.5.2 Systems architecture – hardware and software	236
	8.5.3 Choosing an architecture	237
	8.5.4 Hardware	238
	8.5.5 Software	243
8.6	Procedural control	243
	8.6.1 Introduction	243
	8.6.2 Identifying procedures	244
	8.6.3 Specifying procedures	244
8.7	Acknowledgements	251
References		251

9 Hazards from chemical reactions and flammable materials in batch reactor operations
R. ROGERS
253

9.1	Introduction	253
9.2	Hazard identification	254
	9.2.1 The defined procedure	255
9.3	Chemical reaction hazards	257
	9.3.1 Thermal explosions	259
	9.3.2 Characterization of the desired reaction	262
	9.3.3 Characterization of exothermic decomposition reactions	263
	9.3.4 Selection of safety measures	267

9.4	Fire and explosion hazards	270
	9.4.1 Flammability characteristics of materials	271
	9.4.2 Sources of ignition	280
	9.4.3 Assessment of hazards and definition of appropriate safety measures	283
9.4	Conclusions	287
	References	

10 Environmental protection and waste minimization   289
C. JONES

10.1	Introduction	289
10.2	Batch reactor waste minimization	291
	10.2.1 Process chemistry	291
	10.2.2 Heat effects	291
	10.2.3 Mixing and contacting pattern	292
10.3	Equipment for the production of solid products	292
	10.3.1 Crystallization processes	294
	10.3.2 Precipitation processes	294
	10.3.3 Solid–liquid separation	294
	10.3.4 Batch drying	300
10.4	Fugitive and other minor emissions	301
10.5	Cleaning wastes	303
10.6	Waste treatment and solvent management	304
10.7	Environmental protection	305
10.8	Conclusions	306
	References	

11 Future developments in batch process design and technology   308
P.N. SHARRATT

11.1	Influences for and against change	308
11.2	New technologies	309
	11.2.1 New or enhanced unit operations	309
	11.2.2 New plant designs	310
	11.2.3 Process intensification	311
11.3	New processes	312
11.4	New design methods	312
11.5	New skill requirements	313
	References	313

Index   **315**

List of contributors

J. Bourne	Vine House, Ankerdine Road, Cothridge, Worcester WR6 5LU, UK
K.J. Carpenter	Process Technology Department, Zeneca plc, Huddersfield Works, Huddersfield HD2 1FF, UK
G.S. Joglekar	Batch Process Technologies, 1291E Cumberland Ave., PO Box 2001, W. Lafayette, IN 47906, USA
C. Jones	Fluor Daniel Limited, Process Department, Fluor Daniel Centre, Camberley, Surrey GU15 3YL, UK
J. Mayes	Simon Carves, Sim-Chem House, PO Box 17, Cheadle Hume, Cheshire SK8 5BR, UK
J. Pekny	School of Chemical Engineering, Chemical Engineering Building, Purdue University, West Lafayette, IN 47907, USA
G.V. Reklaitis	School for Chemical Engineering, Chemical Engineering Building, Purdue University, West Lafayette, IN 47907, USA
R. Rogers	Imburex GmbH, Wilhelmstrasse 2, 59067 Hamm, Germany
A. Rushton	Department of Chemical Engineering, UMIST, PO Box 88, Manchester M60 1QD, UK
P. Sawyer	PES Associates, 44 High Street, Chippenham, Wiltshire SN14 8LP, UK

P.N. Sharratt   Environmental Technology Centre, Department of Chemical Engineering, UMIST, PO Box 88, Manchester M60 1QD, UK

M. Sheehan   Department of Chemical Engineering, UMIST, PO Box 88, Manchester M60 1QD, UK

Preface

The use of batch processes for the manufacture of fine and effect organic chemicals is widespread and underpins such industry sectors as pharmaceuticals, dyestuffs and agrochemicals. Despite their great contribution to wealth creation, batch processes have attracted relatively little attention from the academic world and have not been associated with large investments in industrial process technology research.

Batch process design has a number of core techniques and technologies that are common to most industry sectors. Scheduling and sequence control are both significant topics that do not arise in the design of continuous processes. The development and implementation of robust, efficient schedules is essential in both design and operation of batch processes. Mixing and agitation have a central role due to the widespread use of liquid-phase reactions and stirred vessels. Agitators are required for blending, dispersing two or more immiscible phases, promoting high reaction yields and numerous other duties. Incorrect agitation system design can give rise to dramatic yield loss, inefficient processes and possibly even hazardous situations. Batch products are often solids, so solid–liquid separations are important. Batch filtration equipment comes in many forms and selection of the right system is essential to avoid operational difficulties. Solvents are frequently used, and have a major influence on process design through their impact on reaction rates and separations. An understanding of the underlying molecular mechanisms for the interaction of solvent and solute gives a better chance of manipulating processes for enhanced performance. Batch organic chemical processes raise distinctive safety and environmental issues, linked to the use of solvents as well as the reactivity of many of the materials used. Finally, project management and engineering tend to be different in batch processes where 'fast track' projects and equipment reuse are the norm rather than the exception.

The literature on the topics mentioned above is fragmented and often not in a readily accessible form. This handbook brings together design and process analysis methods in the core areas of batch process design. It is intended primarily for practising technologists in the batch process industries, as well as consultants and contractors servicing those industries. It should also be useful to academics and students working in the area, both as a reference and an insight into some of the problems of this sector.

Acknowledgements

The editor would like to acknowledge the patience and good humour of his family, Susan, James and William during the preparation of this book. He would also like to thank his former colleagues in Zeneca plc for introducing him to the challenges of batch processing and inspiring this book.

1 Chemicals manufacture by batch processes
P.N. SHARRATT

1.1 Introduction

A batch process is one in which a series of operations are carried out over a period of time on a separate, identifiable item or parcel of material. It is different from a continuous process, during which all operations occur at the same time and the material being processed is not divided into identifiable portions. This definition of batch processing includes what has been called 'semi-batch' production, during which material is added continuously to a batch over some period.

Batch processing has been a part of man's activities throughout history. Many of the products required for modern life are manufactured by batch processes. Indeed, life itself may be viewed as a set of linked batch processes. Yet until recently, the chemical engineer and the mainstream chemical industry have viewed batch processing as unfashionable or even primitive. This attitude stems from the development of large-scale continuous processes through much of the chemical industry. Such processes give great economies of scale and have produced large profits for the chemical process industries. These successes were accompanied by a rapid development of continuous technologies and supporting design methods in both industry and academia, teaching of skills linked to continuous process design, and a spread of continuous processing to products that had traditionally been produced batchwise.

However, the spread of continuous processing for chemicals production seems to have stopped and in the developed world appears to be in reverse. Numerous chemical process industries retain batch processing as their primary method of manufacture. Products manufactured by batch processes include pharmaceuticals, agrochemicals, dyestuffs, photosensitive materials, food additives, perfumes, vitamins, pigments and many more. Freshwater [1] notes that while profitability in petrochemicals fell by 25% in 1992, pharmaceuticals was the fastest growing area, with an increase of 20% in 1992.

The maturity of the bulk chemicals industry, combined with falling profit margins, changing markets and strong competition from emerging nations (often with cheap raw materials), has changed the face of bulk chemicals production. The technological advantages available to chemicals companies in developed nations no longer overcome the lower wage costs, growing markets and desire for new industry in the developing nations. Such threats to profitability led major chemicals companies to try to move back into the speciality and fine chemicals sector; a sector whose core technologies are primarily batch processes. Growth in the demand for healthcare products has similarly boosted interest in batch processing.

Despite this change of economic emphasis, batch process design is poorly served with design tools and methodologies when compared with continuous processing. Interest in universities has been fragmented and often poorly focused – failing to address issues of importance to manufacturers. Academia has recognized the opportunity to contribute, but seems unclear as to what is required. Industrial research and development is also fragmented.

A key issue is that in some countries, e.g. the UK, the traditional training of the chemist and chemical engineer do not always match the needs of the batch process industries. Process design is often driven by chemistry and the role of the process engineer may merely be to deliver the laboratory process at a bigger scale. In order to make a significant contribution to process design, the chemical engineer needs a much stronger background in physical organic chemistry (to manipulate reaction, solvent and separations) and elementary synthetic chemistry (to be able to communicate with chemists). Conversely, the chemist must be aware of the importance of mass transfer, mixing, heat transfer, control and safety issues.

This chapter examines the chemical process industries that use primarily batch production. Issues arising in process design are discussed and the core technologies identified. The topics of the later chapters are put in the context of batch process design and development.

1.2 Industrial background

Within the chemical process industries, batch processing is focused on the fine and speciality chemicals sectors, while continuous processing is dominant in bulk chemicals production. In order to understand some of the features of batch process design, it is important to understand the nature of the businesses which use it.

1.2.1 Definitions

A useful concept in the classification of chemical products is that of differentiation. Undifferentiated products are sold simply on the basis of their chemical composition and purity. There are likely to be a number of manufacturers of the same product who compete mostly on the basis of price. It is likely that the product has a number of different applications. Differentiated products are sold for what they do, rather than their composition. The chemical formula may well be insufficient to characterize them, i.e. they may be sold as part of a formulated product. They probably have only few, specific applications. Other manufacturers may make similar products, but these will have a different composition and perform differently.

Scale of production is another important feature. Bulk chemicals production using continuous processing has a substantial cost advantage for production levels in excess of a few thousand tonnes per year. Small-scale processes producing tens or hundreds of tonnes per year of product are almost invariably batch.

Table 1.1 shows a classification of chemical products based on scale and differentiation. Fine chemicals are sold on the basis of their chemical composition for use as intermediates in the production of other materials. They often have only one or a few end uses. Speciality (also called specialty or effect chemicals) are purchased because of their effect rather than composition. A good example is a pharmaceutical – the user is not concerned as to the chemical composition so long as the desired effect is obtained.

There is some overlap in the definitions of fine and speciality chemicals. For instance, active ingredients for pharmaceuticals manufactured on a large scale for many users may be thought of as fine chemicals as well as an effect product. There may be several purchasers of the material for use in final products. Formulation of the active material would probably be different between the final products. Examples of products from the fine chemicals and speciality chemicals sectors are listed in Table 1.2.

(a) Speciality/effect chemicals

The bulk of the sales value of speciality and effect chemicals lies in intermediates for the manufacture of pharmaceuticals and pesticides. The number of companies in these sectors is limited, with 10 firms accounting for 75% of pesticide sales and the 15 largest drug companies having 33% of the market [2]. The size of the pharmaceuticals companies arises from the very large development costs associated with the search for new active materials. Manufacture of pharmaceuticals is by either biotechnology or synthetic chemistry, while most agrochemicals are synthesized chemically. The syntheses are generally multistep processes. Companies use in-house manufacturing for some stages, while using custom chemical manufacturers to provide specific raw materials or to carry out certain reaction stages at

Table 1.1 Classification of chemical products

Volume (tonnes year^{-1})	Undifferentiated	Differentiated
High (>1500)	True commodities (e.g. sulphuric acid)	Pseudo-commodities (e.g. engineering plastics)
Low (<1500)	Fine chemicals (e.g. *m*-chloroaniline)	Speciality chemicals (e.g. a pyrethroid insecticide)

Table 1.2 Low-tonnage chemical products

Fine chemicals	Speciality/effect products
Electronic chemicals	Pharmaceuticals
Dye intermediates	Pesticides
Pharmaceutical intermediates	Textile and printing additives
Chiral intermediates	Synthetic perfumes
Perfumes and flavour intermediates	Pigments
Bulk active ingredients for pharmaceuticals	Photosensitive materials
	Speciality surfactants
	Catalysts
	Oils and fats speciality products
	Oil field chemicals
	Dyestuffs
	Perfumes
	Flavourings
	Leather chemicals
	Vitamins

another site [3]. Thus, for a multistage synthesis, different stages may be carried out at different sites and by different manufacturers.

Pharmaceutical companies in particular are driven by the need to identify and develop new products. Patent protection of drugs is of limited duration, and the companies must recoup their high research and development costs over a relatively short time before moving on to new products. Products out of patent are often manufactured by 'generic' drug companies who carry out little if any product research and development and can thus sell products at much lower gross margins.

(b) Fine chemicals

Fine chemicals can be further subdivided into three categories [4]: basic intermediates, high-volume active materials and advanced intermediates (such as chiral materials). Estimates of annual world-wide production of fine chemicals varies. Recent estimates [2, 5, 6] are in the range US$26–64 billion per year. In 1995, it was estimated [2] that the world-wide market for fine chemicals was US$42 billion. Of this, 30% was used as pharmaceutical intermediates, 35% was for pesticides, 23% for flavours and fragrances, and 12% for other uses. Growth is expected mainly from the pharmaceutical sector, at 6% annually, with pesticides growing at only 1%. Over the period 1985–1991 the average rate of growth in fine chemicals production was 5% [7] and continues at such levels [5].

While companies in developed countries have sought to move their production towards fine and speciality chemicals to take advantage of the growing markets, manufacturers in developing countries are also looking to increase their market share. Basic intermediates and high volume active materials are becoming commodities, and producers in Asia are a threat to existing producers [4]. They can take advantage of low labour costs and relaxed safety/environmental controls to undercut producers in industrialized countries [8]. This is forcing companies to look to high-technology products to maintain a competitive advantage.

The production of chiral intermediates is seen as one key growth area. Demand for enantiomeric materials is growing at the expense of their racemic counterparts [6], driven primarily by the pharmaceutical industry. Other capabilities seen as key to serving the pharmaceutical industry are quality assurance, an extensive skill-base in chemical development and production, environmental protection expertise, and a full understanding of the customer's requirements [9].

1.2.2 Product life cycles and regulatory influences

For a product in the speciality chemicals area, there tends to be a product life cycle, which can take the forms shown in Figure 1.1. Product sales are affected by factors in addition to price and efficacy – brand image, competing products, the advantage gained by being first in the market with a new product and marketing expertise can all have a large impact. The result of

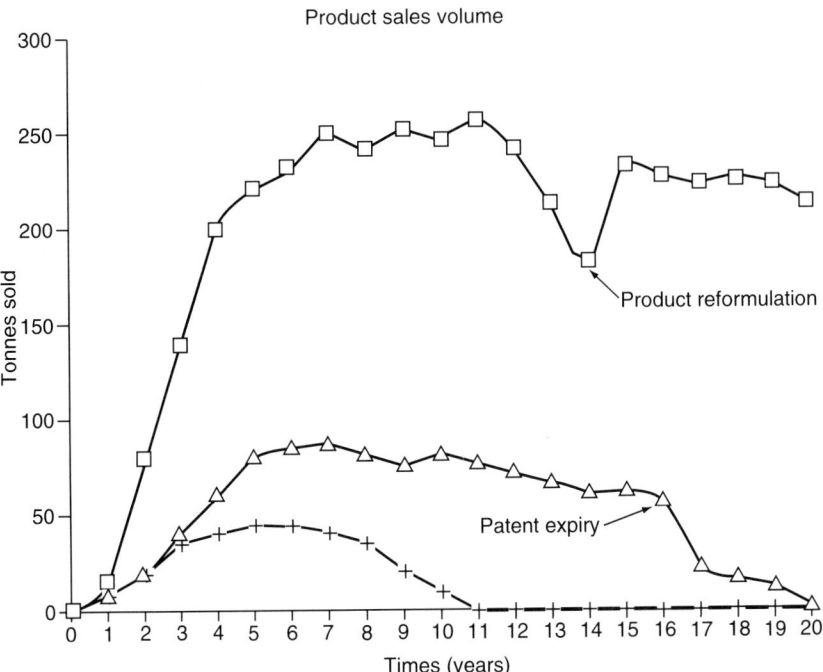

Figure 1.1 Typical product life cycles for speciality/effect chemicals: crosses, poor product; triangles, average product; squares, good product.

this is that prediction of the market performance of a product is rather uncertain. In the early stages of its life, a commercially good product may be indistinguishable from a bad one. There is thus a strong element of risk and, to minimize this, there is a temptation to make a low initial capital investment to produce low quantities of material. If the product sells well, further, staged investments can be made, gradually increasing capacity. Also, note that a rapid growth in demand can only be met by being able to bring new capacity on stream quickly. Choice of the initial capacity will then depend on factors such as:

- Availability of capital.
- Availability of any existing plant which could be used to make the product.
- Use of 'generic' facilities – multipurpose units designed to be flexible enough to manufacture (usually small) quantities of material.
- Economic situation and the degree of risk associated with the product.
- The technology used, particularly such issues as the ease of scale-up or expansion of production.
- The lead time required to expand capacity.
- Company culture and politics.

If the technology is simple, then time to expand capacity will tend to be short and it may be possible to make a relatively small initial investment for a small unit. Another possibility is to use multipurpose or flexible plant to produce market development quantities of materials, or to use toll manufacture by a subcontractor.

The product life cycle is also influenced by regulatory issues. Fine chemicals production will generally be subject to national and international regulations on safety and environmental performance. For example, in the UK, plant will be subject to control by agencies such as the Health and Safety Executive (HSE), the Environment Agency (EA) and local authorities. In addition to these controls on production, national and international controls to ensure product safety are applied to many fine and effect chemicals.

One example of the importance of regulations is in the controls on pharmaceuticals. In the UK, pharmaceutical products are controlled by the Medicines Control Agency (MCA). To obtain a product licence from the MCA a comprehensive account of the method of manufacture of the material must be provided. For synthetic products the synthetic route, reactants, catalysts, conditions, purification methods and details of any isolated intermediates must be supplied. Additionally, quality control procedures must be described for raw materials, processing and final product. Similar, and probably more demanding conditions are imposed by the US Food and Drug Administration (FDA) on any pharmaceuticals that are intended for import to the US.

Further problems exist in respect of patent protection and product registration. Patent protection only gives protection for a limited period (about 17 years in the UK). Many biologically active products require some form of registration (FDA, Ministry of Agriculture, etc.) to demonstrate safety, efficacy and environmental acceptability. This applies for drugs, agrochemicals, pesticides, etc. The process of obtaining such registrations takes up to 7 years in some cases. Thus, the period available for exploitation without direct competition is small – perhaps 10 years. This means that:

(i) Fast lead times between decision to invest and first production are attractive.
(ii) Profit margins need to be large, particularly for drugs – to pay for the high development costs in the short time while patent protection prevents low cost 'generic' copies competing in the market.

It is worth noting that manufacturers are finding new ways to increase the duration of patent protection, e.g. by the patenting of the formulation of a drug. This allows use of the same active ingredient but delivered in a novel way, e.g. in a different physical form.

Another impact of process registration is restriction of process modification. Changes to pharmaceutical processes are strictly controlled under US FDA regulations. Significant changes require re-registration; the cost and effort associated with a change is seldom felt to be warranted by the benefits. This problem can prevent the process development that often takes place in more loosely regulated processes.

1.3 Reasons for the use of batch processing

The economic and technical requirements discussed in Section 1.2 strongly favour the use of batch processing. The following advantages arise.

- The plant used, typically stirred vessels of either stainless or glass-lined steel of $1-50 \text{ m}^3$ in volume, is easily modified for use on new products. Such equipment is very versatile and can be used for reactions, separations, blending heating or cooling operations.
- By use of standard equipment, mechanical design, procurement and fabrication times can be shorter than for the 'one-off' equipment often required for continuous plant.
- A single vessel may be used for several distinct operations during a process. A stirred tank can be used to blend reactants, heat them to reaction temperature, carry out the reaction, cool, distil off solvent and crystallize the product. In a continuous process, each operation would be tend to be carried out in a separate unit. If the capacity of the unit is to be increased, it is possible to add a second vessel in series with the first and carry out some of the operations in that.
- The substantial standing investment in plant that can readily be reused for new processes makes investment in specialized continuous equipment

unattractive. Only a substantial commercial benefit would induce a typical batch manufacturer to consider straying from the use of traditional batch technologies. Even then, the commercial and technical risks in making such a move are unlikely to be taken – particularly with the short development times available for new processes.
- The technology is easily scaled up from experiments carried out in a chemical laboratory. Pilot plants can often be eliminated from the design process, reducing lead times.
- Batch plant is much more robust than continuous plant to inaccurate knowledge about the process, allowing shorter development times.

The issue of robustness to incomplete knowledge is important. Batch processes often have chemistry that is substantially more complex than a typical continuous process. The cost of evaluating kinetics and physical parameters to the accuracy required to use 'standard' chemical engineering design methods (e.g. integrating the reaction kinetics in order to define the reaction time) would be high. However, such detailed analysis is seldom necessary. It is usually sufficient to know how long an operation will take and even that need not be known to excessive accuracy. The use of rules of thumb for scale-up based on identification of the rate-controlling process is a common way of assessing the time taken for an operation on the full scale. Discrepancies between the expected and actual times for each operation tend to average out over the many operations that constitute a single batch. Thus, the impact of inaccurate knowledge is much less serious than for a continuous process whose overall productivity is limited by the capacity of whichever unit has lowest throughput.

Modifications to batch processes in the event of incorrect process design are often less serious than for continuous processes. A batch reaction can simply be left for longer if it is incomplete. Additional reagents can be added if a test result is out of specification. With continuous plant, modification to plant is more likely to be required.

In summary, the robustness of batch allows both reduced lead times (because less rigorous process investigations are required than for a continuous process using the same chemistry) and lets the operator make modifications to the process (like a change in recipe) without significant equipment modifications. These are key virtues in businesses where speed to market is critical.

1.4 Batch process design for fine and speciality chemical production

The design of batch processes for fine/speciality chemicals production differs significantly from the design of large-scale continuous processes. This section introduces some of the important considerations in batch process design, many of which will be dealt with more fully in later chapters.

1.4.1 Process chemistry

Figure 1.2 shows the typical stages of design for manufacture of an organic chemical by a batch process. Such a scheme would be typical of an operator who identifies new product molecules as well as carrying out design and manufacture.

A central feature of batch process design for organic chemicals production is the importance of chemistry. During process development it is not unusual to solve problems by a change in the chemistry to avoid engineering difficulties, e.g. avoiding a difficult separation by choosing an alternative chemical route. This is particularly the case during the early stages of development, the 'route selection' stage. Other important features relating to chemical considerations are the use of solvents as reaction media and the great range of reaction types used. The following sections cover these points in greater detail.

(a) Route selection
The chemical route is the set of chemical reactions which turn the feedstocks of a process into the products. In fine/effect chemicals production, route selection is a major activity of process development, traditionally carried

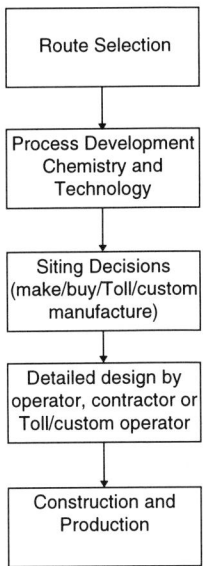

Figure 1.2 Stages of batch process development.

out by chemists with only peripheral involvement from the chemical engineer. Selection of a good route has a major impact on the performance of the final process.

The number of feasible chemical routes depends most strongly on the type of product. Simple bulk chemicals can usually only be manufactured by a small number of chemical reactions and it is sensible only to consider a limited range of feedstocks. At the other end of the scale, pharmaceuticals and agrochemicals can be manufactured by hundreds of different routes from a wide variety of starting materials. Other distinctions are summarized in Table 1.3.

Key features of the route selection decisions are that they are taken *early* in a project on the basis of *limited information*. Usually, particularly for new processes or products, kinetic and yield information is likely to be sketchy, and based on either literature values or most likely laboratory-scale tests. Information on the hazard potential of the substances involved may well be unavailable or incomplete. Separation of the product may not be proven – yields being found by techniques such as chromatography that cannot always be scaled up economically.

Despite the poor quality of information, decisions must be made, primarily to reduce the number of routes for continued development. The decisions are further complicated by the multiple decision criteria. Not only must a process be as cheap as possible, but it must be safe, compatible with the environment,

Table 1.3 Features of the route selection problem for bulk and speciality chemicals

Feature	Bulk chemicals	Speciality chemicals
Number of feasible routes	Few	Many – possibly hundreds
Number of feasible feedstocks	Few	Many
Number of reaction steps required to make product	Usually one or two	Several (three or four typical, 20 for some pharmaceuticals)
Effort in route selection as a proportion of total design effort	Low	High
Impact of poor route selection on profitability	High	Moderate
Patent protection	On processes or technologies	On process chemistry – often used to block competitors' use of similar routes
Competing technology	One route usually best	Competitors may use a variety of routes
Skills used	Chemical engineering and some physical chemistry	Synthetic chemistry, with some chemical engineering
Focus of design	Driven by cost	Driven by yield
Scale-up	Pilot plants and simulation tools commonly used	Use of rules of thumb to scale laboratory experiments is prevalent

operable and represent a good commercial risk. It is unlikely that a company can afford to develop all possible routes through to the flowsheet level, or that pilot-plant trials could be considered for more than one or two possible routes. In order to make the best use of limited information, companies often turn to tools such as structured decision making to assist with the selection of one or a few routes to take through to further development.

The problems of route selection are complicated by the need to take account of potential safety and environmental problems with the final process. Consideration of these matters needs to be woven into the route selection and design process to ensure that:

- Appropriate tests are carried out to identify potential problems and to provide data for decision making.
- Safety and environmental matters are appropriately weighted as criteria when key process decisions are being taken.
- Unavailable or weak data are highlighted, allowing a judgement to be made on the risk associated with any decision.
- Resource requirements to resolve risk and uncertainty can be identified as early as possible.

A method of integrating health, safety and environmental considerations into process design has been presented by Jones [10]. At these early stages of design, decisions have to be based on fairly simple criteria. For example, the use of amounts of waste generated as a measure of likely environmental performance is common. For example, 3M have used a waste ratio [11] – essentially the proportion of material leaving the plant that is not product – as a means of comparing and controlling environmental performance. Safety and environmental matters are addressed further in Section 1.4.2, and Chapters 9 and 10.

(b) Use of solvents
Most batch chemical production processes involve the use of solvents. They are necessary for a number of reasons, including the following.

- Many of the materials processes are solids and cannot be liquefied or evaporated without decomposition. Solvent is necessary to mobilize them and facilitate processing.
- The solvent provides a suitable environment for the reaction to occur. The selection of the right solvent is often key to obtaining high product yields.
- The thermal mass of the solvent absorbs heat, limiting the potential adiabatic temperature rise and thus influencing safety.
- Operation under solvent reflux allows accurate temperature control. Also, the latent heat of the evaporating solvent can be used as an efficient way of removing heat from a reactor.
- Solutions (and even slurries) are much easier to transport from vessel to vessel than solids.

The extensive use of solvents is a major contributor to the character of batch processing, influencing features from the unit operations used to safety and environmental issues. For example, the major environmental problems associated with batch organic manufacture include solvent wastes, solvent contaminated aqueous wastes and volatile organic compound (VOC) emissions. Various aspects of the use of solvents and their impact on process performance are dealt with in Chapter 3.

(c) Chemical route development

Once a route has been selected, it will still require significant development to maximize yields and other performance measures. This work will usually be carried out by development chemists. The final result will be a recipe for production of the product at full scale. A recipe would include process conditions, durations for each operation, and the amounts of materials charged and discharged and their source/destination. In essence the recipe is equivalent to the mass balance and the flowsheet of a continuous process. The recipe is conveniently presented in tabular form. An example is given in Table 1.4.

Within a company and even at a specific site, particular reaction types may be favoured (or disliked). This bias may be due to special expertise in a given area of chemistry, because of the types of product made or for historical reasons. However, one reason for not using some reactions centres on the way in which development work is carried out. Typically, process chemistry

Table 1.4 Typical recipe format (manufacture of sodium t-butoxide)

Operation	Mass added (kg)	Volume added (l)	Cumulative mass (kg)	Cumulative volume (l)	Duration (h)	Cumulative duration (h)
1. Check pan empty	0	0	0	0	0.2	0.2
2. Charge toluene to vessel R100	3000	3464	3000	3464	0.8	1.0
3. Charge sodium sticks to R100	230	237	3230	3701	2.0	3.0
4. Heat to 105°C	0	0	3230	3701	2.0	5.0
5. Start agitator when vessel temperature reaches 105°C	0	0	3230	3701	0.1	5.1
6. Add t-butyl alcohol slowly (10% excess); vent hydrogen to scrubber S101	81	100	3310	3801	2.5	7.6
7. Cool to 35°C	0	0	3310	3801	2.2	9.8
8. Discharge to storage T102	−3310	−3801	0	0	2.2	12.0
					Cycle time	12.0

development is carried out in glass vessels of volume about 1–5 l. Agitation is used and heating/cooling provided via a jacket or isomantle. Using such equipment makes it difficult to carry out reactions that require particularly fast mixing, high heat transfer rates, etc. In such cases, the skills needed to deliver a successful process may lie with the process engineer rather than the chemist. Development laboratories may not have the appropriate tools to investigate these reactions in a way that leads to efficient processes. Also, one way to deal with these reactions is to dilute the reactants with solvent to reduce reaction rates and act as a heat sink. This leads to processes that generate substantial quantities of effluent, use much larger plant than is necessary and have higher processing costs than could otherwise be achieved. Clearly, this problem can be addressed both by the provision of more appropriate equipment and by ensuring that process engineering expertise is available and used during this development stage.

1.4.2 Process design

In many batch process industries it is debatable whether any distinct stage of process design exists in the way it does in the design of large, continuous processes. By the time the chemical route is developed and sufficient data are available for traditional process engineering activities to take place, the process is so well defined that little remains to do other than specifying equipment. Methodologies that work well in continuous process design cannot be transferred directly to batch processing. The use of formal process synthesis methods in batch process design is discussed in Section 1.4.2(a) while the opportunities for thermal integration are considered in Section 1.4.2(b).

The published literature on batch process design, particularly that originating from the academic world, is seldom of any use in practical situations. In general, the focus of batch process design research has been on matters that are readily handled by computing. Work in scheduling and the simulation of some unit operations is widely published, as are attempts to automate aspects of batch process design by the use of expert systems. While the issues addressed are of some importance, the approaches often fail in real applications. The problems that arise include:

- Inaccurate or missing chemical, physical and process data, combined with insufficient time or resource to obtain them.
- The uncertainty in the times taken for process operations.
- The time taken to become familiar with the proposed techniques is too long.
- The techniques attempt to transfer continuous process thinking to batch systems without full consideration of the constraints under which batch design operates.
- The techniques do not fit into the usual development procedures adopted for batch process design.

One activity that does fall into the broad classification of process design is scheduling. Scheduling is vital to predicting and maximizing the production rate of a process. It is necessary to consider not only the size of equipment, but also the time taken to carry out the desired operations. Production rate is then the amount of material produced per batch (a function of the equipment size) divided by the time taken. Inappropriate schedules that leave equipment lying idle for long periods are potentially costly. Scheduling is amenable to various types of analysis and modelling, and will be discussed more fully in Chapter 2.

Simulation is another tool that does find a number of useful applications in design of batch processes. Various common operations can readily be modelled by standard process engineering design packages, e.g. condenser design can be carried out effectively using the standard packages used by continuous process designers. To do this, the condenser has to be assumed to operate at a pseudo-steady state. This assumption is usually good enough. Where time-dependent variables are important, dynamic simulation should be considered to replace or augment experimental studies. This can be used to assist in the analysis and sizing of specific equipment (discussed further in Section 1.4.2.c) and may be extended to whole-process modelling (as discussed in Chapter 2).

(a) Process synthesis methods

The main shortcoming with systematic methods for conceptual design and flowsheeting of batch processes has been the weak link with the process chemistry. Most batch organic process development is carried out by chemists and is primarily experimental. In order to influence and interact with development work of this type, process design techniques must be tailored to work with the data that are likely to be available and at a time when the results obtained can be used to benefit. This said, a range of process synthesis approaches have been tried and some of these may be useful in particular cases.

Douglas [12] presents a hierarchical method that considers different parts of the process in order. He recommends that the decision as to whether a process should be operated batchwise or continuously be taken at the start of process design – its implications are so fundamental to process design. After the batch/continuous decision, the reactor is considered, followed by any recycles, the separation systems and finally utilities requirements. At each stage, short-cut design methods are used to size equipment and identify economic tradeoffs. In principle, this methodology can be applied to batch design, although there are significant problems, e.g. the interaction of process design with scheduling and pieces of equipment that perform multiple duties. A continuous process flowsheet has a much simpler logical structure than a batch process.

The use of computers in batch process synthesis has been widely studied. For example, Rippin *et al.* [13] suggested the use of knowledge-based systems to automate some aspects of process design, and to link together process flowsheet development, plant item sizing, discrete event simulation and scheduling. The knowledge base would be combined with optimization methodologies to make decisions. While these more sophisticated methods have the ultimate objective of automatic design of batch processes it is clear that this will not be achieved for some time.

(b) Thermal integration

Thermal integration to reduce process energy consumption has had minimal impact on the batch industries. This can be contrasted with the continuous industries where substantial energy savings combined with capital savings have been identified in a wide range of processes [14]. The underlying problem is that heat sources and sinks tend to be available at different times in the process. Furthermore, energy costs tend to form a rather small proportion of operating costs.

A number of approaches have been developed that address various aspects of the use of heat in batch processes. Two of these have some merit – the manipulation of schedules and the use of regeneration or other means of heat storage.

(i) Scheduling tools can be used to smooth out utility demands and to maximize any process integration opportunities. To do this, constraints on utility use or the requirement for hot and cold streams to be available together can be added to a scheduling model.
(ii) Regeneration involves the storage of heat in an inert material to be transferred to another stream at a later time [15]. A complementary technique (and one that is often cheaper and easier to achieve) is to retain a heel of hot material from one batch to provide the heat for the next.

(c) Simulation

One technique that has wide applicability is dynamic simulation. This involves the construction of a mathematical model based on time-dependent differential equations that represent the changes in a process with time. Examples of applications to individual operations where dynamic simulation can bring benefits are:

- Yield optimization of reactions.
- Investigation of possible control strategies.
- Modelling of hazardous situations such as runaway reactions to provide a design basis for protective equipment.
- Batch distillation.

Because of the cost and time required to collect sufficient data to establish and validate dynamic models they tend only to be used where substantial benefits are expected. The use of techniques such as computer-controlled pilot reactors with calorimetry makes the data easier to collect.

A range of dynamic modelling software exists. Both specialized packages aimed for batch chemical process simulation and general-purpose dynamic simulation packages can be used effectively. The former required less specialized skills for model development, but may be less flexible.

An example of the use of dynamic simulation for process development was presented by Gillanders [16]. The commercial software package BATCHCAD was used to model the complex (but not exceptionally) reaction scheme shown in Figure 1.3. The reaction scheme was based on substantial experimental work and even then was not definitively proven to be correct. The experiments were carried out on the laboratory scale in a reaction calorimeter. The model was validated against tests on a pilot reactor. The validated model was then used to guide development trials, to model hazardous situations such as runaway reaction and in scale-up. A cost saving was identified through a reduction of reactor size compared with that which otherwise would be chosen.

It was also concluded that while simulation was useful, it was important not to waste time by refining the model beyond that required to realize the benefits.

(d) Physical properties modelling and prediction

An issue that frequently causes difficulties in any design or process analysis calculations is the lack of reliable physical properties data. This is particularly difficult for the novel or uncommon materials often encountered in batch processes. Fortunately, with many processes taking place in solvents, many of the properties essential to process design are those of common solvents. A useful tabulation of important properties of the main industrial solvents has been presented by Smallwood [17]. The use of estimation methods provides one means by which properties can be found, or limited data (such as a boiling point determination) can be extrapolated if required [18].

(e) Safety and environment

Hazard and environmental studies used throughout the design process to ensure operational and personnel safety as well as acceptable environmental performance have been described by, for example, Kletz [19], Scott and Crawley [20] and Jones [10]. An important requirement is to have formal,

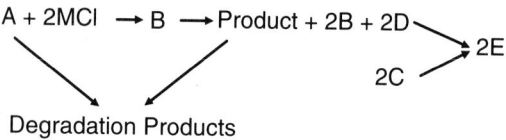

Figure 1.3.

recorded systems that enforce systematic and thorough study of the potential for safety and environmental problems, as well as charting progress to the resolution of these problems. It is particularly useful during these studies (such as hazard and operability (HAZOP) studies) to have operating staff present. The high degree of manual intervention common in batch processing means that operators have substantial knowledge of (and interest in) potential problems.

Alongside process development there may be studies to establish the toxic and environmental effects of intermediates and products. These need to be borne in mind by designers as, for example, the discovery of an unexpected toxic effect could change the degree of containment required and thus significantly increase equipment costs.

A model for the consideration of process safety during process design (i.e. chemical reaction hazards and fire and explosion hazards) is presented in Chapter 9. Environmental considerations during batch process design are more fully explored in Chapter 10.

1.4.3 Unit operations

While the unit operations used in batch processing can all be found in continuous plant, the type of equipment used and the mode of operation are rather different. A key distinction is that batch processes tend to use the same piece of equipment for multiple operations – usually a stirred tank manufactured in a corrosion-resistant material. Thus, a single process vessel might be used to blend reactants, heat them to reaction temperature, cool, perform a liquid–liquid extraction, boil off solvent and crystallize the product. This influences design in a number of ways:

- Design tends more often to be a rating exercise, i.e. predicting how well a given piece of equipment will perform a given duty.
- The performance of stirred vessels in mixing and heat transfer are very important.
- Multiple inlets and outlets are required to process vessels, raising issues of safety and control.
- HAZOP studies are substantially more complex.

Other than the stirred tank, common batch process equipment includes filters and dryers. Solids handling, often involving a significant manual component, is widespread.

(a) Mixing and agitation

An extremely important aspect of fine chemicals processing is mixing. Mixing systems carry out a multiplicity of duties: blending of materials, suspension of solids, dispersion of gases in liquid, dispersion of two immiscible liquid phases, enhancing heat and mass transfer rates, etc.

Because large volumes of liquids are processed, mixing times can be of the orders of minutes or longer in batch stirred vessels. Where the processes occur on a shorter timescale than mixing, or where mixing is insufficient to achieve complete homogenization, the process efficiency can be affected. This problem is most noticeable on scale-up from the laboratory scale. A particular concern is the impact of mixing on the yield of competing reactions. The identification of any potentially mixing-sensitive operations and the specification of mixing equipment to achieve appropriate mixing performance on the production scale are important steps in process development and design. These issues are dealt with in depth in Chapters 4 and 5.

(b) Solids production

Many of the products of the fine chemicals industry are solids – often produced by crystallization, filtration and drying. The formation of solid products by crystallization or precipitation is not described in detail in this handbook. Manipulation of the crystallization process relies on applying knowledge of chemistry, kinetics, mixing and agitation. For a full treatment of the chemical and kinetic aspects of crystallization the reader might consult the work of Mullin [21].

Filtration is a unit operation that frequently causes operational difficulties. The incompletely understood flow properties of slurries, the difficulty in predicting filterability and the potential for blockage of pipework all contribute to this. The topic of batch filtration system design is dealt with in Chapter 6.

Product drying can be carried out in the filter or in a range of types of drier. Tray driers, fluidized bed driers and rotary drum driers are all used. The type of drier chosen will depend on factors such as the properties of the solids, the potential for hazards (both toxicity and fire or explosion), the extent of drying required and the cost of equipment. The selection and design of driers has been dealt with by Perry and Chilton [22].

1.4.4 Process operation and control

Without some form of control a process cannot be operated. Computer control is becoming more and more widespread in the batch industries. However, even in automated plants, batch processing usually has a much greater degree of manual intervention than continuous processes. This may involve charging of reactants, discharge of products, setting of valves and other control elements, starting and stopping drives (pumps and agitators) and sampling of intermediate or final products.

(a) Control

Batch process control seeks to ensure that the correct sequence of operations is carried out. It must also keep process variables such as temperature and pressure within the desired ranges during any given operation. The

CHEMICALS MANUFACTURE BY BATCH PROCESSES 19

common use of multiproduct plant, multistage processes and complex reaction mixtures further complicates the task. Control may be manual, fully automatic or mixed. Manual operation is still common although the cost and quality benefits of computer or PLC (programmable logic controller) control make them increasingly attractive.

Control needs to be considered carefully throughout the design process. Appropriate control is essential to safe and efficient operation, so it is important that control systems (whether they be computer software, manual or other systems) be considered fully during the assessment of health, safety and environmental hazards.

Automatic batch process control is discussed in detail in Chapter 8.

Where any operations are carried out manually, several matters become vital to safe, efficient production including:

- Appropriate operator training.
- Clear, *written*, instructions for carrying out operations.
- Appropriate record keeping for batch progress.
- The provision of safe and efficient means to carry out the required tasks.
- Clear labelling of all valves, vessels and controls.

Good record keeping is essential to minimize the risk of error in carrying out a production process that may extend over many hours – or indeed several working shifts. Often, each batch will be associated with a record sheet that lists the operations to be carried out, and provides space for operators to record the time at which each operation started and finished, to record any problems, and to sign that they have carried out the operation. Results of any measurements or analyses may also be recorded. Without such records it would be much easier, for example, to charge reactant twice by mistake.

The provision of training and validation that the training has been understood are also key to successful operation. Not only does training reduce the risk of error, it also motivates operating staff. Clear labelling and appropriate design of equipment also reduces the risk of error, e.g. the chances of sending material to the wrong vessel is much increased if routeing is carried out by opening and closing valves on a complex, unlabelled manifold.

Efficient, ergonomic design of equipment is important if 'short-cuts' by operators are to be avoided. Dialogue with operators during the design of manual handling and other operations can reduce the risk of producing cumbersome, difficult or unsafe work practices.

(b) Sampling

Sampling of material from batches to check for product quality, reaction completion or process conditions (e.g. pH) is common. A number of problems arise:

- The difficulty in obtaining a representative sample.

- The risk of loss of containment.
- The need to delay processing while the sample is obtained, analysed and any decision taken.
- The need to obtain rapid, accurate analysis.

Obtaining a representative sample from anything other than a homogeneous single-phase system presents substantial difficulties. For example, if solids are present, it is very unlikely even with full agitation that they are homogeneously distributed in the vessel. Even in liquid blending it is possible (particularly with higher viscosity systems) for pockets of high or low concentration to persist for minutes or hours. If there is any hold up in the sample device (e.g. a long line between vessel and sample point), then material that is not representative of the current situation will be obtained.

Samples are obtained in a variety of ways, including:

- Use of a dip-can to remove a sample through an open port in the vessel.
- Use of a sample valve to allow material to drain out or be forced out under pressure.
- Use of double valve arrangements ('block and bleed') to allow enclosed sampling of hazardous materials, or sampling from vessels/lines under high or low pressure.
- Use of gas or pump driven sample loops to circulate liquid past a sample valve so as to obtain a representative sample.

Clearly, it is important to consider carefully the sample system so as to avoid operational problems.

1.4.5 Health, safety and environmental issues during operation

The handling of toxic and flammable materials is commonplace during batch organics processing. Fire and explosion hazards are dealt with more fully in Chapter 9. Environmental problems will be considered in Chapter 10. It is also important to consider the potential human health problems, particularly those arising for operators. The nature of many batch processes, with manual sampling, charging and discharging of hazardous materials, does give rise to a range of potential health problems. It is important that these industrial hygiene issues are appropriately considered during design and effectively managed during process operation.

Regulations governing human exposure to substance, and indeed the management of safety, vary from country to country. However, the basic steps that are required to ensure operator safety tend to be common. These are:

- Identification of the materials which have the potential to cause harm, and the route by which harm can be caused (i.e. inhalation, ingestion or skin absorption).

- Identification of those likely to be exposed to the hazards, and the degree of exposure.
- Consideration of the risk (i.e. the combination of the chance of exposure and its results).
- Definition of appropriate measures to reduce the risk to an acceptable level.
- Implementation of those measures.
- Ongoing checking of the control measures, and where appropriate operator exposure and health.

Such health risk assessments should be built into the design process and the normal operational procedures of processes. In general it is preferable to adopt measures that separate operators (and others) from the possibility of exposure rather than relying solely on protective equipment. Thus, for example, provision of ventilation in solids charging booths would be preferable to the provision of dust masks that may be forgotten or not used.

Other health aspects that need to be considered are the potential for exposure to noise, heat, cold, electrical hazards, falling and moving equipment. Appropriate consideration during design as well as operating systems (e.g. work permit systems for the management of maintenance tasks) can help to reduce risk [19, 20, 23].

1.5 Plant design

The widespread use of stirred vessels and other 'generic' equipment (e.g. pressure filters) has led to significant effort in standardization of mechanical engineering design within batch processing companies. This is reflected by the product ranges of equipment suppliers, e.g. it is possible to buy glass-lined steel reactors, ready-fitted with agitator gearbox and drive, in a range of standard sizes. Commonly used heat exchanger types, such as plate or carbon block devices, again tend to come in fixed sizes. The time and cost benefit of using such standard equipment often outweighs any benefit of having a vessel exactly the right size for the duty. Indeed, as it is common for the batch size to be modified after a period of operation, oversized vessels are more versatile.

Multipurpose and multiproduct plant is designed to be reconfigured quickly for new products. However, it is also common for dedicated plant to be adapted for a new process after its original product is discontinued. Intelligent reuse of process plant can bring savings of 15% on capital cost and 30% on project timescale [24].

As has already been mentioned, a major business objective is to minimize the time between project sanction and production. To this end, the organization of the design process in batch chemicals production tends to be significantly different from other projects. 'Fast track' projects are commonplace – with every effort made to eliminate unnecessary activities, run activities in parallel and to minimize the amount of effort in design. Implementation of fast track projects and use of standard/modular equipment are discussed in more detail in Chapter 7.

1.6 Summary

The design and operation of batch processes for the manufacture of organic chemicals such as is very different to the manufacture of bulk chemicals by continuous process. The process chemistry tends to be more complex, but less well understood than in a continuous process. Reactions are usually carried out in, or suspended in, the liquid phase, often in the presence of solvent, at moderate temperatures and pressures (typically -20 to $140°C$ and atmospheric pressure). Products are often solids. These features mean that mixing and filtration are key operations.

Design of process and production schedules and their implementation through the use of sequence-driven control software or manual control is another central feature. Because of the materials handled, safety and environmental concerns are significant, and require appropriate consideration. The industries using mainly batch processing – particularly the pharmaceutical and agrochemical industries – require short lead times, making rapid process development and plant design essential.

The unique features of batch processing form a distinct, important yet poorly understood area of process engineering. This book seeks to bring together some of the key aspects to provide both a useful source of reference to practising technologists and to promote a clearer understanding of the nature of the technical problems of batch processing.

References

1. Freshwater, D.C. (1994) Why batch? An outline of the importance of batch processing. *IChemE NW Branch Symposium Papers*, 3.
2. Anon (1995) Fine and intermediate chemicals makers emphasize new products and processes. *Chemical and Engineering News*, 17 July, 10.
3. Anon (1995) Custom chemicals. *Chemical and Engineering News*, 13 February, 44.
4. Mullin, R. and Roberts, M. (1994) Fine & custom chemicals: manufacturers head for technology's high ground. *Chemical Week*, 2 February, 20.
5. Mullin, R. and Roberts, M. (1995) Fine chemicals: playing for the top of the big supplier lists. *Chemical Week*, 15 February, 46.
6. Anon (1994) Market, environmental pressures spur change in fine chemicals industry. *Chemical and Engineering News*, 16 May, 10.
7. Anon (1993) European fine chemicals market holds opportunities, obstacles. *Chemical and Engineering News*, 5 April, 17.
8. Anon (1993) Global fine chemicals industry faces rising competition, squeeze on profits. *Chemical and Engineering News*, 19 July, 21.

9. Simons, J.M. (1994) The fine chemicals industry: how to adapt to a changing environment. *Speciality Chemicals*, March, 50.
10. Jones, D.J. (1992) The ICI fine chemicals manufacturing organization process SHE study procedure, *IChemE NW Branch Symposium Papers*, 3, 2.1–2.10.
11. Benforado, D.M., Ridlehoover, G. and Gores, M.D. (1991) Pollution prevention: one firm's experience, *Chemical Engineering*, September.
12. Douglas, J.M. (1988) *Conceptual Design of Chemical Processes*, International Edition. McGraw-Hill, New York.
13. Rippin, D.W.T., Hofmeister, M. and Halasz, L. (1989) 'Batch kit': a knowledge based environment for solving problems of design, planning and operation of batch systems. *IChemE NW Branch Papers*, 2, 6.1–6.20
14. Linnhoff, B., Townsend, D.W., Boland, D., Hewitt, G.F., Thomas, B.E.A., Guy, A.R. and Marsland, R.H. (1994) *User Guide on Process Integration for the Efficient Use of Energy*. IChemE, Rugby.
15. Sadr-Kazemi, N. and Polley, G.T. (1996) Design of energy storage systems for batch plant. *Trans. IChemE Part A*, in press.
16. Gillanders, A.G. (1990) Dynamic simulation case study – a pharmaceutical batch reaction. Presented at *IChemE NW Branch Symposium on Dynamic Simulation*, Manchester.
17. Smallwood, I. (1993) *Solvent Recovery Handbook*. Edward Arnold, London.
18. Walas, S.M. (1982) *Phase Equilibria in Chemical Engineering*. Butterworth, Boston, MA.
19. Kletz, T. (1992) *HAZOP and HAZAN Identifying and Assessing Process Industry Hazards*. IChemE, Rugby.
20. Scott, D. and Crawley, F. (1992) *Process Plant Design and Operation*. IChemE, Rugby.
21. Mullin, J.W. (1993) *Crystallization*. Butterworth–Heinemann, Oxford.
22. Perry, R.H. and Chilton, C.H. (eds) (1984) *Chemical Engineers' Handbook*, 6th edn. McGraw-Hill, New York.
23. Kletz, T. (1994) *An Engineer's View of Human Error*. IChemE, Rugby.
24. Hirst, J.N. (1996) Revamp vs new: guess which is better!, *Fast Tracking Improved Project Efficiency, IChemE NW Branch Papers*, 1.

2 Scheduling and simulation of batch processes
G.V. REKLAITIS, J. PEKNY AND G.S. JOGLEKAR

2.1 Introduction

In this chapter, the batch process scheduling problem is discussed and approaches to its solution are outlined. The approaches considered include approximate techniques, model-based optimization methods and combined continuous/discrete simulation methods. Model-based scheduling techniques and combined simulation methods are complementary decision-support technologies which offer the best prospects for the routine creation and validation of economical and practical schedules. The extension of this methodology to perform design optimization is feasible given available commercial tools. The material for this chapter was drawn from several sources, especially [1–5].

2.2 Batch process features

The manufacture of all chemical products involves three key elements: a process or recipe which describes the set of chemical and physical steps required to make product, a plant which consists of the set of equipment within which these steps are executed, and a market which defines the amounts, timing and qualities of the product required. A distinguishing feature of continuous operations is the one-to-one correspondence between the recipe steps and the plant equipment items: the flowsheet is the physical realization of the recipe and its structure remains fixed in time. In batch plants, the structure of the recipe and the plant equipment network structure are in general distinct. Moreover, the equipment configuration may change each time that a different product is made. Thus, in the batch case there exists an additional engineering decision level: the assignment of recipe steps to equipment items over specific intervals of time. These assignment decisions are inherently discrete in nature, introducing a combinatorial aspect to operational and design problems which is not normally present in the continuous process case.

In the subsequent discussion, the following terminology will be used.

A *resource* is any production input such as raw material, utility or operator time which is required to execute a recipe step. Resources can be *renewable* (restored to original availability level after usage, e.g. steam) or *nonrenewable* (requiring re-supply, e.g. raw material).

A *recipe* is a network of tasks which must be executed to produce a product. Each task consists of a sequence of chemical/physical steps that are executed in the same vessel (Figure 2.1). At the simplest level, each step is described by a processing time, a size factor which defines the capacity required per unit amount of task output, input/output ratios which describe the proportions in which inputs must be supplied and outputs are generated, and task resource requirements consisting of a fixed amount plus an amount proportional to the task output.

A *production line* is a set of equipment assigned to each task of a given recipe. Assuming that the identify of a batch is preserved in the production line, then the *batch size* will be the amount of final product made in one batch. If the production line is used to produce a series of identical batches, it is often convenient to operate the line in a cyclic fashion. The *cycle time* is then the time between the completion of batches.

A *Gantt chart* is an equipment occupation diagram in which time is the ordinate and the abscissa has an entry for each equipment item.

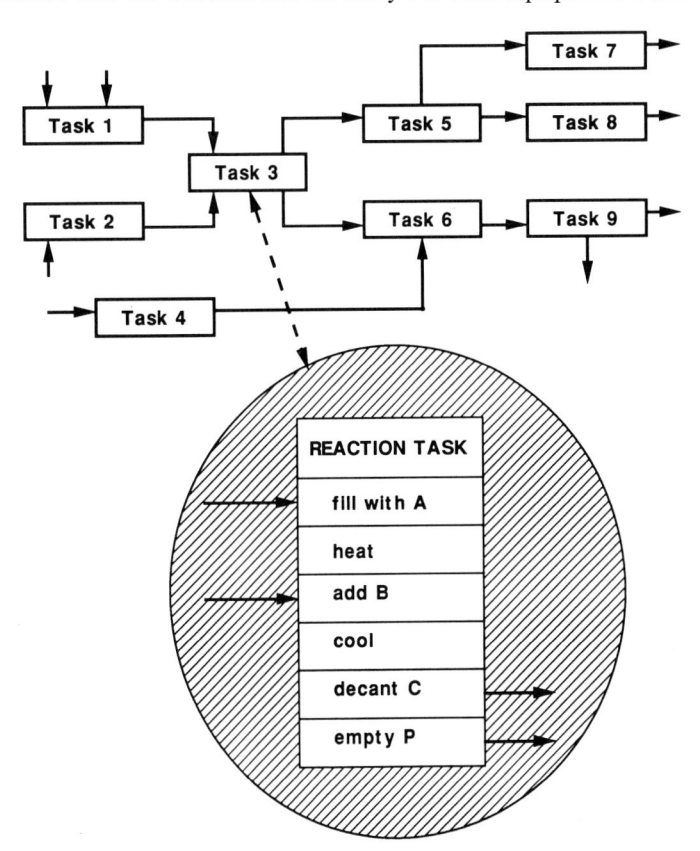

Figure 2.1 Recipe, tasks and subtasks.

A *campaign* is a time interval during which one or more production lines are dedicated to making a specific set of products.

Figure 2.2(a) shows a Gantt chart for a serial four task recipe in which a distinct unit is assigned to each task. Note that the transfer of a task output to the next task in the recipe is denoted by an arrow. The cycle time is 6, corresponding to the maximum of the processing times of the four tasks of the recipe. As is typical, several of the units are idle for a considerable portion of the time but at least one is continuously engaged and becomes cycle time limiting (the so-called 'bottleneck cycle'). In this illustration the campaign consists of three batches.

As noted earlier, a characteristic feature of batch production is the need to specify an assignment of units to tasks. In general this assignment need not be

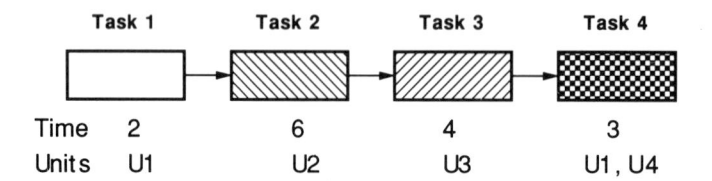

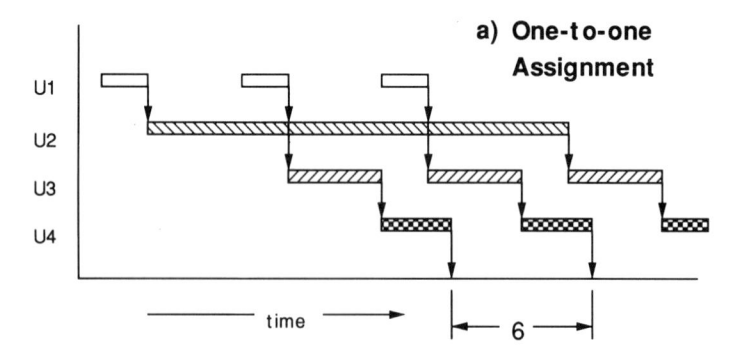

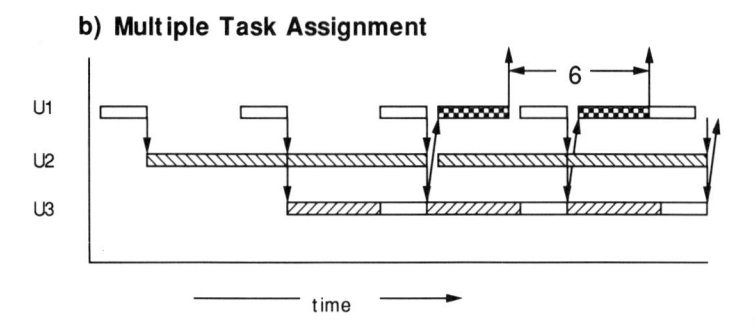

Figure 2.2 One-to-one and many-to-one task to unit assignments.

one-to-one. Rather, multiple tasks can be assigned to the same unit and multiple units can be assigned to execute the same task. For the recipe of Figure 2.2, task 4 can be executed in two different units (U1 and U4). Since these two units are inefficiently utilized in the one-to-one assignment shown in Figure 2.2(a), an improvement in equipment utilization can be achieved by assigning U1 to execute both the first and the fourth task, as shown in Figure 2.2(b), thereby releasing U4 for other uses. Improvements can also be achieved by assigning multiple units to a task which is performance limiting. If a unit assigned to a task is batch size limiting, then assigning another unit which allows the batch at that task to be split and processed in parallel (parallel unit in-phase) will allow an increase in the batch size. (A set of in-phase units assigned to a task is called a *group*.) Alternatively, if the task is cycle time limiting, then adding another unit and alternating the processing of batches at that task (parallel unit out-of-phase) will effectively reduce the task processing and thus the cycle time. As shown in Figure 2.3, the addition of a second U2 unit out-of-phase, reduces the cycle time to 4.

Based on the nature of the product recipes and the allowable task/unit assignments, batch operations can be roughly classified into three basic types: the multiproduct plant, the multipurpose plant under campaign mode and the general multipurpose plant.

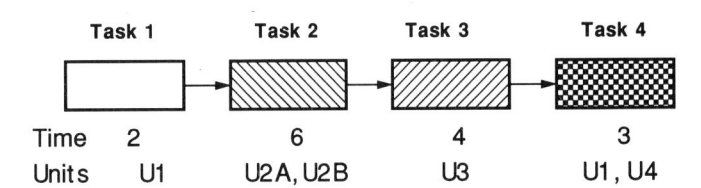

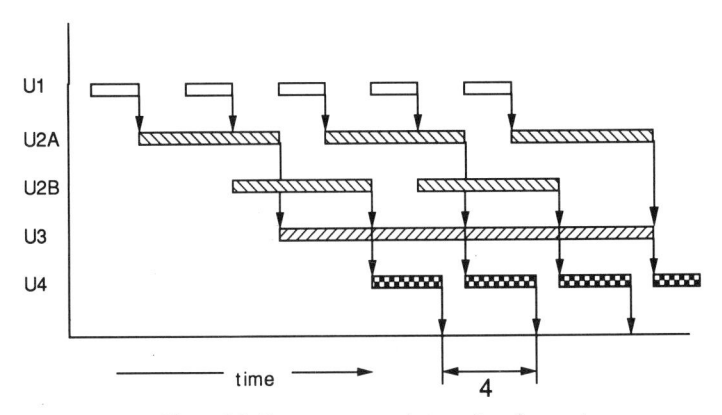

Figure 2.3 One-to-many task to unit assignments.

The *multiproduct plant* is employed for a set of products whose recipe structure is the same (or nearly so), the production line employs fixed many-to-one unit/task assignments, the line is operated cyclically and multiple products are accommodated through serial campaigns. It should be noted that the special case of the multiproduct plant that occurs when campaigns are reduced to single batches is sometimes referred to as a *flowshop*.

The *multipurpose plant under campaign operation* is appropriate for products with dissimilar recipe structures, allows many-to-many unit/task assignments and employs multiple campaigns involving one or more production lines, each operated cyclically.

The *general multipurpose plant* is a multipurpose plant operated with no defined production lines, rather production occurs in an aperiodic fashion involving many-to-many unit/task assignments on an individual batch basis.

The distinction between these operational types is illustrated in Figure 2.4. Two products A and B are to be produced, each involving a two step recipe. Three multipurpose units are available, each capable of accommodating all four tasks. Figure 2.4(a) shows a production line in which U1 is assigned to task A1 and U2 and U3 are assigned out-of-phase to task A2. If the same unit/task assignments were employed for product B, we would have

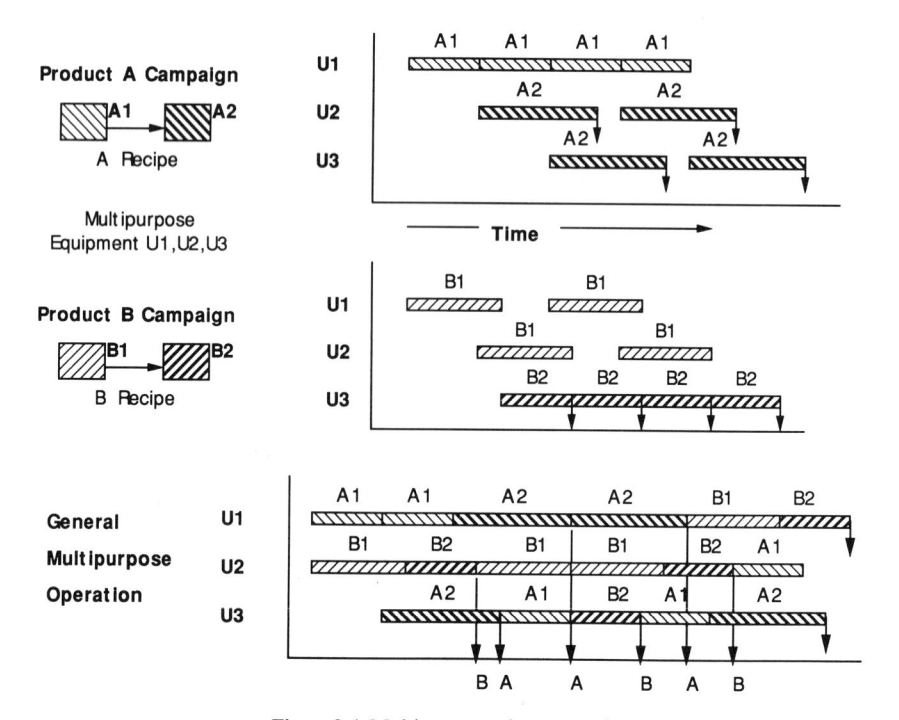

Figure 2.4 Multipurpose plant operation.

a multiproduct operation. In Figure 2.4(b), a different assignment is selected for product B (U1 and U2 are assigned to B1 and U3 to B2). Both lines operate in campaign style with their own characteristic cycle times. For instance, a campaign of four batches of A might be followed by a campaign of three batches of B, followed by another campaign of six batches of A, etc., as required to meet specific product orders. In Figure 2.4(c), production is in the general multipurpose mode, with tasks assigned to units in a flexible fashion, no clearly defined production line, and certainly no cyclic patterns of batch completion. Note that as a result of the imposition of a cyclic production pattern, the equipment utilization in the campaign mode (as evident from the idle time gaps) is in general not as efficient as the utilization obtained when that constraint is relaxed. However, if cross-contamination is a consideration, the flexible, acyclic operation would require more frequent equipment clean-out than in the regular campaign mode where clean-outs may only be required between campaigns.

2.3 The scheduling problem

A key problem which arises in batch operations is the scheduling of the plant to meet specified product requirements. Specifically, given the mode of operation, the product orders, the product recipes, the number and capacity of the various types of existing equipment, the list of equipment types allowed for assignment to each task, any limitations on shared resources (such as utilities or manpower) and any operating or safety restrictions, the scheduling problem is to determine the order in which tasks use equipment and resources, and the detailed timing of the execution of all tasks so as to optimize plant performance.

2.3.1 General features

The scheduling problem involves three closely linked elements:

- Assignment of units and resources to tasks.
- Sequencing of the tasks assigned to specific units.
- Determination of the start and stop times for the execution of all tasks.

For instance, given two reactors (U1 and U2) and six product batches (A–F) which need to be processed, the assignment step might involve allocating A, B and C to U1 and D, E and F to U2. The sequencing step would involve determining the processing order on each unit (e.g. first B, then C, and then A on U1), while the timing step would assign specific start and stop times for each batch on each unit. The above problem elements are shared by scheduling problems arising in a wide range of applications – ranging from machine shops to transportation systems to class room allocation.

Not surprisingly, a large literature, dating to the early 1950s, exists in the operations research domain on solution approaches to scheduling problems. The literature related to batch processing began to grow only in the mid-1970s.

Note that in the above example, the assignment component of the problem involves binary decisions (assign U1 to task A or not) as does the sequencing component (position A first in the sequence or not). The timing component can be a discrete decision problem or not depending upon whether time is treated as a continuum or divided into individual time quanta. It is the binary decisions which provide the challenge to scheduling problem solution. Indeed, a theoretical worst case analysis of computational complexity has shown that even the conceptually simplest forms of scheduling problems (those involving only sequencing considerations such as the sequencing of products on a single process with set-up costs which are dependent on the product processing order) can exhibit exponential growth in computational effort with increasing problem size. For instance, consider a multiproduct plant consisting of a single production line which must process five different polymer grades, A–E, in campaign mode. Suppose that the set-up/clean-out time for each grade depends on the previous grade processed. For instance, in changing from A to B it is 2 time units, while the transition from B to A takes 10 time units, etc. The number of possible grade sequences is 5! or 120. If 10 grades must be processed, the number of grade sequences grows to nearly 4 million; with 20 grades, the number of sequences exceeds 10^{18}. If the campaign sequence which minimizes the set-up time is sought then the search for the best sequence may involve examining most of the candidate sequences – a challenging task for the case of 20 grades, even if each evaluation only takes a nanosecond of computation.

Fortunately, recent research experience has shown that by creative problem representation, clever exploitation of problem-specific structure and effective algorithm design, practical scheduling problems can be solved before 'hitting the wall' of exponential growth. For instance, effective solution approaches are available for the above campaign sequencing problem which can solve problem instances considerably larger than 20 in reasonable computing times. In general, tailored approaches have been proposed for each of the types of operating modes, taking advantage of the occurrence of specific resource constraints types, inventory characteristics and cost structures.

In summary, a single, universal solution approach to all scheduling problems does not exist (contrary to vendor claims) and it is highly unlikely that one will ever be found. Effective, practical solution methods can be devised but require careful engineering to exploit features of specific applications. In the next section we will briefly review the types of basic approaches that are available for use in tailoring solutions to scheduling problems.

2.3.2 Generic solution approaches

The categories of solution algorithms which have been advanced for the solution of scheduling problems include: rule-based dispatching methods, randomized search methods, artificial intelligence related methods, simulation approaches, and model-based optimization methods.

Rule-based methods, which involve the application of one or more sorting or dispatching rules to produce a candidate schedule, are fast and easy to implement. However, their applicability and performance are very case specific. Moreover, they offer no guarantees of quality or even of feasibility of candidate solutions.

Example. As an illustration of a rule-based scheduling approach, consider the scheduling of batches of six polymer grades on three unequal parallel extruders so as to minimize the total tardiness in meeting the promise dates for each of these grades. Data for the problem are given in Table 2.1.

A reasonable but suboptimal schedule might be devised by sequencing the grades by earliest promise date and then assigning the grades to specific extruders based on earliest time to completion. Two rules are applied in this instance: sort orders by earliest promise date and assign extruders based on earliest completion. Based on the first rule, the grades will be assigned in the order C A F E B D. The second rule will assign grades to extruders as follows:

Extruder 1: C F D
Extruder 2: B
Extruder 3: A E

As shown in Figure 2.5(a), where promise dates are denoted by an arrow, B will be early by 10 time units but E will be tardy by 8. The alternative assignment, shown in Figure 2.5(b):

Extruder 2: A B
Extruder 3: E

Table 2.1 Data for rule-based scheduling example

Polymer grade	Processing time			
	Extruder 1	Extruder 2	Extruder 3	Promise date
A	5	10	8	10
B	10	20	12	30
C	5	10	8	8
D	20	40	18	40
E	15	30	25	25
F	15	30	20	20

Figure 2.5 Example schedules: optimal and rule-derived.

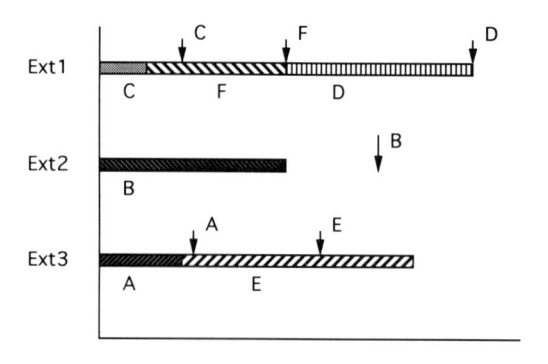

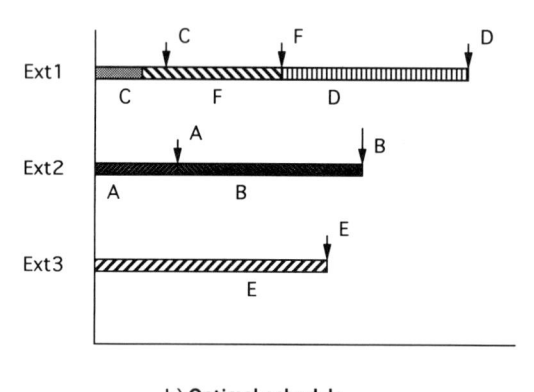

which leads to on-time completion of all grades, would not be found using these dispatching rules. In this instance, a less advantageous assignment of A leads to later advantages in the assignment of E. These types of trade-offs are typically missed by rule-based methods. As a result, rule-based scheduling approaches can offer no guarantee of the quality of the solution obtained.

Search methods involve some form of guided search of the candidate solution space. Typical examples includes genetic algorithms, which are randomized strategies for evolutionary hill-climbing in a combinatorial domain, or simulated annealing, which is a probabilistically based search which allows deviations from the strict hill-climbing discipline. This class of methods is inherently heuristic in nature, requires considerable problem customization and offers no solution quality or computational performance guarantees, except possibly asymptotic convergence.

AI-related methods include constraint-based reasoning and neural network approaches. The former family of methods, dating to the work reported in the early 1980s, conducts enumeration along partial decision trees guided by information about constraint 'stiffness'. The focus is primarily on solution feasibility. Constraint-based reasoning approaches can be computationally quite inefficient as schedules become more tightly constrained and thus nearly exhaustive enumeration must be performed to arrive at a feasible solution. Moreover, such approaches do not take advantage of the mathematical structure of the constraints, which in many instances are linear.

The neural network approach employs the popular network construction in which the weighting parameters of the network are determined from training data. In the context of scheduling applications, such training requires the availability of many very good schedules but suffers from the inherently poor performance of neural nets as extrapolation models.

Simulation methods, as the name implies, use a discrete simulation model of the application as a surrogate to the system being scheduled. Having constructed the simulation, the user is left to his own devices to propose and test alternative schedules, and thus to arrive at good schedules by trial and error. As a variation, a selection of dispatching rules might be employed to facilitate the generation of candidate schedules. In this fashion rule-based assignment is combined with simulation. Again, the limitation of this essentially empirical approach is that there are no guarantees of solution quality or feasibility and no known systematic methods for converting infeasible schedules to feasible ones.

Model-based optimization employs a mathematical model of the application as the basis for conducting a systematic search of the solution domain using numerical and logical methods, such as linear programming and branch and bound. The advantage of a model is that it offers a rigorous measure of the quality and the feasibility of any solution that is obtained. In the case of linear models, upper and lower bounds on the optimal solution can be obtained and these can serve as good estimates of how close the current solution is to the optimal one if the computation must be terminated prior to achieving the optimum. However, model formulation may require considerable expertise and the optimization process can be quite computationally intensive. Moreover, reliable solution in reasonable time requires careful exploitation of model structure and algorithm engineering.

Example. As an illustration of a model-based scheduling application, consider the production of N products, indexed from $i = 1$ to N on a batch production line over P time periods, where each time period t is of duration H_t. Each product i is characterized by a batch size B_i and a cycle time T_i. At the end of each period t, there is a demand D_{it} which must be met for each product i. The objective is to determine the campaign lengths in terms of number of batches, n_{it}, of each product i to be run within each period t so as to minimize the inventory costs of finished products. To describe this application, in addition to the decision variables n_{it}, a set of variables S_{it} which denote the inventory level of each product i at the end of each time period t must be introduced. The key constraints which any solution must satisfy are: the time availability constraints, the inventory balances, and the demand requirements.

The *time constraints* are that the total time required to produce all batches scheduled within a period cannot exceed the length of the period.

$$\sum_i n_{it} T_i \leq H_i \qquad \text{for all } t \qquad (2.1)$$

Note that this constraint neglects the set-up and transition times which may be incurred between product campaigns.

The *inventory balances* mean that for each period and product, the beginning inventory plus the production during the period minus the demands due at the end of the period must equal the inventory at the end of the period.

$$S_{it-1} + n_{it} B_i - D_{it} = S_{it} \qquad \text{for all } i \text{ and } t \qquad (2.2)$$

The demand constraint states that for each period and product, no back-ordering is allowed. Hence, the end of period inventory must be zero or greater

$$S_{it} \geq 0 \qquad \text{for all } i \text{ and } t \qquad (2.3)$$

while the initial S_{i0} is fixed and known.

Finally, the objective function must be defined – in this case it is to minimize inventory costs.

$$\text{Minimize:} \quad \sum_i CI_i \sum_t H_t [S_{it-1} + S_{it}]/2 \qquad (2.4)$$

where the inventory cost for each period is based on the average of the beginning and ending levels, the length of the period, a carrying charge, CI_i, per unit time and amount of products held.

Assuming that the number of batches n_{it} can be treated as continuous variables, the above model is a linear programming (LP) problem (linear objective, linear constraints), which can be readily solved using any of a number of commercial LP packages.

If set-up/change-over times and costs are significant, then the model must be modified to include appropriate logic variables. Let X_{it} be a set of 0–1 variables each of which takes on the value 1 if $n_{it} > 0$ and 0 otherwise. We ensure that this connection between the variables is maintained by defining the set of constraints

$$MX_{it} \geq n_{it} \qquad (2.5)$$

where M is a sufficiently large constant. If the campaign set-up time is denoted by Δ_i and the set-up cost by C_i, then the revised time constraint becomes

$$\sum_i n_{it} T_i^2 H_t - \sum_i X_{it} \Delta_i \quad \text{for all } t \qquad (2.6)$$

and the revised objective function which includes the set-up costs becomes

$$\text{Minimize:} \quad \sum_i CI_i \sum_t H_i [S_{it-1} + S_{it}]/2 + \sum_i \sum_t C_i X_{it} \qquad (2.7)$$

With this modification, the problem becomes a mixed integer linear program (MILP): a linear program in which both continuous and binary variables are present. Solution of this model form can be readily carried out using commercial MILP solvers, provided that the number of binary variables is not very large.

Note that in the above formulations, since inventory balances and demand requirements are only enforced at the boundaries of the time periods, there is no need to explicitly track the separate tasks occurring within the time periods. In the more general form of the scheduling problem to be considered in the next section, such details are fully treated.

2.3.3 Uniform discretization scheduling approach

The most important feature of the general multipurpose plant operation is that equipment and resource utilization profiles exhibit no regular pattern over time. For instance, as shown in the example given in Figure 2.6, multiple

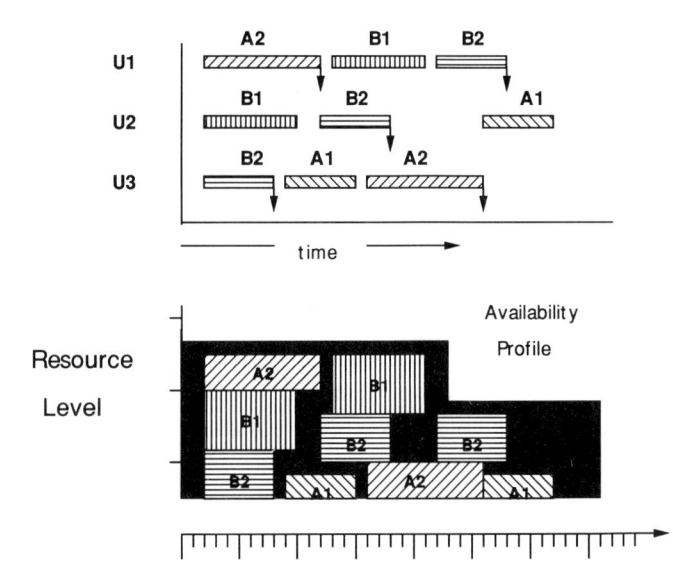

Figure 2.6 General multipurpose plant schedule structure.

tasks of different products are assigned to a given unit in no recurring order and no cyclic resource utilization structure is evident over time.

The difficulty in applying rule or search method-based approaches to this resource-constrained problem, lies in ensuring that resource constraints are met by any proposed unit/task assignment. If the assignment proves infeasible, then correction can only be achieved through brute enumeration of alternative start/finish time adjustments. Constraint-based reasoning approaches can accommodate problem constraints but do not offer optimization capability and do not take advantage of the mathematical nature of the constraints, which in most cases are linear functions of the decision variables. The multipurpose batch plant scheduling problem can, however, be cast into the form of a mathematical scheduling model providing that a suitable device is found for representing equipment and resource utilization over time. Specifically, the scheduling model must contain sets of constraints which will ensure that at each point in time in the production horizon each item of equipment is only assigned to a single task and that the utilization level of each resource which is shared by the simultaneously active tasks does not exceed the available supply.

The classical approach to this problem was proposed in the early days of mathematical programming research and was subsequently elaborated in the resource constrained scheduling context by Pritsker et al. [6] and others. The modelling device employed is to discretize time in some suitable fashion, to introduce assignment variables specific to each time interval and then to write for each time period a constraint set which would ensure that resource restrictions were not exceeded. This approach was first applied in the context of the multipurpose batch plant by Sargent and coworkers [7, 8].

If time is subdivided into suitably small uniform time quanta, then a binary decision variable, W_{ijt}, can be defined for each time quantum, such that, W_{ijt} takes on the value 1 if task i is performed in unit j in time quantum t, and 0 otherwise. Typical model constraints might, for instance, take the form

$$\sum_i W_{ijt} \leq 1 \quad \text{for each } j \text{ and } t \tag{2.8}$$

indicating that in time interval t the specific unit j can be assigned to at most one of the tasks for which it is suitable. Similarly, one can write mass balance constraints on the material k resulting from a given task i, which express the fact that the material available at the start of an interval plus that produced over the interval minus that consumed during the interval as input to other tasks and that allocated to meet demands arising in that interval must be equal to what is available to the next time interval.

$$S_{kt-1} + \sum_i \rho_{ik} \sum_j B_{ijt} - \sum_i \gamma_{ik} \sum_j B_{ijt} - D_{kt} = S_{kt} \tag{2.9}$$

for each material k and time interval t. Variable B_{ijt} denotes the batch size which is used when task i is executed in unit j at time t and parameters ρ_{ik} and γ_{ik} denote, respectively, the amount of material k which is produced when a task i is executed and that which is consumed when a task i is executed.

Similar balances can be written on the utilization of each renewable resource within each time period. Assuming fixed task processing times, resource utilization amounts, and material factors, it is possible to express all of the necessary constraints as linear functions of the binary and continuous variables (batch sizes, material amounts, etc.). The resulting scheduling problem can thus be posed as a mixed integer linear program.

Example. Four products (A to D) are manufactured in a facility consisting of five equipment items following a recipe consisting of three serial tasks. The processing times and feasible unit/task assignments are shown in Table 2.2.

For simplicity the tasks material consumption/production factors are all unity and the unit capacities are all unity. The scheduling objective is to process one batch of each product in the shortest possible time within a maximum time horizon of 15 h, given that all units are initially in an idle and empty state.

Since the shortest task processing time is 1 h, a time discretization interval of 1 h is sufficient to describe the dynamics without further rounding or approximation. With 15 time periods, four products, and the alternate unit assignments for the first and second tasks of each product, a maximum of 300 of the W_{ijt} variables will be required to describe the problem in the UDM formulation. Of course, because of logical schedule start and finish restrictions, e.g. tasks 2 and 3 cannot be initiated before task 1 of the corresponding product is completed, and tasks 1 and 2 must be initiated before the last possible time their successor tasks can be initiated, some 104 of these variables can be excluded from the formulation. The resulting uniform discretization (UDM) formulation, which will involve at most 196 binary variables, can be solved using a suitable MILP package to yield the schedule shown in Figure 2.7. Note that the production can be completed in 10 time units.

Table 2.2 Processing times for products A–D

Products	Processing times (h)		
	Task 1	Task 2	Task 3
A	4	2	1
B	2	4	2
C	3	1	1
D	1	3	1

Feasible units U1, U2, U3, U4, U5

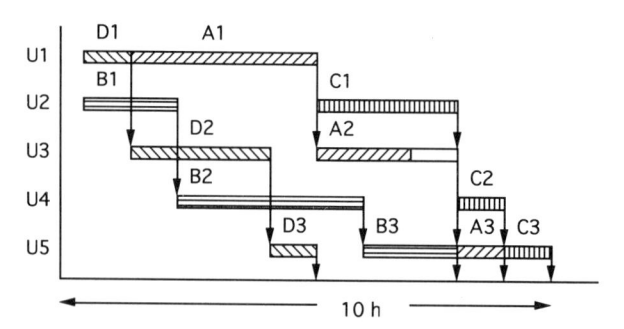

Figure 2.7 UDM-derived schedule.

In general, the MILPs can grow to be quite large. However, as noted in [9], it is possible to formulate the MILP constraints in various ways, some of which provide tighter relaxations and therefore lend themselves to more effective solution than others. Moreover, as shown in that work, it is critically important to develop solution methods that fully exploit the structure and data of these types of problems. Several other UDM formulations have been proposed in the literature [10–12], with various means of representing key problem features such as sequence dependent change-over times and losses. Collectively these various UDM formulations offer the advantages of accommodating complex recipe structures, treating alternative intermediate storage policies and limitations as well as handling multiple task-unit assignments, partial equipment connectivity and batch/lot size selection.

However, all UDM forms share a common limitation, i.e. the approximation of the underlying problem dynamics that may result from the use of time discretization. In general, in order to model rigorously the processing events that will take place, the size of the time quantum must be chosen to equal the shortest duration event. For instance, if task processing times range from 10 to 0.25 h, the latter value must be chosen for the discretization. If the scheduling horizon is 100 h, a problem with 400 intervals is created. To reduce problem size it is often necessary to choose a coarser interval. For instance, instead of 0.25 h, a discretization interval of 1 h might be selected, reducing the horizon to 100 intervals. Of course, the schedule obtained will overestimate the time required for the 0.25 h tasks and may need to be readjusted prior to execution.

2.3.4 Requirements for implementation

While the focus of the above discussion has been on the solution methodology, the implementation of computing-based scheduling technology requires consideration of a number of other implementation issues. These include:

- Electronic access to process and order information.

- Model development facilities.
- User interface to scheduling solutions.
- Linkage to control and planning systems.

Scheduling applications are very information intensive, requiring a wide range of information, including up-to-date status information for process equipment and storage vessels, accurate recipe specifications, resource availabilities and raw materials inventories, timely order information, and current cost and pricing data. These various items are usually located in different plant and business data bases, which must be linked to the scheduling application for routine, reliable and accurate access.

The scheduling system must provide model access facilities at several different levels of depth. The data level for plant scheduling personnel, the language level for plant technical support staff, the formulation level for the application specialist and the algorithmic level for the technical specialist. For routine scheduling runs, only the data level view should be required. However, the system should provide tools for model revision and solution method tuning. While several general purpose mathematical programming packages do provide model equation generation capabilities, they generally do not provide access to the algorithmic level nor do they offer the various data views appropriate for non-specialist use.

Scheduling applications are not only input data intensive but they are also data intensive on output. Accordingly, facilities are required for case management and the display of individual solutions. These displays should include schedule animation, traditional graphics such as Gantt charts, inventory and resource utilization profiles, and order time-lines. Since scheduling solutions may require tuning and editing to meet qualitative considerations, graphical tools for schedule editing and constraint verification and adjustment must be provided. Finally, solution analysis and interpretation capabilities are highly desirable. Insights into the resources which limit the solution, sensitivity to model parameter variations, and the impact of specific order parameters on schedule efficiency can facilitate continuous improvement of the manufacturing operation.

Once a schedule is generated, it must of course be implemented on the plant, and its implications for materials requirement and logistical plans disseminated. The former implies linkage to process control systems and the latter communication to the appropriate planning applications. Both should ideally be executed in electronic form for speed, accuracy and reliability. Furthermore, the degree of compatibility of the actual schedule with production planning targets should be communicated so that mismatches between the two levels can be reconciled.

Although several vendors do provide mathematical programming-based scheduling tools, at present no commercial system provides all of these features. Several, including Superior++ [13], provide a significant fraction.

A potential client is well advised to evaluate vendor capabilities in all of these implementation aspects before making acquisition decisions.

The topic of process control is further considered in Chapter 8. The reader interested in a more detailed consideration of implementation aspects may wish to consult [14].

2.4 Batch process simulation

Process simulation is a well established tool for the analysis of continuous chemical processes. Both steady-state simulators, the so-called flowsheeting systems, as well as dynamic simulation systems are widely used as surrogates for the actual process. They allow effects of changes in flowsheet and unit design and operating parameters to be investigated efficiently. Simulation is also a highly effective tool for batch process systems engineering. However, because batch processes are inherently not operated at a steady-state and their performance is governed by task-resource assignment and sequencing decisions, batch process simulation requires specialized features. Some of these features can be borrowed from the well-developed discrete event simulation domain. In this portion of the chapter the role, essential elements and capabilities of batch process simulators are summarized.

2.4.1 The role of simulation

The model-based approach to scheduling offers the clear advantages of a global, simultaneous view of all assignment, sequencing and timing decisions; the possibility of obtaining a guaranteed feasible and optimal solution (within the limits of model accuracy); as well as the availability of bounds on the optimal solution if the computations are terminated prior to convergence. However, to attack problems of practical scope, mathematical programming models generally require that modelling simplifications be made to keep problem dimensionality within reasonable limits. These include the use of simplified dynamics (a coarse time discretization and constant parameter values over time intervals), linear and state invariant task representation, fixed resource utilization profiles, and, of course, deterministic problem data. By contrast, as will be shown in the subsequent development, plant simulations allow arbitrary recipe complexity, including detailed step level representation; general nonlinear models of the step dynamics; stochastic parameter variations, and state-dependent processing logic (such as execution of alternative recipe branches if quality specifications are not met) to be accommodated. However, these advantages are obtained at the price of a local view of assignment, sequencing and timing decisions.

In the following sections we summarize the features required of a simulator in order adequately to represent batch operations. The mechanisms that are used to address these features are outlined.

2.4.2 Simulation model features

If a simulation model is to serve as a surrogate for a batch plant, then it must be able to accommodate the following features of this mode of operation:

- *Processing dynamics.* Batch and semicontinuous processes are unsteady state and thus the state of the plant as described through a set of variables undergoes changes with time. Some of these variables, such as the level of material in a vessel, change continuously with time. Other variables may change instantaneously as the result of the start or finish of operations or as a result of specific operating decisions, e.g. a reaction step may be terminated when the concentration of a particular species reaches the desired value, or an operator may be released after completing an assignment.
- *Flexible processing network structure.* In batch processes, multiple products may be made over time, each according to a unique recipe, each having specific equipment and resource requirements.
- *Operating decision logic and dynamics.* To represent a batch process it is not only necessary to represent the dynamics of the recipe tasks but also to model the decision logic under which task execution order and resource assignments are made. Indeed, the operating decisions concerned with the assignment of equipment to an operation and the assignment of materials and limited, shared resources as and when needed during the course of the operation will in general dominate the overall performance of a batch plant.
- *Parameter uncertainty.* Batch processes are generally subject to uncertainties in recipe parameters, equipment availability, product requirements and resource availability. Thus, it is often necessary to model these uncertainties in order to predict expected plant performance.

Some of these features are accommodated in conventional discrete event simulation methodology (event calendar, next event time advance, task dispatching logic and Monte Carlo sampling) while other features require continuous dynamic simulation techniques (state event processing, solution of differential algebraic models). The resulting technology is called *combined continuous/discrete* simulation.

2.4.3 Elements of combined simulation

Central to a combined discrete/continuous simulator is the scheme according to which the process being modelled is advanced through time. The four key components of the *time advance mechanism* are:

- The event calendar and its manipulation.
- Solution of the differential/algebraic system of equations.
- Detection and processing of discrete changes.
- Implementation of operating decisions.

(a) The event calendar

The events during a simulation represent discrete changes to the state of the process. The events are of two types: time events or state events.

A *time event* is a discrete change of which the time of occurrence is known *a priori*, e.g. if the recipe for a particular step requires the contents of a vessel to be mixed for 1.0 h, then the end-of-mixing event can be scheduled 1.0 h after the start of mixing.

A *state event* occurs when a state variable or some function of the state variables crosses a value, the *threshold*, in a certain direction. An example of a state event in which the function is linear is given in Figure 2.8. The exact time of occurrence of a state event is not known *a priori*, e.g. suppose the recipe for a particular step requires the contents of a vessel to be heated to 400 K. The time at which the contents of the vessel in which this step is performed reaches 400 K may not be predicted precisely when the heating is initiated because it depends on the charge size, composition, and initial temperature, utility availability, current value of overall heat transfer coefficient, etc. As a result, during simulation it is necessary constantly to check for the possible occurrence of state events.

The *event calendar* is a list of time events ordered on their scheduled time of occurrence. Associated with each event is an *event code* and additional descriptors which determine the set of actions, called the *event logic*, which are implemented when that event occurs. In the example given

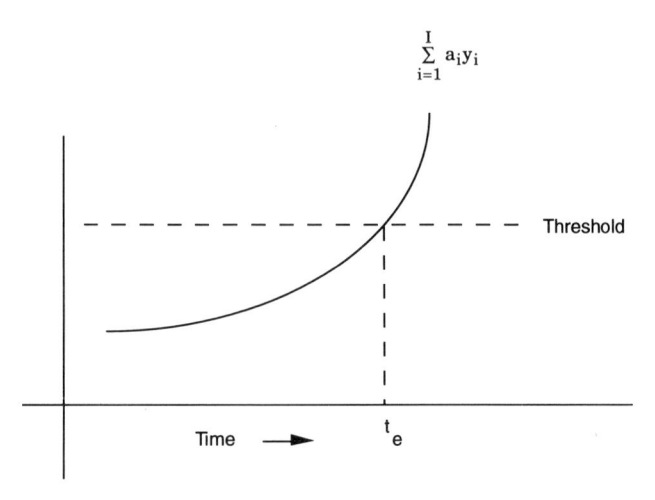

Figure 2.8 Example of state event.

above, the end-of-mixing event may initiate a heating step in the vessel, or may initiate the addition of another ingredient to the vessel. The event calendar is always in a state of transition. The events are removed from the calendar when they occur, and new events are added to the calendar as they are scheduled by the event logic and other components of the time advance mechanism. Several time events on the event calendar may have the same time of occurrence.

(b) Differential/algebraic equation processing

As in conventional dynamic process modelling, the continuous changes in the state of batch plant can be represented by a system of non-linear differential/algebraic equations, the *state equations*, of the following form:

$$F(y, \dot{y}, t) = 0 \tag{2.10}$$

where y is the vector of dependent variables, the state variables and $\dot{y} = dy/dt$. The initial conditions, $y(0)$ and $\dot{y}(0)$ are assumed to be given. In general, the state vector should comprise only those variables that are associated with active tasks and thus could potentially change with time, e.g. if the material is heated using a fluid in a jacketed vessel and there are no material inputs and outputs, and no phase change, the composition and the total amount do not change. Only the temperature, volume, enthalpy and heat duly change with time. Therefore, there is no need to integrate the individual species balance equations nor the total mass equation. The state equations must be solved using a suitable integrator, such as DASSL [15] which uses backward difference formulas (BDF) with a variable step predictor–corrector algorithm to solve the initial value problem.

(c) Detection and processing of discrete changes

Since the exact time of occurrence of a state event is not known *a priori*, after each integration step the time advance mechanism must invoke an algorithm to detect whether a state event occurred during that integration step. To detect the occurrence of a state event the values of the desired linear combination of state variables before and after the integration step are compared with the threshold. For example, suppose a heating step is ended when the temperature of a vessel becomes higher than 400 K. If the temperature before an integration step is less than 400 K and after that step it is greater than or equal to 400 K, then a state event is detected during that step. The direction for crossing in this case is 'crossing the threshold of 400 K from below'. Thus, if the temperature before an integration step is higher than 400 K and after that step it is lower than 400 K then a state event as described above is not detected.

The next step after detecting a state event is to determine its exact time of occurrence, the *state event time*. The state event time is the root of the equation:

$$y_a - thr_a = 0 \qquad (2.11)$$

where y_a is the linear combination of variables for which the state event was detected and thr_a is the threshold. The upper and lower bounds on the root are well defined, i.e. the time before and after the current integration step. The DASSL integrator maintains a polynomial for each state variable as part of its predictor–corrector algorithm. As a result, a Newton interpolation which has a second-order convergence can be used for determining the state event time [16].

The processing of an event consists of implementing the actions associated with the specified event logic, e.g. ending a filling step may result in releasing the transfer line used during filling and advancing the vessel to process the next step. For user convenience, the simulator should provide a library of elementary actions, such as shutting off a transfer line, opening a valve, releasing a utility, as well as the capabilities to accommodate user-generated event logic.

(d) Implementation of operating decisions

While the recipe for each product defines the order in which tasks must be executed in order to produce a given product, the initiation and sequencing of tasks are governed by operational decisions. There are two types of decisions which must be implemented. *Pull*-type decisions serve to initiate the lead tasks of a recipe while *push*-type decisions serve to execute the remaining tasks of the recipe which are driven by the completion of predecessor tasks. Pull-type decisions must be initiated externally through user imposed processing sequences while push-type decisions are initiated by upstream tasks when they become ready to send completed material downstream. Typically, the time advance mechanism reviews the processing sequences at each event to check whether a task can be initiated on available equipment. Typically, a prioritized queue discipline is used to process the requests for assignments of equipment generated by intermediate tasks which seek to push material downstream. The queues also are reviewed at each event time by the time advance mechanism to determine whether any requests could be fulfilled. The processing sequences and priorities used in the queue management are specified by the user. Hence, by suitably manipulating the processing sequences and the priorities the user can influence the assignment of equipment to operations and the movement of material in the process. The priorities are also used for resolving the competition for shared resources.

(e) Monte Carlo sampling

The operating conditions in a typical batch process are characterized by random fluctuations, e.g. the duration of an operation may not be the same on every occurrence, but instead vary within a range, or a piece of equipment may break down randomly, forcing delays for repairs. The random fluctuations significantly affect the overall performance of a process and thus their effects should be included in simulation studies.

Process variability is accommodated in combined simulation by providing the capability to sample the appropriate model parameters and to collect the necessary data for statistical analysis. For instance, each stochastically varying parameter is associated with a random number stream which is given an initial seed. Successive specific values of the parameter are obtained by sampling the specific distribution function which describes the variation of that parameter. During the execution of the simulation, each time a new value of the parameter (e.g. the duration of a specific task) is required, it is generated by sampling. In order to draw statistically valid conclusions from simulations involving sampled parameters, it is necessary to replicate simulation runs using different random number seeds so as to compute the means, minima, maxima and standard deviations of key performance variables. These in turn are used to construct confidence intervals and to conduct hypothesis testing. In general, the simulation run length, data truncation to reduce the bias introduced by the initial transients and the number of experiments are some of the important factors which must be considered for reliable statistical analysis. Monte Carlo sampling methodology is well established in the discrete event simulation literature and is explained in texts such as [17].

Example. In this example the combined simulation methodology is illustrated using a simple batch process example drawn from [3].

Consider a plant consisting of three pieces of equipment: two mixing tanks, MIX_1 and MIX_2, and a filter, FLTR1. The equipment connections are shown in Figure 2.9. Transfer lines 1 and 2 are available for charging raw materials A and B, respectively, into the mixing tanks. Transfer line 3 feeds material to the filter from either mix tank A or B. Transfer lines 4 and 5 are used for removing material from the two outputs of the filter.

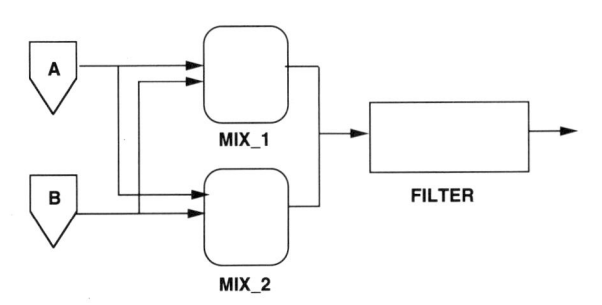

Figure 2.9 Equipment network of example process.

The recipe for product PR_1, consists of two operations, MIX and FILTER with details given in Table 2.3.

The MIX operation can be performed in either MIX_1 or MIX_2, while the FILTER operation can only be performed in FILTER. A processing sequence to initiate two batches of operation MIX is specified. The simulation run is to terminate at 100 h.

At the beginning of the simulation the event calendar has two events: 'START SIMULATION' at time 0.0 and 'STOP SIMULATION' at time 100.0. The review of the processing sequence at time 0.0 results in the assignment of MIX_1 to perform the MIX operation, beginning with the FILL-A elementary step. Since there is only one transfer line available to transfer raw material A, the MIX operation cannot be initiated in MIX_2. Also, the FILTER operation cannot be initiated because neither of the mixers is yet ready to send material downstream. Since the FILL-A elementary step requires 1 h, a time event is scheduled at time 1.0. The implementation of the filling step in MIX_A entails solving the differential equations representing the species mass balance and the total amount equations. Figure 2.10(a) shows the event calendar after all the events at 0.0 are processed. Since FILL-A requires the solution of differential equations, the simulator starts marching in time by integrating the state equations. The goal of the simulator is to advance the process up to the next event time. Since there are no state event conditions active during FILL-A, the integration halts at time 1.0 because of the time event.

At time 1.0, the FILL-A step is completed and FILL-B is started in MIX_1. Also, since transfer line 1 is released after the completion of FILL-A in MIX_1, the MIX operation is initiated in MIX_2. A new event to end FILL-A in MIX_2 is scheduled at time 2.0. Figure 2.10(b) shows the event calendar after all the events at 1.0 are processed. As per the recipe in Table 2.3, FILL-B ends based on a state event (MIX_1 becoming full), hence there is no event on the event calendar to end FILL-B in MIX_1. The FILTER still cannot be initiated at 1.0. The new set of equations to be integrated consist of the species and total mass equations for both MIX_1 and MIX_2. Since, there is one active state event condition to end FILL-B in MIX_1, the integrator checks for a state event after each

Table 2.3 Recipe for product PR_1

Mix	Filter
1. Fill 20 kg raw material A in 1.0 h (FILL-A) 2. Fill raw material B at 40 kg h^{-1} until the vessel becomes full (FILL-B) 3. Mix for 3 ± 0.2 h (STIR) 4. Filter the contents (FEED-FLTR)	1. Filter the contents of either MIX_1 or MIX_2 at the rate of 60 kg h^{-1}; 90% of the material coming in leaves as a waste stream; stop filtering when 20 kg is accumulated in FLTR1 2. Clean FLTR1 in 0.5 h (CLEAN)

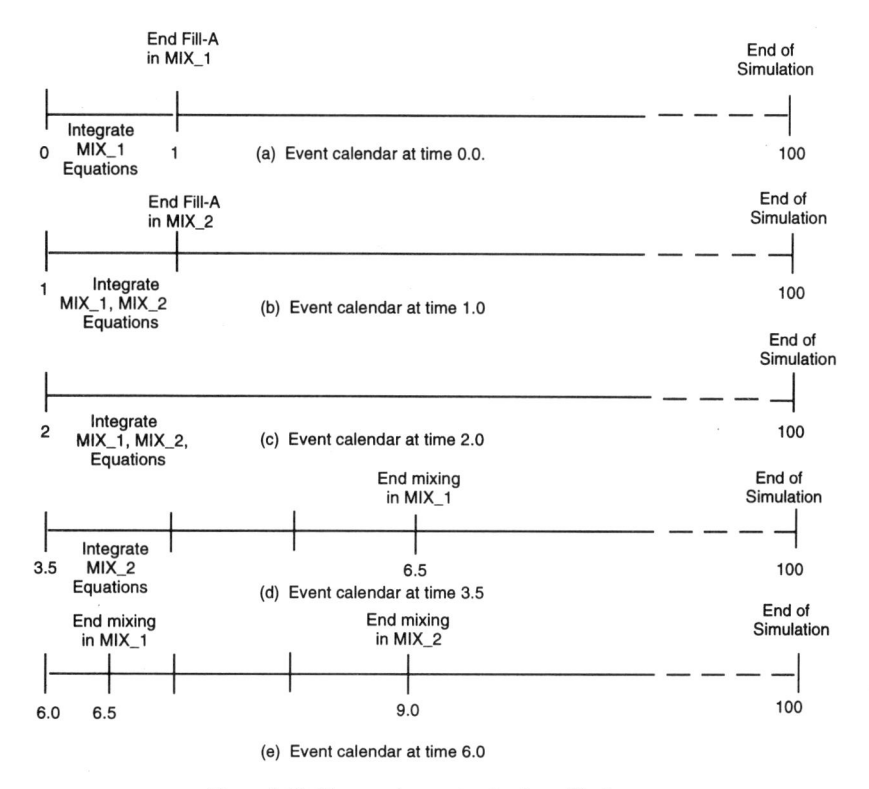

SCHEDULING AND SIMULATION OF BATCH PROCESSES 47

Figure 2.10 Changes in event calendar with time.

integration step. The next known event is at time 2.0 when FILL-A is completed in MIX_2.

Suppose no state event is detected and the integration halts at time 2.0. After completing FILL-A, MIX_2 has to wait because FILL-B is still in progress in MIX_1 and there is only one transfer line available for transferring raw material B. Therefore, after processing the events at time 2.0, there is only one event on the calendar as shown in Figure 2.10(c). The state vector consists of the species and total mass variables for FILL-B in MIX_1.

Suppose a state event is detected at time 3.5 because MIX_1 becomes full. The integration is halted and the appropriate event logic is executed. Transfer line 2 is released and MIX_1 is advanced to the STIR elementary step which has no state equations. Suppose the duration of the first batch of STIR is 2.9 h. An event is scheduled at time 6.4 to mark the end of the STIR elementary step in MIX_1. Also, FILL-B is initiated in MIX_2. Figure 2.10(d) shows the event calendar after all the events at 3.5 are processed. The state vector consists of the species balance and total mass variables for FILL-B in MIX_2. One state event condition is active, i.e. end FILL-B when MIX_2 becomes full. The goal of the simulator is to advance the process to 6.4. However, a state event is detected at 6.0, and the integration is halted.

At 6.0, MIX_2 is advanced to the STIR elementary step which has no state equations. Suppose the duration of the STIR elementary step for the second batch is 3.1 h. As a result, a time event is scheduled at 9.1 to mark the end of the STIR elementary step in MIX_2. Figure 2.10(e) shows the event calendar after all the events at 6.0 are processed. Since both MIX_1 and MIX_2 are processing the STIR elementary step there are no state equations to be integrated and the time is advanced to 6.4.

At 6.4, the FEED-FLTR elementary step is started in MIX_1 and the FILTER operation is initiated in FILTER, beginning with the SEPARATE elementary step. The state variables consist of the species and total mass variables for MIX_1 and FILTER. Two state events are active when integration is resumed, one to mark the end of the FEED-FLTR elementary step when MIX_1 becomes empty and one to stop filtering when the total mass in FILTER reaches 20.0 kg.

MIX_1 becomes empty at 8.4. As a result, the flow into FILTER is stopped, and MIX_1 is released. Since there are no more batches to be made, MIX_1 remains idle for the rest of the simulation. 12 kg is accumulated in the FILTER. After processing all events at 8.4, the time is advanced to 9.1 since there are no state equations to be integrated.

At 9.1, the FEED-FLTR elementary step is started in MIX_2 and the SEPARATE step is resumed in FILTER. The state vector consists of the species and total mass variables for MIX_2 and FILTER. Two state events are active when integration is resumed, and there is only one event on the event calendar, i.e. end simulation at 100.0.

At 10.433, the SEPARATE step is ended because 20 kg is accumulated in FILTER. Therefore, the flow to FILTER is stopped, with 40 kg still left in MIX_2. The CLEAN step is initiated in FILTER and an event is scheduled at 10.933 to mark its end. Since there are no equations to be integrated the time is advanced to 10.933.

At 10.933, FILTER is reassigned to the FILTER operation and the filtration of the rest of the material in MIX_2 is resumed. At 11.6 MIX_2 becomes empty and the filtration is halted with 4.0 kg left in FILTER. Since there are no more batches to be made and no more material to be processed the time is advanced to 100.0, thus ending the simulation.

2.4.4 Limitations of discrete simulation languages

The discrete or combined continuous/discrete simulation packages such as GPSS [18] and SLAM [17], in principle, provide the basic discrete simulation methodology necessary for modelling batch processes. However, since these packages are general purpose simulation languages, they do require considerable expertise and effort for simulating batch processes. In general the discrete simulators have not been widely accepted as effective tools for simulating batch processes because of limitations in process representation, control mechanisms, treatment of process dynamics and model implementation. However, it must be noted that because these simulators are written in procedural languages, such as FORTRAN, and/or provide a mechanism to incorporate user written code, there is no inherent limit on adapting them to satisfy a particular need, provided the user has the time and the expertise required for developing the customized code.

(a) Process representation
In discrete simulators the manufacturing 'activities' are performed on 'entities' by 'servers'. The servers and activities are equivalent to process equipment and elementary processing steps in a recipe, respectively. An entity is a widget which can be very loosely compared to a batch of material. A widget keeps its identity and its movement can be tracked in the process. Also, when a server completes an activity, modelled as a time delay, the entity is released and the server can be assigned to process another entity. In a batch process, during a task several elementary steps are performed in the assigned piece of equipment. Also, not all elementary steps can be modelled as pure time delays, e.g. a step can end based on a state event, or it can end based on interactions with other pieces of equipment. During a task, material from several upstream steps may be transferred into a piece of equipment and several downstream steps may withdraw material from it. Thus, a batch of material may constantly change its identity. Also, whenever there is a transfer of material between elementary steps, pieces of equipment must be available simultaneously from several processing stages, e.g. in order to filter material from a feed tank and store the mother liquor in a storage tank, three pieces of equipment ('servers'), one from each stage, must be available simultaneously.

(b) Operating decisions
The key mechanism for the movement of entities assumed in discrete simulators is 'push entities downstream', i.e. whenever a server finishes an activity it releases the entity and the entity waits in a queue for the assignment of a server to perform the next activity. However, as noted in the previous section, in batch processes, both the 'push' and the 'pull' mechanisms for the movement of material are required.

(c) Process dynamics
Discrete simulators with continuous variable capabilities use explicit integration algorithms such as the Runge–Kutta method for solving the system of differential equations describing the processing steps. Such explicit methods cannot solve a system of differential/algebraic equations (DAEs) very effectively and therefore the differential equations must be stated in

explicit form (that is with derivatives as the left hand sides of the equations). Furthermore, explicit methods are not recommended for solving stiff equations. Since even a simple dynamic process model such as filling a tank requires the solution of DAEs, the discrete simulators have a serious limitation. The problem is compounded by the fact that the more complex models such as reactor and evaporation have implicit DAEs and are generally stiff.

In discrete simulators, the differential equations and the state events must be defined prior to a simulation run. In the context of multiproduct batch process simulation, this requires the definition of the equations for all feasible combinations of steps and equipment items, and the associated state event conditions, along with the logic to set the values and derivatives of the variables which would be active at a given time, prior to a simulation run. While not impossible, this is an overwhelming requirement for the team generating and maintaining the model. Also, defining all of the possible combinations in the state vector and checking the state event conditions at the end of every integration step would result in a tremendous computational overhead considering that at any time at most one combination could be 'active' per equipment item.

(d) Code-driven structure

In discrete simulators, since the event logic, dynamic process models and operating characteristics are implemented through user-written code, the models tend to be very problem specific. Therefore any 'what if...' which is outside the scope of the existing code can be studied only after suitably revising the code. Since the main objective of a simulation study is to evaluate various alternatives, the prospect of having to change the software code in order to evaluate the impact of a change restricts its use to experts. The code-driven feature has been mitigated in part in versions with graphical user interfaces which provide icons representing standard model building primitives, such as queues. However, the need to build the model of a step from these elements remains.

2.4.5 Features of a process-oriented combined simulator

Over the past 15 years, several process modelling systems have been reported in the literature, e.g. gPROMS [19], UNIBATCH [20], BOSS [16] and BATCHES [21], which incorporate the combined discrete/continuous simulation methodology necessary for simulation of batch/semicontinuous processes. Of these, BATCHES is the first commercial system to address fully the special requirements of batch processes from the process representation, data management and analysis standpoint. For a more detailed presentation of BATCHES capabilities, the reader is directed to [4]. Here, the main modelling constructs, decision and control mechanisms, and input/output features provided by the BATCHES simulator are summarized.

(a) Modelling constructs

The BATCHES simulator provides two constructs, the *equipment network* and the *recipe network*, to model the many to many relationship between equipment items and tasks, and to model the recipes of various products.

The *process equipment network* represents the physical layout and connectivity of the equipment items in the given process. The equipment parameters describe the physical characteristics of each equipment item such as volume or heat transfer area. Also, any physical connectivity constraints are specified in the equipment network through the use of transfer lines, e.g. some pieces of equipment from a stage may be connectable to only a few pieces of equipment from another stage. Similarly, to transfer material between two stages, a manifold may be available which allows only one active transfer at a time.

Each product in a batch process is represented by a *recipe network*. In a recipe network, an operation is represented by a task, while an elementary processing step is represented by a subtask. The appropriate choice of task details and subtask models and parameters allows the user to represent the recipe to as fine a detail as desired.

Subtask model. The most important subtask descriptor is the model used to represent the corresponding elementary processing step. Each subtask model in the model library represents a specific physical/chemical operation, such as filling, emptying, reaction under adiabatic conditions, continuous separation of material and so forth. Each model, whether a simple holding tank model or a complex batch reactor model, will in general consist of a set of simultaneous DAEs. If the models in the library do not meet the requirements of a particular elementary processing step, new models can be added to the library.

Subtask parameters. The key subtask parameters are: subtask duration, state event description, operator and utility requirements. Typically, the available resources are shared by the entire process and the instantaneous rate of consumption of a resource cannot exceed a specified value at any time. Individual tasks compete for the resources required for executing their subtasks. Thus, before the execution of a subtask the following conditions must be satisfied:

- The required amount of input material must be available upstream.
- Equipment items must be available downstream for every output.
- A transfer line must be available for each material transfer.
- Operators and utilities must be available in the specified numbers or amounts.

Link between equipment and recipe networks. For every task in a recipe an ordered list of equipment items suitable to process that task is specified. If at a given time there is a choice, the piece of equipment best suited to perform the task is selected. A particular piece of equipment suitable for several tasks appears on the lists associated with the corresponding tasks. The suitable equipment lists provide the link between the equipment and recipe networks.

(b) Decision and control logic

In BATCHES the operational decisions are controlled through processing sequences, equipment and resource assignment queue disciplines, and conditional branching constructs.

- *Processing sequences.* This 'pull'-type construct allows user specification of the number of times or amounts of specific tasks which are to be executed. Typically, these will be the lead tasks of a recipe. The processing sequences merely specify the desired sequence: the actual start and end times of operations will be determined by the simulator as it executes the required tasks.
- *Equipment and resource queues.* The 'push'-type decisions for assigning equipment items and other resources to tasks are handled through queues and search options. The primary criterion for ordering the requests is user specified priorities, and the secondary criterion is First In First Out (FIFO). Search options dictate the order in which equipment and resource needs of specific tasks are to be identified and reserved.
- *Conditional branching.* Often, during an operation a different set of subtasks or tasks may be executed based on quality control considerations or the state of the material at that particular time, e.g. if a certain concentration is exceeded, a reprocessing task must be performed to bring the material within allowed specifications. This can be accomplished through conditional branching information which is associated with specific recipe subtasks.

(c) Input/output capabilities

Since BATCHES simulation models are data driven, a simulation model is built by specifying appropriate values of various parameters associated with the modelling constructs. The system provides interactive graphical network builders, menus, and forms to build, catalogue and alter simulation models. Output options include animation, time series output, textual summaries and graphical summaries of key statistics. An animation is a dynamic presentation of the changes in the state of a system or model over time displayed on a facility model which is constructed using icons. Time series outputs consist of values of user-selected variables that are recorded over the course of the simulation run. Examples include the amount of material in a specific piece of equipment, the process inventory, the cumulative production, utility usage, etc. Textual reports may include mass balance summaries, as well as equipment and resource utilization statistics.

Additionally, the cycle time summary for each piece of equipment, along with a breakdown of the time lost due to waiting for material and resources, is reported.

Inherently, a dynamic simulator must march through time in a sequential fashion. As infeasibilities are encountered in executing the assignment and sequencing decisions imposed on the simulation, the simulation executive only has at its disposal the timing and reassignment variables which are associated with those tasks that are active at that point in time or with units that are idle and thus can be assigned. It has no predictive capability to look ahead over several assignments before making assignments and cannot readjust the execution of tasks already completed or initiated. Moreover, it can neither provide bounds on the optimal scheduling problem solution nor initiate procedures to search for the optimum schedule. Thus, while a simulation can incorporate a high level of recipe and process detail, it has very limited and myopic decision-making capabilities. However, the high degree of process fidelity allows the simulation to serve as an effective surrogate for actual plan operation under a specified set of assignment and sequencing decisions. It can be used, for instance, to verify or establish processing capacity under these decisions, to study the effects of uncertainties via Monte Carlo experiments, and to refine the timing of individual steps within tasks.

The strengths and limitations of simulation tools in addressing scheduling problems are in marked contrast to those of optimization-based scheduling approaches. Although it uses approximate and aggregated models of batch process operation, the optimization approach inherently has a global view of all of the assignment and sequencing decisions which must be made. Thus, the two methodologies are to a large extent complementary. A useful approach to arriving at feasible, robust and/or optimal schedules is to attempt to combine these two methodologies. Indeed, this has been done in a number of studies (e.g. see [22–26]). In these applications, the simulation model was coupled in an *ad hoc* fashion with a simplified algebraic or conceptual model which was solved using mathematical programming or rule-based methods. The latter served to generate the necessary assignment decisions which could then be evaluated by the simulation. The limitations of these studies were that the generation of and linking between the levels of representation were *ad hoc*, and there was no structured feedback of information between the levels which could assure an approach to overall optimality. To systematize the linkage between simulation and optimization, a strategy must be used which consists of deriving the simulation model and mathematical programming formulation employed for scheduling starting with a common source of information, and using the two in an iterative loop. A schematic representation of such a strategy is shown in Figure 2.11.

As the first step, a subtask level representation of the underlying process must be developed from the detailed recipe information. The

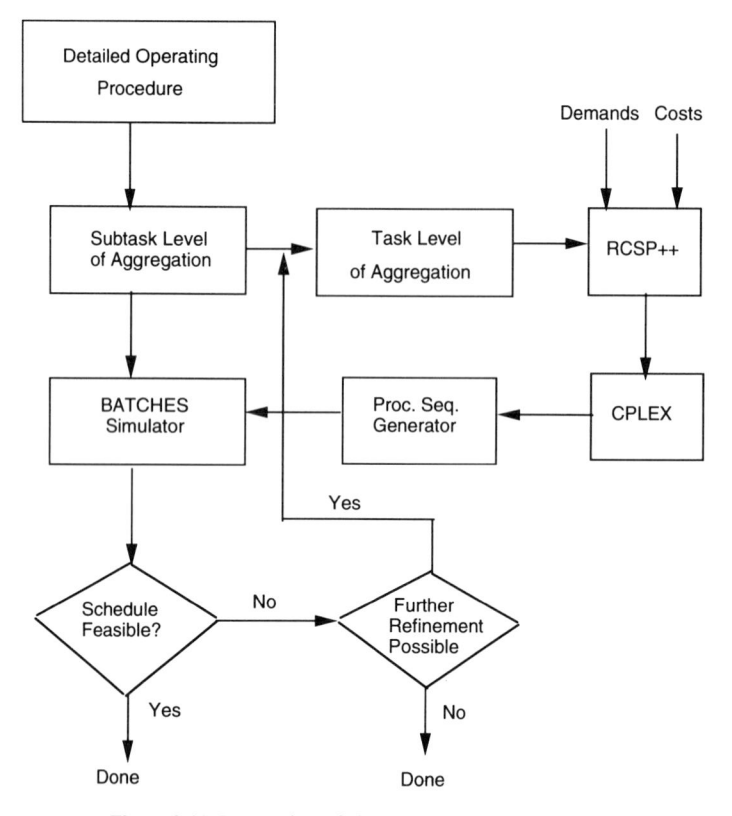

Figure 2.11 Integration of simulation and optimization.

recipe information could be in an electronic form as detailed operating procedures in a batch control system or in the form of a standard operating procedure manual. The derivation of a subtask level representation requires user interaction to decide the level of detail. A simulation model can then be automatically generated from the subtask level representation. Similarly, the parameters describing the task level aggregation can in principle be derived automatically from the subtask level representation. The generation of the mathematical programming formulation requires task descriptors, costing information, selection of performance criteria and demand information.

The iterative solution loop consists of firstly solving the scheduling problem using either rule-based or mathematical programming methods. The schedule generated by the scheduler is translated for use by the simulator. The detailed simulation model exercises the schedule on the surrogate plant and determines its feasibility or robustness. If the simulator determines the schedule to be infeasible or to lack adequate flexibility, it provides insights into the dynamics of interactions between the various process elements, thus providing clues about the probable causes for the infeasibility.

In general, if the simulator detects infeasibilities, it is because the tasks require more time than the lumped durations used in the scheduling formulation due to subtask level interactions. To reflect this fact, the simplest recourse is to change the task durations used in the scheduling problem. The task level descriptors can then be adjusted to approximate the low level interactions at an aggregate level and the scheduling problem is resolved using the adjusted values. The iterative loop is executed until the simulation model verifies the feasibility or desired level of robustness of solutions generated by the scheduler.

Example. To illustrate the integration of simulation and optimization, we consider the operation of a plastics packaging plant [27]. The objective of this study was to determine robust schedules in the presence of processing time uncertainties. The BATCHES simulator was used as the simulation tool. The RCSP++ package [12] was used for generating the mathematical programming formulation of the scheduling problem based on uniform time discretization. The CPLEX package was used for solving the resulting MILP. A schedule translator converted the schedule generated by CPLEX into processing sequences used by the simulator. The simulation model of the recipe was developed from the recipe description. The task level aggregate descriptors, used as an input to RCSP++, were also manually generated from the recipe description. The iterative loop was manually implemented by analysing the results of the simulation, deciding the changes in the task durations and changing the input data file used by RCSP++.

The plant structure for the plastics packaging plant is shown in Figure 2.12. A batch of resin is prepared in S and is transferred to a blending unit, B1 or B2. Each blender may store the material prior to and after blending. Material from B1 is transferred to a packaging line, LB1 or LB2, through an intermediate task V, requiring additional processing time. Material from

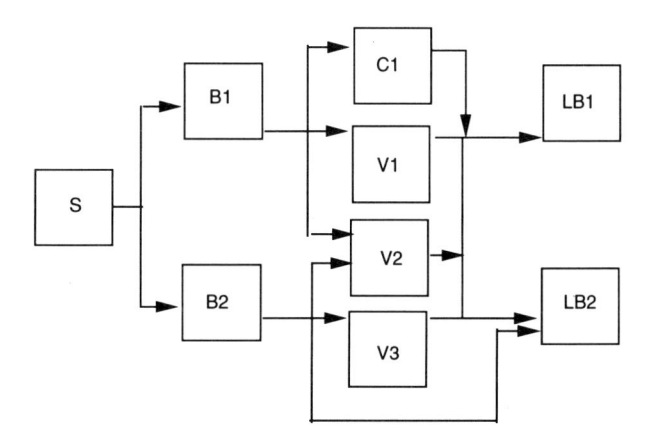

Figure 2.12 Equipment network of plastics manufacturing facility.

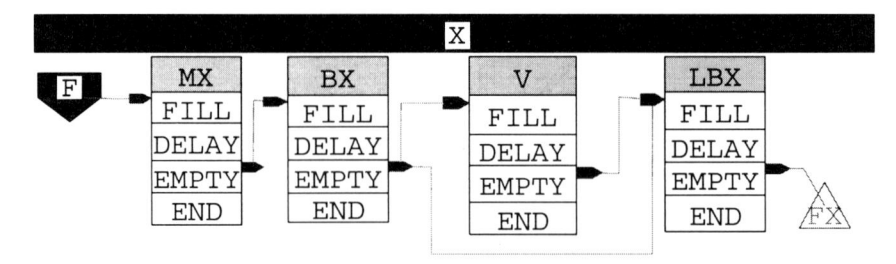

```
┌─────┬─────┬──────X──────┬─────┐
│ MX  │ BX  │     V       │ LBX │
│FILL │FILL │   FILL      │FILL │
│DELAY│DELAY│   DELAY     │DELAY│
│EMPTY│EMPTY│   EMPTY     │EMPTY│
│ END │ END │    END      │ END │
└─────┴─────┴─────────────┴─────┘
```

Figure 2.13 Recipe network for a plastics product.

B2 can be fed directly to packaging line LB2 or through an intermediate task V to packaging line LB1. During the actual process, the material changes state only once, when it is blended. All other steps are involved with transporting and storing material so that it can be packaged in time to meet customer demands.

Two finished products, X and Y, are produced in this facility. The recipe network for product X is given in Figure 2.13. For both products, the key processing times are as follows: the preparation task requires 1.0 h, the blending task requires 2.0 h, the transfer task requires 1.0 h and the packaging task requires 1.0 h. The demand for each product is given in Table 2.4.

The objective of the scheduler was to minimize the total inventory cost, levied at $10/kg per unit time, while meeting order due dates in the presence of processing time uncertainties. The Gantt chart for the optimal schedule for the base case is given in Figure 2.14.

A series of simulation runs was made with the durations of the preparation, blending and packaging steps uniformly distributed between around 20% of the average. These simulations showed that, on average, three orders out of a total of seven missed their due dates, clearly indicating that the optimal schedule generated in the base case is not robust. In the next iteration, two scheduling problems were solved. In one the processing times of the preparation, blending and packaging steps were increased by 10%, and in the other by 20%.

Simulation runs were made with the same sampling strategy as in the first case. As expected, there were no late orders for the schedule which used 20% longer processing times. However, the schedule which used 10% longer

Table 2.4 Demand for a plastics-producing facility

Product	Amount	Due date	Product	Amount	Due date
X	30	7	Y	25	8
X	90	12	Y	60	11
X	60	18	Y	25	15
			Y	30	16

BASE_1.GNTTASK

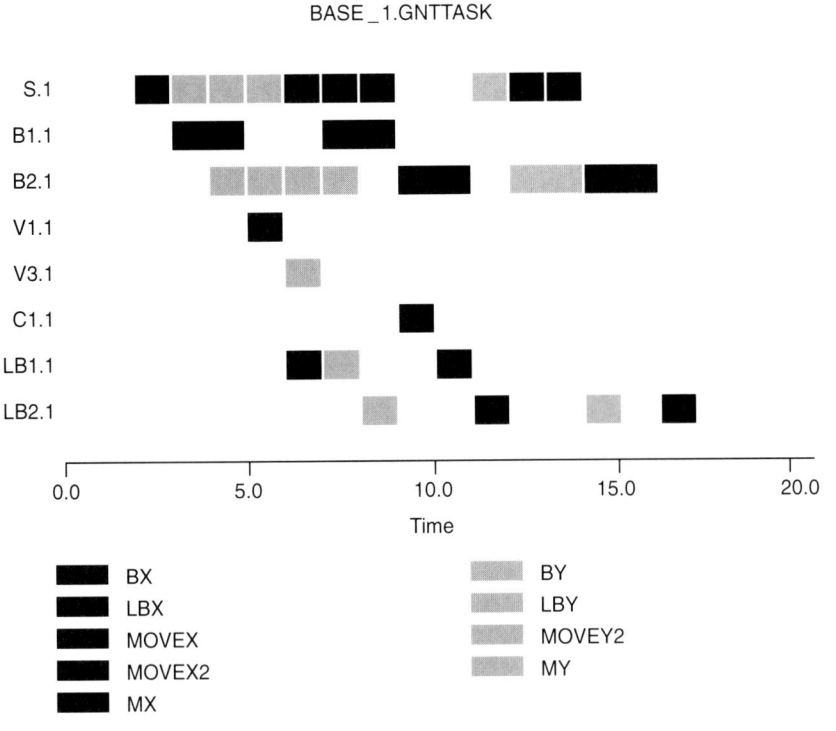

Figure 2.14 Gantt chart for plastics example.

processing times missed 0.3 orders on average, much fewer than the base case. An increase of 20% in processing time certainly created a robust schedule for the level of uncertainty under consideration, but it reduced the time available on equipment items for processing additional batches. This example demonstrates the effective use of simulation and optimization technologies to investigate the obvious tradeoff between robustness and productivity.

2.5 Design implications

While the scheduling problem focuses on effective utilization of existing production resources to meet product requirements, the batch process design problem involves the determination of what the optimal level of those production resources should be. Thus, given the mode of operation, the product orders, the product recipes, the list of equipment types allowed for assignment to each task, any limitations on shared resources, such as utilities or manpower, and any operating or safety restrictions, the conceptual design problem is to determine the required number and capacity of the various types of equipment, the order in which tasks use equipment and resources, and the timing of the execution of all tasks so as to optimize plant annualized cost.

Evidently, the principal difference between the definition of the scheduling problem and the above statement of the design problem lies in the relaxation of the equipment number and capacity from the status of problem parameters to optimization variables. Indeed, as the way the plant is scheduled will determine its capacity, the design problem can be viewed as an upper level decision problem which has imbedded in it the scheduling problem. Thus, to solve the former we must necessarily also solve the latter. Of course, there are differences in the timescales which must be considered: at the design stage product demands are not known at the level of individual orders and instead might be aggregated at quarterly, seasonal, or annual requirements. Moreover, because of differences in the level of certainty of the demand requirements (long range forecasts in the design case versus concrete orders in the scheduling case) the scheduling subproblem solutions required in the design case may be less rigorous.

In principle, in defining the design problem one should also include the choice of mode of operation as one of the design optimization variables. After all, the selection of operating mode, e.g. cyclic or acyclic, multiproduct or multipurpose, is at root dictated by economic considerations: cost of inventory, change-overs, complexity (measured in labour and automation costs) and off-specification production. Indeed, since the general multipurpose operational mode can be viewed to encompass the other two limiting modes as special cases, the mode-specific design problems can in principle be subsumed by that of the general multipurpose plant. However, direct optimization over the operational mode proves impractical, firstly because all of the mode-dependent costs are difficult to quantify and, second, because more effective solution methods can and have been devised for mode-specific formulations.

A detailed discussion of optimization approaches to the solution of various mode-specific forms of the design problem is beyond the scope of this chapter. The interested reader is referred to references [1] and [28] for discussions of these developments. Simulation tools can of course be employed to support design studies. However, this can only be achieved through case studies in which equipment sizes and number are systematically examined to arrive at successive design improvements.

References

1. Reklaitis, G.V. (1995) Computer aided design and operation of batch processes. *Chem. Engng Educ.*, **2**, 76–85.
2. Reklaitis, G.V. (1995) Scheduling approaches for the batch process industries. *ISA Trans.*, **34**, 349–358.

3. Clark, S.M. and Joglekar, G.S. (1996) Features of discrete event simulation. In *Batch Processing Systems Engineering* (G.V. Reklaitis, D.W.T. Rippin, P. Hortacsu and A.K. Sunol, eds), Springer Verlag, NATO ASI Series F.
4. Clark, S.M. and Joglekar, G.S. (1996) Simulation software for batch process engineering. In *Batch Processing Systems Engineering* (G.V. Reklaitis, D.W.T. Rippin, P. Hortacsu and A.K. Sunol, eds), Springer Verlag, NATO ASI Series F.
5. Reklaitis, G.V. (1996) Overview of scheduling and planning of batch process operations. In *Batch Processing Systems Engineering* (G.V. Reklaitis, D.W.T. Rippin, P. Hortacsu and A.K. Sunol, eds), Springer Verlag, NATO ASI Series F.
6. Pritsker, A.A.B., Waters, L.J. and Wolfe, P.M. (1969) Multiproject scheduling with limited resources: a zero-one programming approach. *Management Sci.*, **16**, 93–108.
7. Kondili, E., Pantelides, C.C. and Sargent, R.W.H. (1993) A general algorithm for short-term scheduling of batch operations – I. MILP Formulation. *Comput. Chem. Engng*, **17**, 211–223 (earlier version presented at PSE'88, Sydney, Australia).
8. Shah, N., Pantelides, C.C. and Sargent, R.W.H. (1993) A general algorithm for short-term scheduling of batch operations – II. Computational issues. *Comput. Chem. Engng*, **17**, 224.
9. Pekny, J.F. and Zentner, M.G. (1994) Learning to solve process scheduling problems: the role of rigorous knowledge acquisition frameworks. In *Foundations of Computer Aided Process Operations* (D.W.T. Rippin, J.C. Hale and J.F. Davis, eds), CACHE, Austin, Tx, pp. 275–309.
10. Zentner, M.G., Pekny, J.F., Reklaitis, G.V. and Gupta, J.N.D. (1994) Practical considerations in using model based optimization for the scheduling and planning of batch/semicontinuous processes. *J. Process Control*, special issue on Batch Processing, **5**(4), 259–280.
11. Pantelides, C.C. (1994) Unified frameworks for optimal process planning and scheduling. In *Foundations of Computer Aided Process Operations* (D.W.T. Rippin, J.C. Hale and J.F. Davis, eds), CACHE, Austin, Tx, pp. 253–274.
12. Zentner, M.G., Pekny, J.F., Miller D.L. and Reklaitis, G.V. (1994) RCSP++: a scheduling system for the chemical process industry. In *Proceedings of Process Systems Engineering Symposium*, Kyongju, Korea.
13. SUPERIOR++, Registered trade-mark of Advanced Process Combinatorics, Inc., West Lafayette, IN 47906.
14. Bodington, C.E. (ed.) (1995) *Planning, Scheduling, and Control Integration in the Process Industries*. McGraw-Hill, New York.
15. Petzold, L.R. (1982) A description of DASSL: a differential/algebraic system solver. In *IMACS World Congress*, Montreal, Canada.
16. Joglekar, G.S. and Reklaitis, G.V. (1984) A simulator for batch and semicontinuous processes. *Comput. Chem. Engng.*, **8**, 315–327.
17. Pritsker, A.A.B. (1986) *Introduction to Simulation and SLAM II*. Systems Publishing, W. Lafayette, IN.
18. Schreiber, T. (1974) *Simulation using GPSS*. John Wiley, New York.
19. Barton, P.I. and Pantelides, C.C. (1991) The modelling and simulation of combined discrete/continuous processes. In *International Symposium of Process Systems Engineering*, Montebello, Canada.
20. Czulek, A.J. (1988) An experimental simulator for batch chemical processes. *Comp. Chem. Engng*, **12**, 253–259.
21. BATCHES, registered trademark of Batch Process Technologies, Inc., West Lafayette, IN 47906.
22. Overturf, B., Reklaitis, G.V. and Woods, J.M. (1978) GASP-IV and the simulation of batch/semicontinuous operations: single train process. *I&EC Proc. Des. Dev.*, **17**, 161.
23. Kuriyan, K., Reklaitis, G.V. and Joglekar, G.S. (1987) Multiproduct plant scheduling studies using BOSS. *I&EC Proc. Des. Dev.*, **26**, 1551–1558.
24. Young, R.A. and Reklaitis, G.V. (1989) Capacity expansion study of a batch production line. *I&EC Res.*, **6**, 771–777.
25. Mignon, D.J. (1996) 'Retrofit design and energy integration of brewery operations. In *Batch Processing Systems Engineering* (G.V. Reklaitis, D.W.T. Rippin, P. Hortacsu and A.K. Sunol, eds), Springer Verlag, NATO ASI Series F.

26. Clark, S.M., Harper, P., Kavuri, S. and Joglekar, G.S. (1993) An integrated system for the operation of beverage manufacturing processes. In *Proceedings of FOCAPO-94*, Crested Butte, Colorado.
27. Joglekar, G.S., Clark, S.M., Pekny, J.F. and Reklaitis, G.V. (1995) Towards the integration of simulation and optimization for batch process engineering. In *Workshop on Analysis and Design of Event Driven Operations in Process Systems*, Imperial College, UK.
28. Subrahmanyam, S., Pekny J.F. and Reklaitis, G.V. (1996) Decomposition approaches to batch design and scheduling. *I&EC Res.*, **35**, 1866–1876.

3 Solvents in chemicals production
M. SHEEHAN

3.1 Introduction

Solvents are some of the most widely used and important chemicals, accounting for world-wide sales of almost $US3 billion in 1992 [1]. Since the industrial revolution thousands of different solvents have been synthesized by the chemical industry. Their widespread use as cleaning and drying agents and as reaction media is related to their ability to dissolve and transport solids and other liquids. The role of solvents in the fine chemical industry has been dominated by their use as a reaction medium. They influence many aspects of reactions and other processes through physical and chemical interactions with solutes.

Recently there has been an increase in the understanding of the environmental effects of many of the more common solvents. This has resulted in increased public awareness and increased legislation restricting the use and release of solvents, and has led to much research and discussion on the role solvents play in industry and society. An understanding and exploitation of ways to reduce the detrimental effects of solvents, and ways to deliver the desirable effects solvents play in physical and chemical processes is critical to quality of life and the future of industries that rely on solvents.

There are many aspects to the uses and life cycles of solvents in the fine chemicals industry. The basis of solvent selection is currently being modified by increased understanding of the nature and potential effects of solvents. The use of any given solvent in a process can no longer be based purely on precedent. However, complementing the developing list of criteria that may restrict use of a solvent on health, safety and environmental grounds is growing knowledge in the area of solvent selection that can help to find a solvent that may be particularly suitable for a specific purpose. The decision-making process for choosing a suitable reaction solvent must consider both the chemical effects on the reaction and reactants themselves as well as the recovery and recycle of the solvent back into the process. This chapter focuses on the physico-chemical behaviour of solvents that influences both the reaction outcome and the subsequent recovery of solvents.

Section 3.2 characterizes the use of solvents in terms of physical and chemical properties as well as briefly introducing the environmental and health and safety concerns that are associated with the use of solvents. Section 3.3 looks at the solvation process with an emphasis placed on intermolecular forces between solvents and solutes. As well as highlighting the energetics of the solvation process, this section demonstrates the importance of intermolecular forces through a description of different types of solvation process and their characteristic behaviour. Section 3.4 draws on this understanding of the solvation process and the intermolecular forces between molecules to explain the effects of solvents on reactions. The approach here is primarily qualitative due to the lack of quantitative, predictive rules. The energy transfers that occur during reactions in solution are discussed and illustrated with examples from single-phase and two-phase systems.

The final two sections are process oriented, concentrating on recovery of solvents and destruction of solvent-laden wastes. Although predominantly focusing on specific techniques for recovery of solvents, the relationship between the choice of solvent, its chemical behaviour in a reacting system and the options for recovery are demonstrated. Techniques for the treatment and destruction of solvent laden wastes and some potential problems are discussed in Section 3.6.

The main objective of this chapter is to present an understanding of some features affecting the choice of solvents for use as reaction media in organic reactions and the various other factors (economic, environmental, health, safety) that affect this choice.

3.2 Solvent properties

Industrial solvents encompass a wide range of compounds with widely varying behaviour and properties. To classify all solvents unambiguously into discrete groups is an impossible undertaking. However, solvents can be classified according to specific properties from their chemical and physical characteristics, as well as environmental and health and safety considerations. Of course, many of these classifications are interconnected and overlapping.

3.2.1 Chemical properties

Solvents can be broadly classified according to their chemical bonding. As well as organic solvents, water and supercritical carbon dioxide, which are primarily covalently bonded, there are ionic and metallic solvents. The consideration of metallic solvents is outside the scope of this chapter and ionic solvents will be considered only briefly.

Molten salts are ionic solvents that can be used to dissolve salts and metals. They are becoming increasingly important in both organic and inorganic reactions. Their physical properties, such as high thermal conductivity, permit rapid dispersal of heat in a reacting system. Other properties such as high thermal stability, low viscosity, low vapour pressure and wide liquid range make them very useful as reaction media. Ionic solvents behave as strong dissociating agents due to their ability to exchange with solute ions of the same charge. An example of the use of these solvents in organic reactions is the synthesis of vinyl chloride from acetylene and 1,1-dichloroethane using $ZnCl_2/KCl/HgCl_2$.

Solvents have also been classified according to their specific and non-specific solute–solvent intermolecular interactions, such as hydrogen bonding abilities and dipolarity. A commonly used classification involves three main groups: dipolar aprotic (possessing a dipole but without hydrogen bonding ability), protic (with hydrogen bonding ability) and non-polar aprotic solvents. Solvents selection is also influenced by the reactions that may occur between solvents and solutes, e.g. hydrolysis reactions should not be carried out in carboxylic esters, amides or nitriles.

3.2.2 Physical properties

Physical properties such as boiling point, melting point, density, viscosity (and many others) are important descriptors of the processing behaviour of materials. In what follows, the properties related to solubility, solvation, suitability as reaction media, solvent recoverability as well as safety and environmental matters are emphasized.

(a) Phase transition properties

Phase transitional properties include the boiling point (T_{boil}), freezing point (T_{freeze}) and more recently, with the emerging use of supercritical fluids, the critical point (T_{crit} and P_{crit}). When choosing a solvent as a reaction medium it is advisable to have as large a possible working range between the melting and boiling points, as well as thermal stability over this range. Phase transition parameters are useful as general guidelines for choosing solvents that will be easy to recover by distillation (favoured by a high boiling point difference) and choosing solvents that will not restrict the choice of reaction conditions (a wide operating temperature and pressure range). The freezing point is important for consideration of cooling crystallization, the avoidance of blockage by solidified solvents, etc.

(b) Rheological properties

Properties such as the viscosity (μ) density (ρ), liquid expansion coefficient and the surface tension are important ways of determining the transport characteristics of solvents and predicting the phase characteristics of the solvated reaction system. As well as being critical to the design of mixing equipment, they are important to choosing and designing solvent recovery operations. Good partitioning between immiscible phases is favoured by low viscosity, large density differences and sufficient interfacial tension.

The density and thermal expansion coefficient of a solvent are relevant to design of storage tanks, pipelines and process vessels.

(c) Thermodynamic parameters

There are many useful thermodynamic properties that describe solvents, including the latent heat of vaporization (ΔH_{vap}), fusion (ΔH_{fus}) and the specific heat capacity (c_p). These can be related to both the chemical and physical performance of solvents. The very common use of distillation for solvent recovery highlights the importance of terms such as the specific heat (c_p) and latent heat of vaporization. The specific heat and to some extent the thermal conductivity (k) of solvents are indicators of the effectiveness of using solvents as a reaction thermal control medium. The heat of fusion (ΔH_{fus}) can be of use when considering crystallization processes for reactor design or recovery units. Finally, the heat of combustion is of interest when no recovery of the solvent is possible and incineration or use as fuel in cement-kilns are the chosen methods of solvent disposal. An excellent collection of the properties of solvents is given by Smallwood [2].

The latent heat of vaporization is a measure of the degree of association between molecules and is used in the Hilderbrand solubility parameter (Section 3.2.2.e).

(d) Phase equilibrium

Parameters that characterize the equilibrium distribution of solvents and solutes between immiscible phases are important to choosing solvents for effective solvation, characterizing two-phase reaction processes and the designing of recovery operations such as distillation and extraction processes. For gas–liquid systems the important parameters include pure component vapour pressure data and vapour–liquid equilibrium (VLE) data for mixtures, including azeotropic properties. (An azeotrope is a mixture whose vapour at the boiling point has the same composition as the liquid, and thus cannot be separated by simple distillation.) Such data are critical to the design of distillations. One common way of presenting VLE data is through the relative volatilities (α_{ij}) of components in a mixture. Many more complex representations of VLE relationships are available [2–4], of which the non-random two liquid equation (NRTL), the universal quasi-chemical equation (UNIQUAC) and the Wilson equations are among the more popular. Tabulated parameters for a variety of binary (and more complex mixtures) are available, and by insertion into the equations can give a good representation of VLE. There are also methods available for the *a priori* estimation of equilibrium data such as the group contribution method (UNIFAC).

Equally important phase equilibrium properties include liquid–liquid miscibility and solubility. These can help in process design (single phase or two phase). Liquid miscibility is further discussed in Chapter 4 (Section 4.5.1).

(e) Intermolecular interaction parameters
The parameters that characterize solvents in terms of intermolecular forces can be considered the most important and fundamental of all solvent parameters. Intermolecular forces determine a solvent's solvation ability as well as the magnitude of any effect the solvent may have on reaction kinetics, mechanism and equilibria. They are also linked to the possible recovery of solvents – indeed the properties already discussed (phase transition and equilibrium) arise from intermolecular interactions. A more detailed examination of individual parameters that characterize aspects of intermolecular interactions is given below.

The *dipole moment* (m) is a molecular property that can be either permanent or induced through molecular movements or electric fields. Molecules which possess a centre of symmetry in all possible conformations cannot possess a permanent dipole. However, most common organic solvents possess dipole moments. In the absence of specific solvent–solute interactions, the dipole moment dictates the orientation of the solvent about the solute. This can be of great importance in the determination of reaction mechanisms and hence reaction rates and selectivity changes. Although of some qualitative use, it is not a quantitative measure of the solvating ability of a solvent as it does not describe either the magnitude or the accessibility of electric charge. However, when considered alongside the dielectric constant it is a more effective measure of solvent polarity.

The *dielectric constant* (e) represents the ability of a solvent to separate charge and orient its dipoles (either induced or permanent). Unfortunately, since the ability to orient solvent dipoles can be more restricted in the solvent shell surrounding the solute than in the bulk solvent, the dielectric constant has little success in the quantitative prediction of effects on reaction rates. However, it has been used as a general indicator of the solvation ability and chemical reactivity of inorganic salts. It is particularly useful as an indicator for ion solvation because this process involves the separation of ions.

A solvent's *refractive index* (n) has an important relationship to its optical polarizability. Dispersion forces are forces of molecular interaction related to polarizability and it is observed that high refractive indices usually correspond to high dispersion forces. Polarizability also leads to a good solvation ability for large, polarizable anions. All aromatic compounds possess high refractive indices.

An important form of interaction in some solutions is that arising from the interaction of an electron pair donor and an electron pair acceptor. The *donor number* (DN) developed by Gutmann [5] is a quantitative measure of the solvent's ability to act as an electron pair donor. Similarly, it is used as a measure (semi-quantitative) of the nucleophilicity of electron pair donor (EPD) solvents. The higher the solvent donor number, the stronger the interaction between solvent and acceptor. The formal definition of the donor number is the negative enthalpy of formation of the 1:1 adduct between the solvent, S, and antimony pentachloride in 1,2-dichloroethane (as shown in equation 3.1).

$$S + SbCl_5 \rightarrow S.SbCl_5 \qquad (3.1)$$

The donor number has been proven useful in co-ordination chemistry where it has been correlated [6] with the thermodynamic, kinetic and spectroscopic behaviour of reacting systems. As might be expected, the donor number is a good measure of the solvating ability towards Lewis acids. Some noteworthy solvents with high donor numbers are the amide solvents and dimethylsulphoxide (DMSO).

Complementary to the donor number, the *acceptor number* (AC) is related to the hydrogen bonding ability of solvents. The acceptor number is a measure of the ability of a solvent to accept an electron pair. They are derived from the P-NMR chemical shifts of the solvent in triethylphosphine oxide. The greater the P-NMR shift of the solvent, the greater the acceptor number. Acceptor numbers give some measure of the solvent ability to stabilize anions as well as being a scale on which to compare the protic behaviour of different solvents.

The *Hilderbrand solubility parameter* arises from the idea that the attraction between the solute and solvent can be partially quantified by looking at the energy needed to separate the solvent molecules to create a cavity for the solute. This is termed the cohesive energy density (CED), defined as the energy of vaporization per unit volume ($\Delta U/V_m$). Observing its relationship to the mutual solubility of components, Hilderbrand termed the square root of the CED the solubility parameter, δ.

$$\delta = \sqrt{\Delta U/V_m} = \sqrt{(\Delta H_{vap} - RT)/V_m} \qquad (3.2)$$

The solubility parameter gives a good indication whether the solvent will strongly solvate any given non-electrolyte solute. Although limited, it is also possible to use δ to predict effects on reaction rates. For example, a correlation has been found between δ and the rate constants for the dimerization of cyclopentadiene and the dissociation of hexaphenylethane in various solvents [7].

Unfortunately δ is only of use (on its own) for reactions between neutral non-polar molecules in non-polar solvents. However, solvent polarity reflects the ability of the solvent to interact with the solute (to some extent) and the CED is the energy required to create a hole in the solvent into which the solute can accommodate. Thus these two terms can be considered complementary and reaction kinetics will be related to the values of both terms.

The most popular, comprehensive and possibly most useful empirical measure of *solvent polarity* has been the *Et(30)* scale developed by Dimroth and coworkers [8]. Et(30) values are defined as the transition energy of the dissolved betaine dye (29a) in a given solvent and are strongly dependent on the electrophilic solvation power of the solvent in question. Hence, Et(30) values are highest for protic solvents. The colour of the betaine (29a) dye depends on the solvent in which it is dissolved, and because of the polarity difference between its ground state and excited state, this colour is dependent on the solvent polarity. The betaine (29a) dye shows a large change in its solvatochromic adsorption band (greater than 350 nm, 28 kcal mol^{-1} from diphenylether to water) making it very sensitive to solvent polarity. It has been found that there are good correlations between the Et(30) scale and other empirical solvent polarity parameters. A list of the above parameters for some common solvents is given in Table 3.1.

(f) Environmental considerations

The fine chemicals industry is one of the largest consumers of volatile organic compounds (VOCs) and other potentially environmentally damaging compounds. The difficulty in managing these compounds arises from their volatility and the complex mixtures with other materials often used in processing. Often, large quantities of solvent laden waste are created and sent to be reprocessed or destroyed. The potentially serious effect on the environment caused by solvent wastes has led to tighter regulations, designed to increase responsibility and minimize pollution. Due to this new legislation, public pressure and changes in attitude within the process industries, there has been a move towards the recovery and reuse of solvents, and the phasing out of environmentally troublesome materials.

As well as some potential impacts to ground water, surface water and soil, the major concerns currently with solvents involve their contributions to the

Table 3.1 Physical properties of some important organic solvents, arranged by decreasing values of Et(30) [8, 9]

Solvent	$m \times 10^{30}$ C m	e at 25°C	n_D^{20}	BP (°C)	Et(30) (kcal mol^{-1})
Water	6.07	78.39	1.3330	100	63.1
Formamide	11.24	111.0 (20°C)	1.4475	210.5	56.6
MeOH	5.67	32.70	1.3284	64.7	55.5
Ethanol	5.77	24.55	1.3614	78.3	51.9
Acetic acid	5.6	6.15 (20°C)	1.3719	117.9	51.2
DMSO	13.0	46.68	1.4783	189.0	45.0
DMF	12.88	37.0	1.4269 (25°C)	152.3	43.8
Acetone	9.54	20.70	1.3587	56.3	42.2
Nitrobenzene	13.44	34.82	1.55 (25°C)	210.8	42.0
Pyridine	7.91	12.4	1.5102	115.3	40.2
Ethylacetate	6.27	6.02	1.3724	77.1	38.1
CS$_2$	0.0	2.64 (20°C)	1.6280	46.2	32.6
CCl$_4$	0.0	2.24 (20°C)	1.4574 (25°C)	76.8	32.5
Cyclohexane	0.0	2.02 (20°C)	1.4262	80.7	31.2

load of VOCs released to the atmosphere. Solvents contributed approximately 40% of total VOCs in Western Europe in 1990 [2]. The potential effects include toxic effects on plants and animals, the formation of ground-level ozone, contribution to global warming, and for some materials the depletion of stratospheric ozone. The role VOCs play in the creation of photochemical smog and acid deposition has been known since the 1930s. When VOCs mix with nitrous oxides (NO_x) and are irradiated by UV light a chain of reactions occur. This leads to the production of many photochemical pollutants such as ozone, peroxyacetyl nitrate (PAN), organic and inorganic acids. The production of ground-level ozone is considered as one of the most critical problems facing many of the world's industrialized countries. This is because of its high recorded concentration in the atmosphere (some cities much higher than 200 p.p.b.v.) [10], and its detrimental effects on health (respiratory problems), plant life (reduced biomass, growth and yield of vegetation), materials (attacks elastomers, fibres and dyes) and climate (global warming).

The creation of ozone by volatilized VOCs has been expressed in terms of photochemical ozone creation potential units (POCP). The POCPs for many organic compounds are given in Table 3.2 [10], calculated with respect to the creation potential of ethylene (POCP = 100). Photochemical PAN creation potentials (PPCP) are also available for many VOCs. A comprehensive guide to reducing environmental impact (photochemical reactivity) can be found in the Los Angeles Rule 66 [11].

The two major concerns over changes in the climate of the earth – global warming and ozone layer destruction – can both be linked to the emission of commonly used solvents. Of the three main types of solvents (hydrocarbons, chlorohydrocarbons and oxygenates), chlorinated solvents are under the greatest pressure for replacement.

Compounds such as CO_2, CH_4, N_2O, ozone and fluorinated hydrocarbons (HFCs) all contribute to global warming. It has been suggested [10] that halogenated VOCs are likely to be the second most important source of thermal trapping after CO_2. This is due to the their strong ability to absorb IR

Table 3.2 Photochemical ozone creation potentials (POCP) of common VOCs calculated for the British Isles [10, 12].

VOC	POCP	VOC	POCP	VOC	POCP
Methane	1	Propylene	105	Formaldehyde	40
Ethane	10	1,3-Butadiene	105	Acetaldehyde	55
n-Hexane	50	Acetylene	15	Acetone	20
Cyclopentane	50	Benzene	20	Methanol	10
Mecyclopentane	50	Toluene	55	Ethanol	25
Ethylene	100	Ethylbenzene	60	Methylene chloride	1
1-Butene	95	m-Xylene	105	Chloroform	1
2-Butene	100	1,2,4-TriMebenz	120	Methyl chloroform	0

radiation. In 1993 they are believed to have accounted for 20% of the observed change in radiative forcing.

The problem of ozone depletion is being addressed through bans on the production and dispersive use of a range of halogenated materials. Two solvents that are responsible for a significant proportion of the ozone destruction are 1,1,1-trichloroethane and 1,1,2-trichlorotrifluoroethane (CFC 113) with ozone depletion potentials (ODPs) of 0.15 and 0.8, respectively. Both these solvents had a 1 January 1996 production and import ban under the Montreal Protocol. Other banned materials include halons and carbon tetrachloride. It is likely that the bans will extend to include other materials, and already perfluorinated materials are under pressure because of their high Global Warming Potentials [12].

Many of the materials produced and used by the chemical industry have had limited investigation into their long-term ecological and biological effects. Often chemicals that are initially considered safe are later investigated and found to be hazardous (e.g. DDT, asbestos). This delay can be due either to a long induction period for the effect to be manifested in an organism or because the material gradually accumulates in organisms until it reaches a harmful level. The potential for long-term environmental harm is now an important consideration in process assessment. It is important to note that concerns of this type continue to be raised. For example, grave concerns [13] have been expressed over oestrogen mimicking compounds such as PCBs, phthalates and bisphenol-A. These compounds have been found to be biological accumulators and it has been suggested they can increase the risk of testicular diseases and infertility in males and breast cancer in females.

With many traditional solvents having poor environmental records, large numbers of solvents need to be and are being targeted for replacement [1]. Solvents such as ethylene glycol ethers and hazardous air pollutants such as methyl ethyl ketones and methyl isobutyl ketones are slowly being phased out. Paralleling this decline in destructive solvents is a growth in other, friendlier solvents. These include solvents such as esters, hybrid aqueous/non-aqueous solvents, and blends such as oxygenates and terpenes. More selective, niche solvents are being favoured over multipurpose solvents such as the highly volatile acetone.

Water is anticipated to be the solvent of the future and with the gains in surfactant and emulsion research, aqueous solutions are steadily becoming more effective. Unfortunately this move is more prevalent in the cleaning of equipment and the cleaning industries than the fine chemical industry. Cleaning of vessels involves a substantial use of organic solvent and chemical cleaning. This can often be replaced by an aqueous solvent and increased mechanical cleaning. Water, however, has numerous drawbacks as a reaction solvent and may of today's synthetic reactions simply cannot be carried out in the presence of water.

In assessing the potential for environmental impact of processes, a structured approach is important, as discussed in Chapter 1.

(g) Health and safety parameters

Safety issues arising from the use of solvents include flammability and toxic effects. Flammability issues are dealt with at length in Chapter 9. The problems of toxicity are dealt with briefly below, though for a more comprehensive treatment, the reader should refer to other works [14, 15].

The volatility and widespread use of solvents means that human exposure is likely to occur during chemical process operations. The groups potentially exposed include process operators, other staff at operating sites and those living or working nearby. The potential toxic effects are many and varied – ranging from acute toxicity to health effects brought on by long-term exposure. Clearly, no significant health risk is tolerable, so the emphasis in handling of solvents must be the control of exposure to reduce risk to an acceptable level.

In determining suitable measures to control exposure, there are a number of issues that must be addressed. The operations involved must be defined in order to identify those likely to be exposed to the chemicals, as well as the exposure route (i.e. inhalation, skin contact or ingestion). The latter is important because it relates to possible control methods, as well as allowing a clearer definition of the health hazard. The potential for exposure can then be assessed and compared with data on tolerable exposure levels. In the UK, one such source of data is the HSE's EH40 list, which gives allowable exposure levels for a wide range of compounds [15]. Some data for common solvents are reproduced in Table 3.3. Once the potential for exposure and associated health hazards are identified, control measures

Table 3.3 Occupational exposure standards (p.p.m.) for common solvents (OELs: Criteria Document Summaries, HMSO, London, 1992)

Solvent	8 h time-weighted average	10 min reference period
Acetone	750	1500
Benzene	5	
Butan-1-ol		50 (notation skin)
2-Butoxy ethanol	25 (notation skin)	
Carbon tetrachloride	2 (notation skin)	
Chlorobenzene	50	
2-Chloro ethanol		1 (notation skin)
1,2-Dichlorobenzene		50
2-Ethoxy ethanol	10 (notation skin)	
n-Hexane	20	
2-Methoxyethyl acetate	5 (notation skin)	
Propane-1,2-diol	150 (vapour and particulate)	
Toluene	50	150 (notation skin)
Trichloroethylene	100	150 (notation skin)
Xylene	100	150 (notation skin)

can be defined. These can take the form of protective clothing, extraction systems to remove fumes, containment, etc.

The legislation applicable to control of solvent-related and other health risks will vary from country to country. A model for identification and control of such risks is the UK COSHH (Control of Substances Hazardous to Health) Regulations [16].

3.3 Solubility

The oldest known rule for the solubility of compounds is that 'like dissolves like'. This approximation holds true for many systems but is not absolute. The interactions that contribute to solubility of one compound in another are many and varied, and will be discussed in some detail in this section.

The process of solvation refers to the surrounding of each dissolved molecule by a shell of tightly bound solvent molecules as shown in Figure 3.1. During a chemical reaction the consecutive solvation of reactants, activated states and products of reaction take place.

The solvent shell is a result of intermolecular forces of attraction and repulsion between the solvent and solute. There are three important aspects characterizing this solvation process and the solvent shell. These are the co-ordination number and solvation numbers, the fine structure of the solvent shell, and the interactions between the outer shell and the bulk solvent.

The co-ordination number is the number of solvent molecules in close contact or within bonding distance to the solute such that no other solvent

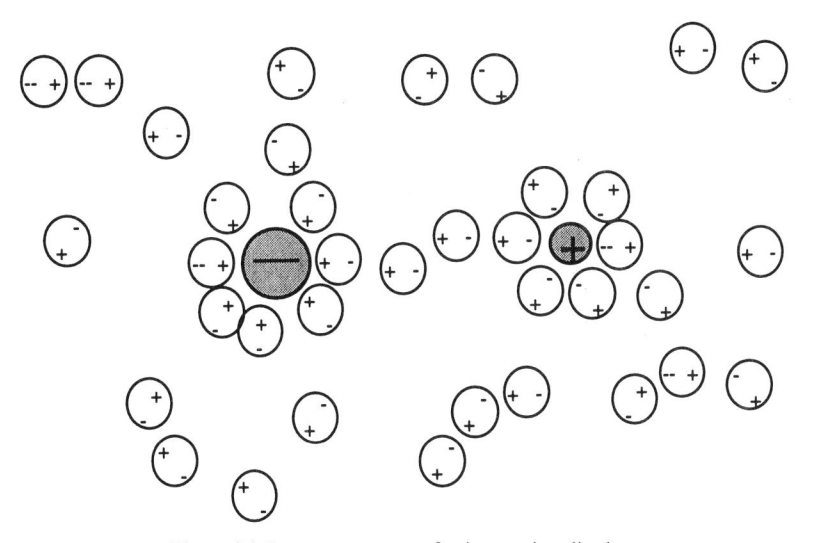

Figure 3.1 Low energy states for interacting dipoles.

molecules lie within this shell. The solvent number is the number of molecules or ions that remain attached to the given solute long enough to experience translational movements with the solute. The solvent shell itself is a dynamic quantity, and as such there is significant exchange and interchange of the solvent shell molecules. The internal solvent shell described by the solvation number and the co-ordination number are often referred to as primary or chemical solvation. The outer shell consisting of partially ordered solvent molecules and the bulk solvent is often termed secondary or physical solvation.

It is important to understand the structure of the solvent shell in a qualitative way to manipulate the local environment for the activated state, and thus manipulate equilibrium and kinetic properties for reaction systems. The fine structure of the solvent shell is unique to various solvents and is most significant when water, a highly ordered solvent, is used. This structuring of the solvent around the solute can have undesired effects on reaction kinetics. For ion solvation, in which more dilute solutions exhibit greater solvation, there exist two different models for this ordering process. In non-ordered solvents such as hydrocarbons there is an internal structure dictated by the shell steadily decreasing in structure towards the disordered bulk solvent. In highly ordered solvents, such as water, there exists a structured region around the ion, then an unstructured and highly mobile region surrounded by a differently but again highly structured water region around this.

The solvent structure and the solvation ability of a particular solvent for a solute can be related to the free energy of solvation, ΔG°_{solv}. This arises cumulatively from the following energy terms.

(i) Cavitation energy to create the hole in the solvent occupied by the solute.
(ii) Orientation energy of the solvent around the solute.
(iii) Isotropic interaction energy arising from long-range intermolecular forces such as electrostatic and polarization forces.
(iv) Anisotropic interaction energy forces arising from specific interactions such as hydrogen bonding and electron proton donor and acceptor bonding.

In order to dissolve a solid, both the interaction energy between the solid molecules and between the solvent molecules must be overcome by the interaction energy of solute and solvent.

When considering the interactions between solvents and solutes, two broad categories of intermolecular forces arise. These are the non-specific, directional, induction and dispersion forces such as those arising from dipole moments and the specific, directional forces which include hydrogen bonding and charge transfer. An understanding of these forces is key to an understanding of the effect of solvents on reaction kinetics, as the individual solvation of each compound contributes to the free energy profile that the reaction follows. The following sections describe the important types of molecular interaction.

3.3.1 Non-specific intermolecular forces

Non-specific forces are forces that cannot be completely saturated, and are electrostatic interactions which act to orientate molecules in the electric field created by an ion, by a dipole moment or by an induced dipole moment. These are of particular importance to solvation as most common organic solvents possess dipole moments of up to 18.5×10^{-30} C m. Only a few hydrocarbons and some symmetric compounds have no permanent dipole moments.

(a) Ion–dipole force
A molecule with a permanent dipole moment will orient itself in the field created by a neighbouring ion (Figure 3.1). This orientation energy is inversely proportional to the distance squared between the ion and the centre of the dipole.

(b) Dipole–dipole force
This important and relatively common force occurs when two neighbouring dipolar molecules orientate themselves in a minimum energy position. This is usually the end-to-end arrangement as shown Figure 3.2, but when the dipoles are extremely voluminous the minimum energy position becomes the side-by-side arrangement. Dipole orientations leading to attraction are statistically favoured and are also temperature dependent. Dipole–dipole interactions can only exist when the attractive energy is larger than the thermal energy.

(c) Dipole-induced dipole force
The dipole moment of molecule has the ability to induce a dipole moment in a neighbouring apolar molecule. When this occurs, the induced dipole will tend to align itself in the direction of the dipolar molecule. This temperature-independent effect will be larger for molecules with high polarizability. There will also be an induced dipole when an ion neighbours an apolar

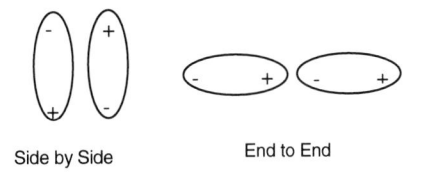

Figure 3.2 Orientation of dipoles in the field created by an ion – side-by-side and end-to-end configurations.

molecule. These induced dipole effects are limited to mixtures of dipolar or ionic compounds and non-polar molecules.

(d) Dispersion force

The continuous movement of molecules that possess no permanent dipole moment results in the creation of a small, fluctuating dipole moment which can in turn induce a dipole moment in a neighbouring apolar molecule. This effect is termed a dispersion force. As well as being universal, this is the most important force in the aggregation of molecules with neither free charges nor dipole moments. Dispersion forces increase rapidly with molecular volume and the number of polarizable electrons. Although dispersion forces are extremely short ranged (being proportional to r^{-6}), solvents with large refractive indices (high optical polarizability) are capable of exhibiting strong dispersion forces. Due to the high polarizability of p-electrons, strong dispersion forces exist between molecules with conjugated p-electron systems such as aromatics.

3.3.2 Specific intermolecular forces

Specific intermolecular forces are those directional forces that can be saturated. In the case of hydrogen bonding and electron pair donor–acceptor interactions they can be very strong in comparison to the non-specific forces of attraction.

(a) Hydrogen bonding

When a covalently bound hydrogen atom forms a second bond to another atom this bond is called a hydrogen bond. The importance of the hydrogen bond has given rise to the classification of compounds containing hydrogen bonding proton donor groups as protic and those without as aprotic. The hydrogen bond is to either an electronegative, different molecule (hetero-intermolecular) or between the same molecules (homo-intermolecular) such as is the case with water, alcohols carboxylic acids and amides. The strength of a hydrogen bond is in the order of 10 times stronger than other non-specific interactions. Hydrogen bonding plays a particularly important role in the interactions between protic solvents and anions. As a result of the small size of the hydrogen atom, small anions are better solvated by protic solvents than larger anions.

(b) Electron pair donor–acceptor interactions

An additional bonding interaction between two valency saturated molecules that involves an electron transfer from the donor to the acceptor molecule is termed an electron pair donor–acceptor (EPD/EPA) interaction. These interaction energies can be very strong (up to $-188\,\text{kJ}\,\text{mol}^{-1}$) and are usually larger than dispersion forces. The necessary condition for the formation of this bond is the presence of an occupied high energy orbital in the EPD molecule and a sufficiently low energy unoccupied orbital in the EPA molecule. This dependence on the orbital type allows further classification of EPD/EPA structures into nine distinct groups. EPD and EPA interactions have also been likened to Lewis acid–base interactions.

Solvents can behave as a donor and acceptor simultaneously. Water, for example, can act as a donor through the oxygen atom as well as an acceptor by forming hydrogen bonds. This gives an indication as to the complicated and ordered structure and behaviour of water as a solvent. The importance of EPD and EPA interactions has led to the development of the scales mentioned in Section 3.2.2 to quantify the strength of various solvents as EPA and EPD.

(c) Solvophobic interactions
When hydrocarbons are dissolved into water there is an increase in the free energy of the system. There is a decrease in the entropy of the system, a reflection of the increase in the order of the system and water's highly ordered structure. However, when two different saturated hydrocarbons come into close contact there can be a partial reconstruction of the original water structure (Figure 3.3). This reconstruction increases the entropy of the hydrocarbon–water interactions and the free energy decreases. This effect is known as a hydrophobic interaction but can be generalized to other solvents such as ethanol and glycerol (solvophobic interaction).

3.3.3 Energy of solvation

The effect of a change in solvent on the reaction kinetics is directly related to changes in the free energy profile of the reaction brought about by different energies of solvation (of reactants, intermediates and products) in different solvents. This observation emphasizes the importance of the energy of solvation in any discussions on solvent effects on reaction rates.

The total solvation energy is extremely sensitive to the size of ion and thus the charge density. The smaller the ion the greater the charge density and the more highly solvated it is. It is found that the smaller anions show a much

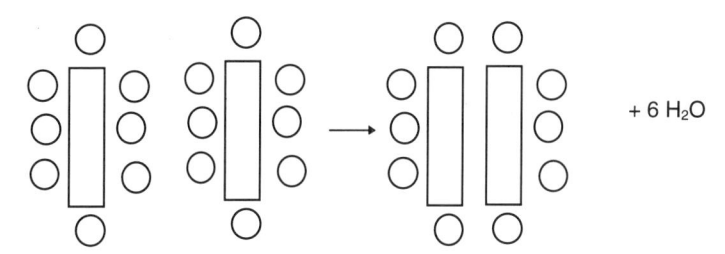

Figure 3.3 Solvophobic interactions.

bigger increase in the free energy than large ions when transferred from protic solvents (i.e. H$_2$O, MeOH) to aprotic solvents (i.e. DMF, Me$_2$SO$_2$, MeCN). While high free energy is linked to high activity, it is important to remember that high free energy also leads to low solubility.

Cations show more complicated behaviour due to the specific bonding and ligand field effects. Cations tend to be more strongly solvated by Lewis bases (i.e. DMF, Me$_2$SO$_2$) which is in turn demonstrated by the high donor numbers for these solvents.

Non-electrolytes show little changes in free energy behaviour between various non-aqueous solvents but tend to exhibit lower solubility in aqueous solvents.

3.3.4 Selective solvation

It has been found in binary solvent mixtures that the ratio of solvent molecules in the solvent shell can be different to the bulk solution as the solute surrounds itself with the component or components leading to a more negative free energy of solvation. This is termed selective solvation and is divided up into homoselective and heteroselective aspects.

In homoselective solvation, both ions of a binary salt are preferentially solvated by the same solvent as can be seen in Figure 3.4. The solvation of calcium chloride (Ca^{2+}, Cl$^-$) is largely by water in a water–methanol solvent mixture. In heteroselective solvation (Figure 3.4) the anion is solvated by one of the solvents and the cation is solvated by the other such as the solvation of silver nitrate in an acetonitrile–water mixture (NO$_3^-$ solvated by water, Ag$^+$ by acetonitrile).

A number of experimental methods are available to examine these effects including conductance and Hittorf transference measurements and spectroscopic techniques such as NMR [17], UV/Vis [18] and IR [19].

3.3.5 Solvent mixtures

In reacting systems it is often possible and advantageous to use a mixture of solvents. Chen [20] developed a method to determine whether or not a solute

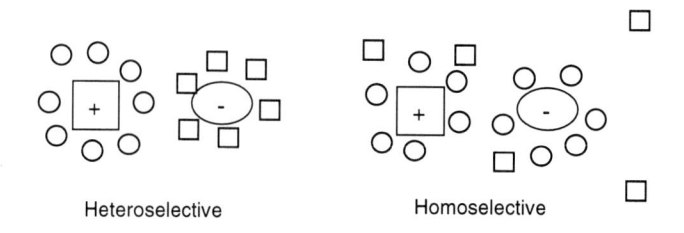

Figure 3.4 Homo- and heteroselective solvation in a binary solvent mixture.

will be miscible in a mixture of solvents. This was by assuming the lattice theory is adequate and taking a 'single liquid' [21] approach to mixtures.

It was found using binary mixtures and optimizing for the lowest interaction between solute and solvent mixture that a mixed solvent may have a stronger solvating power that either solvent on its own. It was found that although solvents 1 and 2 may be non-solvents, a mixture of the two could be a solvent for the solute in question. Similarly a mixture of 1 and 2 may have a higher solvating power than solvent 1 alone although 2 may be a non-solvent. An example of this is a mixture of 80% acetone and 20% hexane which dissolves polymethylmethacrylate at 25°C better than acetone alone, although hexane is a non-solvent.

Care must be taken with mixtures of components that may individually solvate a solute but may be unsuitable as a mixture, as in the case of the dissolution of cellulose acetate in which dissolution is possible in both aniline and acetic acid but not in the mixture of the two solvents. This type of behaviour is common to macromolecular compounds where certain parts of the large molecule are selectively solvated.

3.3.6 *Supercritical solvents*

Recently there has been an increase in the research of the solvation process of supercritical (SC) solvents and the use of supercritical solvents as reaction media. Central to the solute–solvent interactions of SC solvents is the idea of clustering about the solute molecule with respect to the bulk solvent conditions. This clustering about the solute molecule is a common observation, and very large and negative partial molar volumes, typically two orders of magnitude larger than in the bulk solution, have been found about organic solutes in SC solvents. The use of such solvents can enhance solubilization, leading to improved reaction kinetics in some systems [22].

Many interesting differences exist between liquid solvents and SC solvents. These can be advantageous to both reaction kinetics and mechanisms and solvation of the solute in question. SC fluids have a density of about 30% of normal liquids enabling good solvation capabilities and yet low enough for high diffusivity and rapid mass transfer. Some difficulties may arise due to the thermal properties of such systems, e.g. with highly exothermic reactions.

3.4 Solvent effects on reactions

The underlying assumptions leading to description of solvent effects on reactions are that the mechanism passes through at least one intermediate state (the activated or transition state) which subsequently transforms into the reaction products.

3.4.1 Rules of thumb – Hughes–Ingold rules

Hughes and Ingold [7, 23], looking at pure electrostatic interactions between dipolar molecules and solvents in initial and transitional states, summarized the effect of solvents on reactions of different charge types. This theory is of broad importance since it is possible to classify all elimination and substitution reactions as involving different charge types. Reasonable assumptions underlie the Hughes–Ingold rules:

(i) An increase in the magnitude of charge will increase the solvation of the solute.
(ii) An increase in the dispersion of the charge will decrease the solvation of the solute.
(iii) A destruction of charge will decrease the solvation of the solute more than an increase in the dispersion of the charge.

Based on these assumptions, the following rules for the effect of solvents on reactions with different charge types were developed:

(i) For reactions with greater charge density in the activated state than in the initial reactants, an increase in the solvent polarity will result in an increase in reaction rate.
(ii) For reactions with lower charge density in the activated state than in the initial reactants, an increase in the solvent polarity will result in a decrease in reaction rate.
(iii) For reactions involving little or no change in the charge density on going from the reactants to the activated state, a change in solvent polarity will result in little change in the reaction rate.

Thus, a change to a more polar solvent will result in an increase or decrease in the reaction rate depending on whether the transitional state is more or less dipolar than the reactants. This emphasizes the importance of solvent polarity on the prediction of solvent effects on reactions.

As a consequence of the complicated and wide range of solvent/solute interactions, there is also a possibility of a change in the reaction mechanism. This is illustrated by looking at the difference in the charge dispersal of elimination (E) and nucleophilic substitution (SN) reactions. For example, in E_2 and SN_2 reactions the charge is spread over more atoms and thus less dense than in E_1 and SN_1 reactions. Hence an increase in the solvent polarity will favour the E_1 and SN_1 mechanisms. Since the stereochemical outcome of a reaction is strongly dependent on the reaction mechanism, the Hughes–Ingold rules can be applied to predict the product distribution of stereospecific reactions.

(a) Limitations of the Hughes–Ingold rules

One of the limitations of the Hughes–Ingold rules is the treatment of the solvent as a dielectric continuum, characterized by its dipole moment and dielectric constant. This ignores the importance of other intermolecular forces such as hydrogen bonding and EPD/EPA complexing. However, a thorough understanding of the reaction mechanism and analysis of the expected solvent interactions will aid prediction in a qualitative way.

A further limitation arises from the assumption that the effect of entropy changes on the free energy of activation are negligible. In most reactions this assumption is reasonable, although in reactions such as the decomposition of tert-butyl peroxide, in which the decrease in entropy due to the highly ordered solvation of the transition state counteracts the effect of increased solvation of this state. This is of particular importance in highly ordered solvents such as water (see Section 3.3.2).

3.4.2 The transition state approach

The theory of absolute reaction rates, also known as the transition state theory, provides an effective means of evaluating how a change in solvent may influence a reaction. The rate of a chemical reaction is dependent on the free energy profile for a reaction system and in particular, provided that the transition state actually becomes product, on the free energy of activation, $\Delta G^{\#}$. The free energy of activation is defined as the difference between the free energies of the reacting species and the free energy of the transition state as seen in the following diagram (Figure 3.5).

The change in free energy (ΔG) is related to both the change in enthalpy (ΔH) and the change in entropy (ΔS) as shown in equation (3.3).

$$\Delta G = \Delta H - T\Delta S \qquad (3.3)$$

All other things being equal, increasing the solvation of the transition state will decrease the free energy of activation, increasing the reaction rate. Similarly, increasing the solvation of the reactants will reduce their free

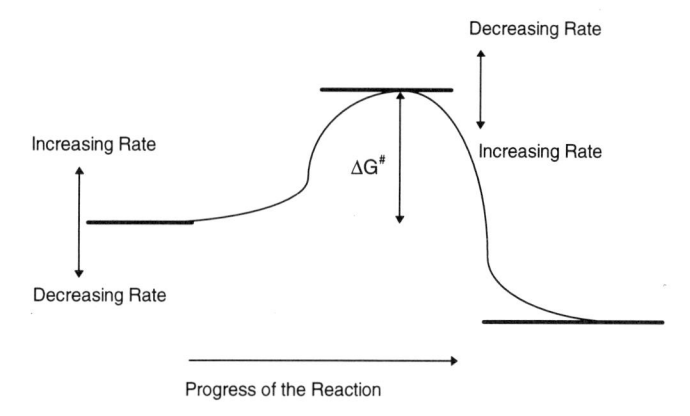

Figure 3.5 Free energy diagram illustrating solvent effects on reaction rate.

energy and increase the free energy required of activation, thus decreasing the reaction rate. These effects are illustrated in Figure 3.6. To quantify this effect, Cox [24] states that a net change of 5.7 kJ mol^{-1} in the solvation energy corresponds to an order of magnitude change in the rate constant at 25°C. The transition state theory can be expressed in terms of changes in the free energy or the equivalent activity coefficient analysis can be used.

The change in free energy between different solvent media (1 and 2) for the solvated reactants (ΔG_{rxt}) and the solvated transition state ($\Delta G^{\#}$) is given by equations (3.4) and (3.5).

$$\Delta G_{rxt} = G_{rxt2} - G_{rxt1} \quad (3.4)$$

$$\Delta G^{\#} = G_2^{\#} - G_1^{\#} \quad (3.5)$$

This leads to quantifying the effect of solvents on reactions through the change in the free energy of activation between the solvated reactions ($\Delta\Delta G^{\#}$).

$$\Delta\Delta G^{\#} = \Delta G_{rxt} - \Delta G^{\#} \quad (3.6)$$

Using the fundamental assumption of the transition state theory – that there is equilibrium between the reactants and the transition state – we can write for a bimolecular reaction A + B → products

$$A + B \leftrightarrow X^{\#} \quad (3.7)$$

Thus, the reaction rate is given by the concentration of the transition state ($X^{\#} = ABK^{\#}$) multiplied by the frequency of the barrier crossings.

$$\text{Rate} = K^{\#} AB(k_B T/\hbar) \quad (3.8)$$

Incorporating the change in free energy occurring on solution in terms of activity coefficients the rate constant can be written (taking $\gamma_i = 1$ for the reference solvent).

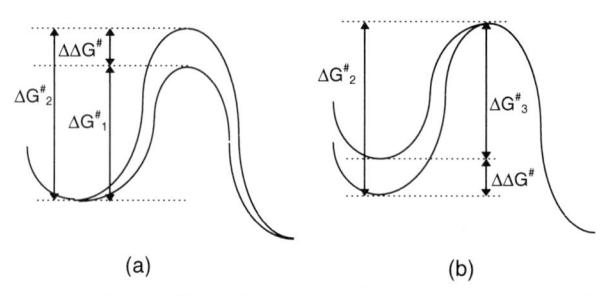

$$k_2 = (k_B T/\hbar) K^{\#} (\gamma_A \gamma_B / \gamma_{\#}) \quad (3.9)$$

Figure 3.6. Free energy diagram illustrating changes in reactant solvation (a) and transition state solvation (b).

Equation (3.10) shows the effect on the reaction rate of a change in solvent and can be derived for any number of reactants and different reaction kinetics

$$-\ln(k_2/k_1) = \Delta\Delta G^{\#}/RT \qquad (3.10)$$

Difficulties using the transition state approach are often encountered when mechanisms have multiple steps. Such a mechanism will have multiple transition states, as depicted in Figure 3.7.

It can be seen that an understanding of the chemistry of the activated state and reactants (i.e. dipolarity, charge dispersal, steric factors) and their solvation is of utmost importance to the prediction and manipulation of the solvent effects on reaction kinetics, equilibria and mechanism. Many authors have attempted to quantify these effects either empirically or theoretically by developing relationships between $\Delta G^{\#}$ or reaction rate and solvent parameters such as the dipole moment, dielectric constant, refractive index, solubility parameter and solvent polarity parameters such as Et(30). A thorough analysis of these relationships is beyond the scope of this chapter. Reichardt [9] gives an excellent summary of many methods. It is, however, important to appreciate that it is very difficult to attain accurate correlations between any one parameter and solvent effects on reactions. In general, most solvent sensitive processes cannot be correlated to a single parameter since polarity is a function of many different interactions, both specific and non-specific. Multiparameter approaches towards the quantification of solvent effects on reaction and solvation are far more accurate although they are also system specific.

3.4.3 Solvation dynamics

The simplest description of solvent interactions is based on the bulk properties m and e. These properties and their effect on solvation has led to free

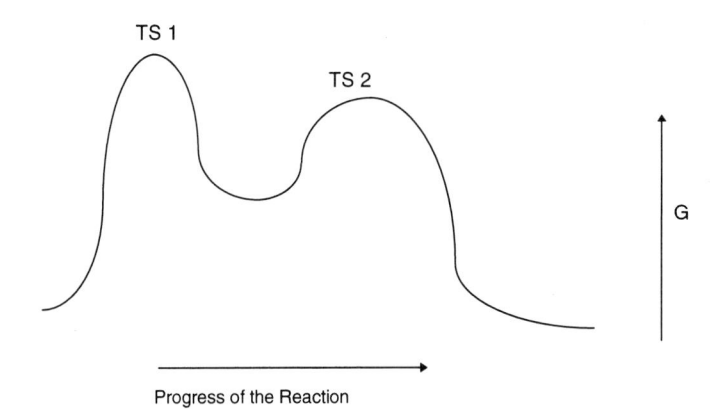

Figure 3.7 Free energy diagram for a reaction with two transition states.

energy relationships between the activated state $\Delta G^{\#}$ and the overall process. This solvent continuum picture is appealing; however, Rossky [25] suggests that the equilibrium-based concept of free energy becomes suspect in reactions where the barrier crossing through the activated state occurs faster than most intermolecular interactions between solvent and solute.

Under these fast conditions the transitional state forms more rapidly than the nuclear rearrangement of the solvent molecules and the transition (excited) state molecule will be in the solvent configuration corresponding to the ground state (reactants). The fate of the reactant encounter can be determined by the instantaneous solvent force acting on the solute in the transition state geometry. In some situations this may oppose charge migration and the activated state may reverse, as found in simulations of the textbook SN_2 reaction between halides and methylhalides [25].

It has been shown [26] that the first inertial response of the solvent, involving tilting and twisting rather than the slower diffusive reorientation motions, dominate aqueous solution dynamics. This further highlights the importance of understanding the role water takes as a solvent medium.

3.4.4 Reaction examples

It is very difficult to completely summarize the effects of solvents on all reactions. Generalizations are often difficult as reactions show a wide range of behaviour and solvent–solute effects are system specific. However, most exhibit similar trends when analysed at the free energy/transition state level.

(a) Isopolar transition states

Reactions between neutral molecules to form relatively isopolar (non-polar, non-radical) transition states usually display very subtle solvent effects. This is related to the similarities in the charge dispersion on going from reactants to products. Cycloadditions, sigmatropic and electrocyclic reactions can be grouped into this field. These represent members of the class of pericyclic reactions. Changing the medium or substituent usually has little or no effect on the rates of pericyclic reactions.

Diels–Alder additions are good examples of pericyclic reactions exhibiting small solvent effects. The dimerization of cyclopentadiene to endo-dicyclopentadiene typifies this reaction (equation 3.11).

(3.11)

Table 3.4 Solvent effects on the rate constant for the dimerization of cyclopentadiene at 20°C [9, 24]

Medium	Gas phase	Ethanol	Nitrobenzene	Paraffin oil
Rate constant ($M^{-1} s^{-1} \times 10^7$)	6.9	19	13	9.8
Medium	carbon disulphide	carbon tetrachloride	benzene	neat liquid
Rate constant ($M^{-1} s^{-1} \times 10^7$)	9.3	7.9	6.6	5.2

The second-order (bimolecular) rate constant for this reaction has been measured in solution and in the gas phase. The rates vary over a factor of about 3, as can be seen in Table 3.4.

This subtle rate effect is to be expected of reactions that occur without change in mechanism between solution and the gas phase, and reflects the extremely weak intermolecular interactions between solvent and reactant. The very low dipole moment of cyclopentadiene (1.4×10^{-30} C m) also indicates that a more polar solvent will result in little change.

(b) Dipolar transition state

Some reactions have considerable differences in charge distribution between their neutral reactants and dipolar transition state. This classification covers a wide range of common reactions such as elimination and substitution reactions and electrophilic additions.

Since the transition state for these reactions is more dipolar than the reactants, an increase in polarity of the solvent is expected to solvate preferentially the activated state over the reactants. Hence, a lowering of the free energy of activation and increased reaction rate should be observed.

These effects can be extremely large. For example a rate acceleration of 10^{10} is observed [24] in the bromine addition to pent-1-ene on changes in solvent polarity and solvent hydrogen bonding ability (from CCl_4 to H_2O). A less dramatic but similar solvent effect is found in ion-forming reactions such as the widely studied trimethylamine, *p*-nitrobenzyl chloride reaction [27] (equation 3.12). As shown in Table 3.5, more polar

Table 3.5 Solvent effects on the rate constant for the reaction between trimethylamine and *p*-nitrobenzyl chloride at 25°C [9]

Solvent	Hexane	Benzene	MeOH	Me$_2$CO	MeCN
Relative rate*	1	17.4	208	494	2954

* Relative to hexane.

solvents will favour the charged product, thus increasing the reaction rate.

$$Me_3N + p\text{-}NO_2C_6H_4CH_2Cl \rightarrow p\text{-}NO_2C_6H_4CH_2NMe_3^+ Cl^- \qquad (3.12)$$

(c) Free radical transition state
Many reactions pass through free radical activated complexes formed during either radical pair formation or atom transfer reactions. Free radicals can be created by oxidation, reduction or by homolytic cleavage of covalent bonds such as C–C bonds. Radical forming activated complexes normally do not exhibit charge separation and, as such, radical forming reactions are considered insensitive to solvent changes. However, it has been found that if specific solvation is not a dominant effect, atom transfer reactions show little solvent effect and electron transfer reactions between neutral species show large solvent effects.

Consider the halogen abstraction (bromine) by the radical 1-ethyl-4-carbo methoxypyridinyl, in which the first atom transfer is the rate limiting step [9] (equation 3.13).

$$(3.13)$$

Very little solvent effect is found on the rate of this reaction, indicating the bromine atom transfer may have the same charge separation as the reactants. Considering that the pyridinyl radical has an expected dipole moment of between 0 and 10^{-29} C m and dibromomethane has a dipole moment of 5×10^{-30} C m, it can be inferred from the absence of any strong solvent effects and the geometric orientation of the transition state, that the dipole moment of the complex is also in the range 0–10^{-29} C m.

In contrast to the above reaction, a large solvent effect is observed with the use of 4-nitrobenzylhalides instead of dibromomethane (a factor of 14 800 change in rate from 2-methyl-tetrahydrofuran to acetonitrile) (equation 3.14). This increase indicates [28] a change in the reaction mechanism to an electron transfer process and the creation of an ion pair from a neutral pair of molecules.

$$(Py^\bullet \text{ radical}) \qquad \qquad \text{Ion pair} \tag{3.14}$$

(d) Reactivity of ambident anions
The use of substrate with more than one reactive site is common in the fine chemicals industry. For example, ambident anions can form more than one type of product in nucleophilic reactions, the proportions of which are very dependent on the solvent medium. The well documented reaction [9, 26, 29] between phenoxide ions and allyl or benzyl halides given in equation (3.15).

$$\tag{3.15}$$

As the solvent becomes a stronger proton donor (protic) (i.e. $CH_3CH_2OH < H_2O < CF_3CH_2OH < C_6H_5OH$), carbon alkylation competes more strongly with the oxygen alkylation. The mechanism of specific solvation through hydrogen bonding explains this behaviour. Through the process of hydrogen bonding, protic solvents selectively solvate the more electronegative reactive site, which in this case is the oxygen site. Thus the accessibility of this centre is reduced, allowing other reaction centres (carbon atoms) to compete for alkylation.

(e) Equilibrium effects
Displacements in equilibrium can occur when going from a gas-phase reaction to the same reaction in solution and on changes of solvent medium. This is a result of the differences in solvation of the starting

and finishing products and the resultant differences of the free energy of solvation. Solvents can effect many different types of equilibria, including association/dissociation, tautomeric, rotational, conformational and isomerization equilibria.

The association equilibria of free ions to form ion pairs is dependent on the solvent medium and to some extent on the relative electrical permittivity of the solvent. Kakabadse [30] states that if the relative permittivity of the solvent in the solution of a 1:1 salt is lowered, a greater proportion of the free ions will associate to form ion pairs. Similar solvent effects can be found in the equilibria of dimerization, trimerization and thus micelle formation.

(f) Mechanism/selectivity effects

Alternative routes of a reaction may only be separated by small changes in free energy. A reaction's sensitivity to this energy can be exploited through the appropriate choice of solvent. The mechanism, selectivity and to a lesser extent the rate of reaction of cycloaddition reactions may be strongly affected by the solvent medium. Consider the reaction of dimethyl ketone and the enamine N-isobutenylpyrrolidine occurring via the concerted and non-concerted reaction pathways shown below (equation 3.16).

(3.16)

As expected from the charge dispersion of the zwitterionic intermediate, an increase in solvent polarity favours the non-concerted reaction mechanism. It has been reported [9] that in the non-polar solvent cyclohexane, 92% of the enamine goes through the concerted reaction whereas in the more polar solvent acetonitrile, 57% of the enamine goes through the zwitterionic intermediate to give the non-concerted product.

A similar example showing solvent effect on selectivity is the Diels–Alder cycloaddition reaction of methylacrylate and cyclopentadiene to produce the endo- and exo-cycloaddition products and the dimerization product dicyclopentadiene (equations 3.17 and 3.11).

(3.17)

The endo/exo selectivity is dependent on the polarity of the solvent medium. The endo/exo product ratio was shown [31] to decrease from 74:26 in non-polar triethyl amine to 88:12 in the polar solvent methanol (30°C). These observations have led to the suggested dipole moment representation given in the reaction scheme. In this scheme the superimposition of the dipole moments in the activated state suggests a more dipolar endo activated state. The parallel dimerization reaction of cyclopentadiene was discussed under (a) above.

3.4.5 Heterogeneous reactions

Reactions that occur across phase boundaries or involve transport between boundaries are becoming increasingly important. Examples of these heterogeneous reaction processes can be found in many areas of the fine chemicals industry as well as other industries such as the biochemical processing and the physiological industry. Two-phase systems are at an advantage over single-phase systems by greatly simplifying solvent recovery.

The desired reaction usually occurs in one phase, but the concentrations of the reactants, production of by-products and the medium conditions will be affected by the presence of the second phase. Thus reaction rates, equilibrium and selectivity will in most cases be dependent on the use of solvents as transport and reaction media. As such, solvent effects on kinetics and mass transfer are critical in heterogeneous reaction systems.

(a) Extractive reaction

Extractive reaction is probably the commonest of all heterogeneous reaction techniques. The process involves the partitioning of the (usually organic) reactant into a separate (usually aqueous) reacting phase. Depending on the rate controlling aspects of the reaction system, this reaction may occur in a diffusional film or in the bulk reacting phase. The reaction rate will depend partially on the solubility of the reactant in the reacting phase.

The reacting phase solvent medium may affect the rate through charge-related solvation of the activated state, as in one-phase systems, but also through its effect on the solubility. For example, consider the hydrolysis of *n*-hexylformate. This has low aqueous solubility, so its reaction with sodium hydroxide can be a two-phase reaction system. Increasing the concentration of sodium hydroxide in an aqueous reaction phase decreases the extraction rate by decreasing the solubility of the organic reactant. Thus, higher hydroxide concentrations reduce the reaction rate [32]. Similarly, the addition of a water immiscible solvent will decrease reactant partitioning into the reacting phase, thus decreasing the reaction rate.

In order to understand the role solvents play in extractive reactions it is necessary to evaluate whether the rate is diffusion or reaction controlled. The latter may be solvent sensitive. From this it is then possible to determine the local conditions under which the reaction occurs, and treat this as a single-phase system. An optimization procedure balancing the delivery of the species to the reactive site (solubility) and the most effective medium in which to undertake the reaction (polarity) can then be used. Design of the system will be further complicated by the consideration of the solvent recovery stage, although the two-phase reaction system greatly simplifies product removal and recovery of solvents.

(b) Phase transfer catalysis

The introduction of phase transfer catalysis (PTC), through the works of Makosza, Brandstrom and Starks in the 1960s, has led to a great interest in and the extensive industrial use of this catalytic reaction method for the synthesis of organic chemicals. It is often the method of choice for reactions involving anionic species either added as alkali salt or generated by a base. Excellent reviews of PTC and its industrial applications can be found in papers by Freedman [33] and Taverner and Clarke [34].

PTC offers many advantages over conventional reaction techniques. It is particularly useful for reactions between two mutually immiscible reactants such as sodium salts and alkyl halides. Industrially, its advantages include increased reaction rates and product selectivity as well as low energy requirements and inexpensive catalysts, bases and oxidants.

PTC can allow the use of inexpensive, non-toxic, recoverable solvents. There is also the possibility to use the liquid reactant as solvent. Overall, the use of two phases greatly simplifies the solvent recovery and product removal stages. Similar observations can also be made about inverse PTC, where the reactant is transferred from the aqueous phase to the organic phase for reaction rather than the organic to aqueous as outlined below.

The reaction between benzoylchloride (PhCOCl) and sodium acetate (CH$_3$COONa) using the catalyst pyridine 1-oxide (PNO) (equation 3.18) illustrates PTC reactions [35].

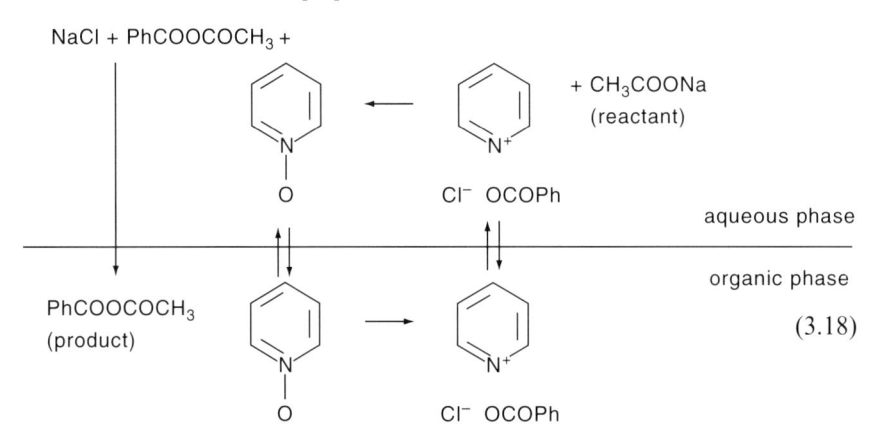

$$\text{(equation 3.18)}$$

As a result of the insolubility of CH$_3$COONa in dichloromethane, the component existing in the organic phase and in equilibrium with the sodium acetate in the aqueous phase is the product of hydrolysis, acetic acid (CH$_3$COOH). This by-product affects the rate and yield of the reaction by reacting with the PNO in the organic and aqueous phases. However, the polar acetic acid affects the polarity of the organic phase, thereby increasing the rate. Thus, the rate of this PTC reaction is strongly affected by both the increased solvation by CH$_3$COOH and by the protonation of CH$_3$COOH. The reaction is sensitive to the polarity of the solvent used. As the dipole moment of the organic solvent increases (polarity increase), the conversion of the PhCOCl increases and the formation of by-product PhCOOH obtained from the hydrolysis of benzoyl chloride increases, which in turn favours formation of the anhydride.

(c) Solid catalysis
In solid catalysis with solvated reactants the choice of solvent can be important to the reaction rate. As an example, consider the Diels–Alder reaction between methylacrylate and cyclopentadiene (equation 3.17) catalysed by Zn(II)-doped K10 clay. It was found [36] that a qualitative relationship between the solubility of the organic solvent in water, the reaction rate, reaction yield and the endo/exo selectivity existed. In miscible solvents most of the reaction takes place in the bulk solvent and the clay takes no part in the catalysis.

It is interesting to note that the rates and selectivities of the clay catalysed reaction in organic media are comparable to the rates and selectivities of the reaction in water. However, much greater concentrations of reactants can be used in the organic-based reaction and lower temperatures (giving higher endo selectivity) can also be used. This demonstrates the wide range of reaction solvents and process choices available without incurring a change in yield, rate or selectivity. Emphasis can then be shifted towards such aspects as recoverability, and health, safety and environmental aspects of solvent selection and reaction process design.

3.5 Solvent recovery

All the world's production of solvents is ultimately either destroyed or dispersed into the biosphere. As well as the ethical grounds and increasing political legislation for the recovery of resources from all process outputs, solvent recovery is often justified on economic grounds. The high expense and commitment required by environmentally sound destruction methods often restricts the possibility of solvent destruction or at least makes it economically attractive to recycle as much solvent as possible before the disposal of final residues. Recycling and reuse of the solvents, particularly less toxic and less environmentally problematic ones, can mean that a more expensive solvent is cheaper in the long run. This emphasizes the importance of choosing solvents and reformulating reactions for easy recovery.

With the search for ever improved environmental performance being a major consideration in the many aspects of the fine chemicals industry, solvent recovery will become an increasingly necessary tool to achieving this aim. Many options for solvent waste reduction and solvent recovery are available. These options range from reformulating reactions and processes to remove the solvent altogether or allow easier recovery to in-house recovery operations using techniques such as distillation, fractional crystallization, adsorption, absorption and extraction. The use of a merchant recovery specialist, contract processor or solvent supplier is also a viable option, and is a trend that is developing more and more in the solvents business.

3.5.1 Origin of solvent wastes

Solvents used in industry perform many operations and can influence the physical and chemical aspects of reactions. They may be used in isolation or in a mixed series with other solvents and they may participate directly in the reaction or be used as a reaction medium only. As a result of the complicated processes involved in charging, discharging, transfer, reaction, work-up and separations stages, solvent mixtures requiring further recovery arise in four general classes.

(a) Solvent vapours mixed with gas
Solvent vapours mixed with gases arise from breathing losses, sparging operations, gas-liquid reactions, air stripping as well as from coating processes.

(b) Solvents in wastewaters
This is the most common mixture encountered in solvent recovery operations. It can arise through the use of aqueous media in the reaction process or at some stage of work-up for extraction or washing. This type of liquid stream will become increasingly important as the shift towards aqueous solvents becomes more widespread.

(c) Contaminated liquid solvents
Solvents may be purposefully contaminated through their use to selectively dissolve by-products and impurities from reaction mixtures. Contaminated solvents are also commonly formed during the cleaning and drying of equipment.

(d) Solvent mixtures
The nature of the fine chemical and pharmaceutical industries typically leads to the use of multistage reactions. Each reaction may require a specific solvent or solvent mixture. It is also possible that solvent mixtures may be used as reaction medium to meet process requirements. Solvent mixtures frequently arise from simple batch distillation.

3.5.2 Designing for recovery

The most effective and economical method of recovering process solvents is to make considerations of the effect a choice of solvent will have on the recovery stage. Most reaction processes occur under many different reaction and solvent conditions. Preliminary designs must consider whether a solvent is necessary at all. Detailing the reaction in terms of its mechanism is extremely important, particularly in the early process design when the solvent required and even process required are not fixed. Often, reactions traditionally undertaken in single phase can be manipulated to run in two phases, greatly simplifying recovery.

In general there are three directions a 'recovery-based design' can follow. These are:

- The evaluation of the necessity for a solvent at all.
- The manipulation of the reaction process conditions.

- Finally, and most commonly, the effective use of conventional methods of solvent recovery.

(a) Solvent necessity

One of the most obvious solutions to solvent recovery problems is the analysis of the phase behaviour of the reactants and the possibility of the complete elimination of the solvent. The reaction may be possible in the gas or both reactants may be used as liquids.

Often, rates of reaction can be increased and by-products can be minimized through appropriate solvation using a solvent, but a complicated trade-off must be made as the necessity for recovery can compete with this economic gain. Alternatively, there is the possibility of using catalysts to replace the rate and selectivity enhancing role the solvent may play. The elimination of solvent and replacement with catalyst in the oxidation of ethyl benzene to acetophenone is an example of this rate enhancing role [37].

(b) Mechanistic analysis

When designing reaction processes and choosing the solvents for use in these processes, previous experience has often been the deciding factor. However, many existing processes have been designed with an extremely narrow list of criteria in comparison to the requirements placed on industry today (including legislative, economic, health, safety, environmental and ethical requirements).

A full understanding of the chemistry of the reaction and in particular the transitional state and mechanism is necessary for the designing of reaction process conditions, minimization of waste and the simplification of the related recovery system. As outlined above, this analysis allows a semi-quantitative assessment of the most appropriate solvents for high rates and yields as well as good selectivity (low by-product production). Similarly, creative manipulation of the reaction system to maintain the best environment for the transitional state, thus good reaction performance, may lead to process alternatives such as two-phase reaction. Mechanistic manipulation can also lead to new single-phase systems with less harmful and more easily recoverable solvents.

In general, mechanistic analysis provides a pathway to understanding the properties of solvents that are critical for effective reaction. Mechanisms can give insight into the intermolecular interactions that dominate solvation of reactant, transitional states and products, and thus dominate the free energy pathway of all reactions. Armed with these details, solvents can be arranged according to their solvation effectiveness leading to a preferred reaction outcome and then can be classified according to their recovery potential (i.e. product immiscibility, partition coefficient with product, by-product or other streams, boiling point differences, opportunities for recrystallization).

As a simple example illustrating the relationship between the chemistry of solvation and reaction and the engineering aspects of reaction and recovery consider the acylation reaction given in equation (3.19).

$$\text{(3.19)}$$

Based on the use of extraction techniques for the work-up stages, Brown [38] aimed to design solvents for better recoverability in this reaction scheme. The first compound in equation (3.19) is insoluble in polar solvents and most reactive when M^+ is solvated. This indicates a good cation solvator, such as a high boiling point ether, is required. Unfortunately this has some solubility in water and can easily end up in an aqueous stream on work-up. Toluene, however, is easy to recover but would require more dilution and would lead to slightly lower yields (5% lower).

Another approach was to accept the solvent would end up in the aqueous stream and design the solvent for recovery from this stage. Although difficult to recover, DMF was chosen for its good solvation properties. Attempts were made to alter chemically DMF (by adding a chemical 'handle') so as to retain dipole moment yet increase hydrophobicity and thus recovery from aqueous streams. This presented problems for recovery from the hydrophobic reaction products. The two solvents suggested by Brown were DMF and n-methone piperidine. DMF exhibited better reaction properties while n-methone piperidine gave better recovery using solvent extraction. At each stage of solvent design an analysis of solvent properties (intermolecular interaction parameters) and their effect on both reaction and recovery were made.

(c) Conventional recovery techniques
In all situations except those where the solvents have been designed out of the process or two-phase reaction systems where solvents may not be contaminated and can be collected in a pure state, there will always be a need to separate solvents from mixtures of compounds. These mixtures may be very difficult and expensive to separate, requiring multistage processing

and inventories of new materials or may equally only require simple separation techniques such as the distillation of non-azeotropic liquids with widely separated boiling points. Solvent selection needs to be carried out in the context of the available separation techniques. A full understanding of these techniques and the parameters that characterize their effectiveness is necessary at all stages of process design.

There is a distinct group of unit operations that have been effectively used for solvent recovery. A summary of a broad range of techniques and specific recovery notes for common solvents is given in the *Solvent Recovery Handbook* by Smallwood [2]. In designing a process it is important to appreciate that there is often flexibility in the exact composition of wastes, particularly those arising from washing and solvent extraction operations. This can be exploited to maximize the amount of waste generated with a recoverable concentration of the solvent. Sharratt has presented methods to carry out this optimization [39].

Techniques for the separation of solvents from various waste streams are briefly described in the following sections. The aim is not to provide a design guide (the design of such technologies is widely described elsewhere), but to point out some particular issues of importance related to the treatment of batch fine chemical solvent residues.

3.5.3 Separation of solvents from gaseous wastes

A vapour contaminated with solvents arising from processes such as chemical drying, coating or a pretreatment such as stripping are often in volumes of the order of $70 \, m^3 \, l^{-1}$ of solvent. This concentration is a result of the necessity to keep the solvent concentration (chlorinated solvents excepted) to a maximum of 25–35% of the lower explosive limit to avoid explosions.

Satisfactory discharge gas–air pollution levels vary for each solvent and will depend on local legislation, but in general will be high for oxygenated solvents (with high biodegradability) and low for highly toxic or environmentally damaging materials. A number of techniques are available to treat these gas streams and, depending on the solvent concentrations and the design requirements, a suitable technique for recovery can usually be found.

(a) Adsorption

Adsorption is predominantly a batch process and is a highly efficient method for higher molecular weight solvents. Almost all organic vapours with molecular weights greater than air can be adsorbed. The adsorption of these compounds is inversely proportional to the volatility of the solvent to be removed.

Activated carbon (AC) is the most extensively used adsorbent and provides the most desirable characteristics for the removal of typical solvent molecules. Theoretically, activated carbon retentivity can exceed 30% of its weight, although practically the most economical operation involves loadings of from 5% (low molecular weight solvents) to 20% (high molecular weight solvents). Two-bed operations can theoretically achieve 99% removal, although practical removal efficiencies are of the order of 80–85%. If emission levels force higher requirements, it is possible to use a three bed arrangement (with two beds in series in line at any time, the second bed is for polishing) to achieve 99.7% recovery efficiency. Desorption of the bed is carried out using steam or hot gas recirculation. The retained water then aids the control of bed temperature but unfortunately results in a solvent–water stream following condensation.

A great deal of care must be taken with the heating of AC beds due to the heat of adsorption of solvents onto AC. Ketones, in particular, can undergo reaction with water and release large quantities of heat. Higher ketones can raise the bed temperature to temperatures at which the AC will ignite. Care must also be taken with the recovered solvent – often solvent inhibitors are preferentially removed and the solvent product may require stabilization by further dosing of inhibitor.

(b) Absorption

Absorption into water (scrubbing) is widely used to capture water-soluble organic vapours. The recovery of the captured solvent is almost never cost-effective. Absorption into non-aqueous media (high boiling ethers or hydrocarbons) can also be used and the captured solvent subsequently recovered by vacuum stripping.

(c) Condensation

Condensation has long been used for the capture of solvent vapours. Unfortunately, cooling water or chilled water temperatures are insufficient to achieve adequate abatement of most emissions. Nevertheless, condensation is highly cost-effective for the primary treatment of concentrated emissions.

The use of refrigeration to obtain lower temperatures is expensive; often prohibitively so if to achieve current legislative limits. However, the use of the source of cooling provided by a site nitrogen supply can deliver cost-effective cryogenic condensation [40]. This technology is well suited to application in batch chemicals processes.

3.5.4 Separation of solvents as liquid

Solvents contaminated with water (or vice versa) are a common problem in industry. There are two major reasons for the recovery of solvents from these streams; the recovery and recycle of solvents back into the process (or elsewhere) and the removal of solvents from water to meet safe environmental discharge requirements for water.

Recovery is also becoming more attractive since solvent removal from water by directly evaporating VOCs into the atmosphere or releasing air from stripping are unacceptable. Standards for effluent water purity for discharge to water courses will usually be higher than the design requirements for the recovered solvents to be reused. Thus, consideration of the treatment of wastewater and the interactions solvents have with this system must be made when choosing the solvents. Although less purification of the solvents is often acceptable it must be remembered that contaminated solvents can have a detrimental effect on the solvation of reactants and transitional states. Also, the use of recycled rather than fresh solvent is difficult in pharmaceutical processes subject to FDA regulations.

In all recovery and treatment systems it is important to bear in mind the point source approach (recovery and treatment should be at the source, streams should not be mixed) to avoid further contamination of aqueous streams. A number of conventional recovery techniques are available for separation of solvents and water. These include decanting, membrane separation, reactive methods, adsorption and extraction techniques as well as steam and air stripping and the many variations of distillation.

(a) Decanting

Many industrial solvents are only soluble to a small degree in water and hence form a two-phase aqueous/organic mixture. Although ineffective for true emulsions, gravity decanters are an effective and cheap method for separating out solvents whose density differs significantly from water. The density difference should be greater than $30 \, \text{kg m}^{-3}$ [2] and the key to good separation is droplet coalescence. Further treatment of the aqueous phase is usually required, although this method alone may be used to sufficiently treat aqueous wastes with solvents of solubilities of less than 0.2% in water.

Decanter temperature is critical to both settling speed and the solubility of the solvent in water. Centrifuges can be used to enhance the effects of density difference if gravitation effects are inadequate or equipment size constraints occur. It may be possible to separate emulsions through enhanced droplet coalescence using a high voltage electric field.

(b) Membrane separation

At least two types of membrane separation are useful for water–solvent separation. Pervaporation involves the removal of water (water passes through) from the solvent using hydrophilic membranes. These membranes are unfortunately mostly soluble in aprotic solvents such as DMF, DMSO and glycol ethers. Water contents of 1–15% in the feed and 0.1–1% in the product are typical performance for pervaporation. Pervaporation will break alcohol–water azeotropes. Alternatively, membrane separation removing solvents (solvent passes through) from water using a lyophilic membrane is possible. These membranes can be destroyed by exposure to highly concentrated solvents.

Design and research for new selective membranes is increasing greatly and improved membranes are becoming commercially available. These improvements make this process worth consideration for the recovery of solvents and solvent–water separation.

(c) Adsorption

Adsorption, typically using AC, is often used as a final polishing stage in the separation of solvents from water. It is, however, a very flexible method of treating variable batch process effluent and can be used for the treatment of more concentrated effluent streams (point source). AC is most effective for the removal of high boiling non-polar solvents. Different solvents in a mixture are removed to a different extent.

Laboratory experiments can be used to quantify the take up of solvents from water where AC data are not already available. This take up of solvents by AC is dependent on temperature, pH and the salt content of the aqueous solution. It should be noted that inorganic salts are not removed during regeneration and over time reduce the AC capacity and effectiveness. Regeneration of the AC also leads to poor adsorption of low boiling solvents such as trichloroethylene. Small on-site regeneration units should be considered very carefully. Generally, on-site regeneration is not recommended for operations consuming less than 0.3 tonne day^{-1} of activated carbon.

(d) Air stripping

Most organic solvents can be removed from waste water to a level suitable for discharge by using air stripping. Air stripping is particularly useful for removal of solvents with low solubility in water or high volatility with respect to water. However, the solvent laden air must be further cleaned to allow its safe release into the environment. This can provide major difficulties due to the low concentration of solvent in the leaving air and the requirement for high purity separation. This treatment usually involves the use of adsorption techniques.

Mixed solvents in low concentrations have no effect on each other in air stripping while the presence of inorganic salts improves the effectiveness of air stripping by increasing the activity coefficient of the solvents.

(e) Steam stripping

Steam stripping has many similarities to air stripping. It involves a more elaborate plant for the stripping stage but a simplification of the final collection of the solvent since all of the gas stream is condensable. Steam stripping units are often combined with distillation units for the separation of water miscible solvents. They may also be combined with a decanter for immiscible solvents. The conversion of a stripper to a fractionator is a relatively simple task and the steam costs associated with this technique are reasonable, assuming that good heat exchange and effective exchanger cleaning is available. Steam stripping is not suitable for water miscible solvents with high boiling points and partial vapour pressures lower than that of water.

(f) Solvent extraction

Solvent extraction is a separation technique that involves separating two immiscible phases that are created by the introduction of a foreign substance called the extractant or extraction solvent. The most critical aspect of extractive techniques for the removal of solvents from water is the appropriate choice of the extraction solvent. Selecting an extraction solvent is a difficult and complicated procedure involving many compromises. There are many characteristics that make a desirable extractant. These include low solubility in water but good solubility in the extracted solvent, chemical stability, high availability, low cost and safe handling characteristics.

The extractant must have favourable selectivity to attain a good separation and high solvent capacity to reduce process solvent requirements. There is also a need to aim for as small as possible solvent loss in the raffinate phase. Solvent selectivity, capacity and loss are all related to the activity coefficients of the compounds in the mixture. Empirical methods have been used to determine this selectivity, but recent developments [41] have attempted more fundamental analysis using molecular modelling. Consideration of the further treatment of both the organic and aqueous streams from the extractor must also be made. This involves choosing the extractant for ease of separation from the extracted phase through either a high density difference or suitability for distillation as well as low biological oxygen demand (BOD) for the water removal stage. These are particularly critical when one considers that the loss of extraction solvent is a function of the amount of effluent water.

As a result of the large ranges in water miscibility of solvents, some extremely hydrophobic solvents such as hydrocarbons and chlorinated hydrocarbons can be used in liquid–liquid extraction processes to drive water out of hydrophilic solvents. Unfortunately the aqueous phase will become saturated with the introduced organic solvent and require further treatment.

Aqueous streams containing large amounts of high boiling organic contaminants create problems when using solvent extraction techniques for clean-up. Water insoluble contaminants will either build up in the extraction solvent or accumulate in the equipment. This can lead to the requirement of flashing the extraction solvent every now and then and having difficult-to-handle residues.

Very hydrophilic solvents (large negative $\log P_{oct/w}$) have proven to be very difficult to separate from water using a third solvent as an extractant whereas sparingly water soluble solvents with $\log P_{oct/w} > 1.5$ can usually be easily extracted with a third solvent.

(g) Fractional crystallization
Mixtures which are extremely difficult or impossible to separate by distillation or extraction techniques can often be successfully separated by fractional crystallization. Fractional crystallization is a staged process involving phase transition in melts or solutions and relies on the fact that the desired component or fraction is preferentially distributed between the crystalline phase and the mother liquor.

It is an attractive technique for separating close boiling components or as a replacement for the top rectification stages of very large fractionators with many stages or very high reflux ratios. It can achieve an ultra-pure product using compact equipment with low energy requirements. This is due to the fact that latent heats of crystallization are much lower than heats of vaporization. However, as well as being a young technology, this technique can be very system-specific and usually will not allow much flexibility.

(h) Distillation
Distillation is possibly the most effective and widely used separation technique in industry today. It is certain to be the technique of choice for most separation objectives. Batch distillation is extremely flexible, making it very suitable for use in the fine chemicals industry where solvent batches may vary daily in both quantity and quality. Distillation exploits the differences in volatility of compounds and can be used to remove water from solvents, remove residues from solvents and to separate solvent mixtures. For this reason, rather than concentrate on particular duties it may perform, a brief rundown of the various distillation techniques will be given. Many excellent texts provide an in-depth coverage of distillation techniques [2].

The relative volatility of solvents can give an indication of the possible use of distillation and the column requirements. Any value of $\alpha_{ij} < 2$ will require either a large column or the use of an entrainer. Other separation techniques should also be considered for these tough separations. The temperature and pressure are also important parameters due to their effect on the volatility.

It is often common in the fractionation of solvent laden wastes from the fine chemicals industry to find the distillate contaminated with reaction by-products and products of hydrolysis and solvent decomposition. Ultimately these can form residues in the still, possibly causing problems with heat transfer, corrosion, the accumulation of unstable material and of course a disposal problem. Several variants of distillation can be used to deal with difficult mixtures.

Steam distillation. Steam distillation is commonly associated with the removal of residues from solvents. It avoids the possibility of heat transfer surface fouling because heat transfer is direct. All solvents with boiling points below 100°C and all solvents not completely miscible in all proportions in water can be evaporated by injecting live steam into the liquid solvent. Unfortunately the steam is contaminated by the solvent and requires further separation such as decantation or further fractionation. However, direct steam injection into a still avoids the residue associated problem of fouling of the heat exchanger surfaces and aids in the removal of residues. Furthermore, as a result of the temperatures involved (100°C at 1 atm), exotherms are unlikely to occur. For a solvent not appreciably water miscible, steam distillation can improve relative volatility.

Problems with this technique include:

- Materials of construction problems, particularly with halide salts.
- The formation of azeotropic mixtures with water.
- Foaming.
- Residues handling and disposal.

Vapour distillation. A method similar to steam distillation for treating solvents contaminated with residues while avoiding direct steam injection and eliminating the problems associated with fouled heating surfaces is the Sussmeyer operation [2]. The Sussmeyer process superheats the solvent batch vapour and returns it to the solvent liquid batch. A demister is used to avoid fouling the heat exchanger. Since this process is carried out under vacuum it is possible to handle solvents with boiling points up to 180°C without decomposition. Vapour distillation of components with boiling points up to 120°C such as alcohols, ketones and esters is considered better than pervaporation on cost grounds.

Vapour pressure reduction. Involatile materials may reduce the vapour pressure of solvents in mixtures, as well as altering the relative volatilities. An example of this phenomenon is the addition of salt to break an azeotrope (salting out). It reduces the vapour pressure of the component in which it is more soluble (often water) thus enabling extractive distillation to be run in a solvent of salt solution.

Vacuum distillation. When distillation is run under vacuum there is often an improvement in the relative volatilities of the separated components. This leads to more efficient separation. Through vacuum distillation there is an opportunity to lower the processing temperature leading to less solvent decomposition and a reduction in the potential for exotherms and polymerization.

Vacuum or pressure distillation can be run alongside atmospheric distillation to take advantages of the differences in azeotrope composition with pressure. This is demonstrated in the separation of water and MEK in which the atmospheric azeotrope contains 35% water and the pressure (100 lb in^{-2}) azeotrope is 50% water. The separation process involves taking the azeotropic atmospheric distillate and purifying it in the pressure column.

Azeotropic distillation usually involves the addition of an entrainer which forms an azeotrope with only one of the components. The newly formed azeotrope separates into two layers, one enriched in the entrainer and the other in the target (separated) component. It is critical that the entrainer is easily recoverable from the separated component.

An example of azeotropic distillation is the separation of a DMF–hydrocarbon mixture using the entrainer water. The water forms an azeotrope with the hydrocarbon but not with DMF, resulting in a two-phase water–hydrocarbon mixture at the top of the column.

Extractive distillation. This predominantly continuous distillation technique involves the addition of another component or entrainer that alters the ratio of the relative volatilities of the compounds to enhance separation. The entrainer is removed from the bottom of the column and is then separated and returned. A common example of this type of system is the methanol–acetone system where water is added as entrainer because it has a strong affinity for methanol and thus lowers its relative volatility, allowing it to be retrieved pure from the top of the column.

3.6 Solvent destruction

There are two main options available for the destruction of solvent wastes. These are biological and non-biological destruction. Industry has gained considerable experience in non-biological methods such as incineration and the emerging use of cement-kilns for the disposal of residues. Extensive research has been undertaken in many areas of biodegradation mechanisms and techniques. However, with present engineering technology, the overriding conclusion is that there are no available methods for the completely safe (i.e. with no impact) disposal of hazardous wastes. In many destructive methods, materials are reduced to CO_2 and/or methane which directly contribute to global warming. Serious problems exist in all aspects of solvent waste disposal and very expensive, time consuming measures must be taken to ensure that there are no significant releases of bioaccumulating, toxic or environmentally damaging materials.

Often a combination of biological and non-biological treatments is the most cost effective and environmentally responsible method of destruction. In the determination of the most appropriate destruction method it is necessary to characterize fully the waste to be treated and attempt to keep this waste as consistent as possible.

3.6.1 Non-biological treatment

Non-biological destruction methods include such techniques as incineration and use as cement-kiln fuel. Wet air oxidation is also suitable for the destruction of aqueous organic wastes.

In recent times incineration has come under public scrutiny, particularly for its role in the production of ash borne dioxin compounds (toxic bioaccumulators). Deviations between safe dioxin limits in the UK and in the US emphasize the present degree of uncertainty the scientific community has on the environmental effects of many chemical compounds. Historically, however, the chemical industry has had an extremely poor record on foreseeing environmental problems (DDT, DCE, PCBs, CFCs, halons, dioxins, etc.) and must remain cautious of all released chemicals, regardless of the present stage of their environmental understanding.

Incineration is presently a viable option provided incinerators are operated with very high performance scrubbers to remove toxic ash from the waste gas stream. There is a need to have highly efficient, high temperature (about 1500°C) combustion to ensure the full destruction of all compounds and production of the least harmful combustion products. The cooling (temperature profile) of the waste gases is equally critical due to the production of toxic compounds in this stage.

Other methods such as usage of solvent waste as fuel in cement-kilns are possible. Although an energy efficient method of destruction, strict control measures are placed on the types of contaminants allowed into the cement process, thus restricting the types of waste suitable for destruction by this method. It is interesting to note that the use of cement-kilns and other inexpensive disposal routes for solvent residues have the potential to increase the amount of solvent recycling. By reducing the cost of disposing of the still bottom residues from solvent recovery, the economics of solvent recovery are improved. The issues of solvent recovery, cement-kiln fuels and incineration are discussed in more detail by Sharratt and Sparshott [40].

Wet air oxidation is a technique involving the oxidation of aqueous organic waste streams using air dissolved in water under high pressure. Typically pressures of tens or hundreds of atmospheres and temperatures in the range 100–300°C would be used. The method can provide a useful pre-treatment for concentrated wastes before biological oxidation, but is not a complete treatment in itself [41].

3.6.2 Biological treatment

As well as traditional aqueous wastewater treatments it is appropriate to consider landfills as biological treatment systems.

Historically, landfills have often been poorly managed and many cases of long-term health and environmental problems have been direct results of the subsequent use of contaminated toxic waste sites as residential areas, and the leaching of hazardous waste into soil, rivers and groundwater. Landfill is relatively inexpensive in comparison to all other destruction techniques. It can be an effective destruction method if it is strictly and correctly managed over the entire duration of the biodegradation of all the waste material (this can be hundreds of years).

However, previous and present day long-term environmental and health disasters, realistic concerns over leaking, release of VOCs through evaporation, long-term requirements, and land use limitations have all led to high public awareness and, as a result, resistance to landfill technology.

Although biotreatment has been around for at least one hundred years, new innovative treatment technologies continue to be developed. Major research concentrates on organism activity. Biotreatment is inexpensive in comparison to non-biological treatment due to the low energy requirements. By the nature of the fine chemicals industry, solvent laden wastes can be very difficult to destroy biologically and an enormous amount of care must be taken to ensure there is no release (particularly VOCs) into the environment.

Aerobic treatment is limited to dilute aqueous wastes. There are four main routes for a compound to take in most conventional aerobic biotreatments. The most obvious route and the primary aim of any biological destruction technique is biodegradation. Volatilization directly into the atmosphere through forced aeration or through the top air–water interface is also an extremely common occurrence. The final two routes that may be taken are the pass-through of contaminants into the liquid stream that is discharged directly into the environment, and pollutant sorption onto solids and precipitation out (landfill) leading to the possibility of leaching into soils and ground water.

Volatilization is a particularly important aspect of aerobic treatment performance for the fine chemicals industry due to the high use of VOCs and the relative inability of aerobic treatment to contain these compounds. As an example of this, Table 3.6 gives the routes taken through the Lakeview treatment facility in Ontario, Canada, by various solvents in waste water.

Means to rectify this loss of VOCs in aerated biotreatments have been attempted through the development of gas phase biofiltration methods. This type of technology is still being developed, although Germany and the Netherlands have as many as 500 full scale biofilter operations on line.

While anaerobic treatment provides an opportunity to treat more concentrated wastes, all practicable forms of biological treatment are likely to be limited to the treatment of solvent-contaminated waste water.

Table 3.6 Fate of organic solvents in aerated biotreatment [42]

Solvent	Per cent biodegraded	Per cent volatilized	Per cent passed through
Benzene	87	13	0
Chloroform	62	20	18
Chloromethane	12	55	33
Ethylbenzene	56	43	1
Toluene	74	25	1
Trichloroethylene	50	29	21

3.7 Conclusion

Solvents play an essential role in the manufacture of many of the chemicals on which modern society depends. Effective, safe and environmentally responsible use of solvents requires a clear understanding of their behaviour and interaction with other materials. Solvents have long been known to have a range of effects on chemical reactions. These effects can be changes in rate, selectivity, mechanism and equilibria. Solvents can also affect the separability of the post-reaction mixture. They also influence the transport and thermal properties during processing. However substantial or subtle these effects may be, they are all important considerations in the design and operation of any process that uses solvents.

The chemistry of solvation, the effect of solvents on reactions, phase behaviour and the eventual recovery of solvents from complex mixtures of compounds are related by the molecular interactions that occur between solvent molecules and other solute or solvent molecules. These interactions and their practical representation through the use of free energy diagrams permit solvated reaction systems to be designed at least in a qualitative way. This design can help to achieve objectives such as high recoverability, high rates and yields, and good selectivity. Increased understanding of both the reaction and solvation chemistry leads to the realization of the considerable flexibility inherent in many reaction systems.

While simple heuristics such as the Hughes–Ingold rules are available and the possibility of using molecular modelling as a design tool is being developed, it is still only possible to make qualitative predictions for the effects of solvents on reactions. However, through an understanding of the chemistry of both the solvent and the reaction it may be possible to make semi-quantitative predictions based on other reactions with similar chemistry. To quantify fully the effect of a solvent on a particular reaction or determine the reliability of a particular recovery method for a unique waste stream, semi-quantitative predictions based on precedence, mechanistic analysis and theoretical considerations can only be quantified through experimentation.

The growing understanding of the potential for health and environmental effects arising from the use of solvents has led to new constraints and restrictions being placed on the fine chemicals industry and other solvent users. These constraints have led to solvent containment, recovery and more comprehensive consideration of the environment during design being developed as partial solutions to environmental problems. The understanding is also being reflected by a more ethical stance being taken by industry and its workers and is leading to a re-evaluation of the role of solvents in industry. Nevertheless, there is still much work to be done to reduce the harm associated with the use of solvents while continuing to exploit their useful properties in the fulfilment of human needs.

3.8 Nomenclature

c_p	specific heat capacity (kJ kg^{-1} K)
e	dielectric constant (–)
G_{rxt}	Gibbs free energy of reactant (J mol^{-1})
$G^{\#}$	Gibbs free energy of transition state (J mol^{-1})
ΔH_{fus}	latent heat of fusion (kJ kg^{-1})
ΔH_{vap}	latent heat of vaporization (kJ kg^{-1})
$\hbar$	Planck constant (J s)
k	thermal conductivity (W m^{-1} K^{-1})
k_B	Boltzmann constant (J K^{-1})
m	dipole moment (C m)
n	refractive index (–)
P	pressure (Pa)
$P_{oct/w}$	partition coefficient for a substance between octanol and water (–)
r	intermolecular or interatomic distance (m)
R	gas constant (J mol^{-1} K^{-1})
T	temperature (K)
V_m	molar volume (m^3 mol^{-1})
α_{ij}	relative volatility of components i and j
δ	Hilderbrand solubility parameter (kg$^{0.5}$ m$^{-0.5}$ s^{-1})
γ	activity coefficient (–)
μ	viscosity (N s m^{-2})
ρ	density (kg m^{-3})

References

1. British Broadcasting Corporation, BBC 2 (1996) Horizon, 8.00 pm, 26 February.
2. Smallwood, I. (1993) *Solvent Recovery Handbook*. Edward Arnold, London.
3. Wallas, S.M. (1987) *Phase Equilibria in Chemical Engineering*, Butterworth, Boston.
4. Reid, R.C., Pausnitz, J.M. and Poliky, B.E. (1987) *The Properties of Gases and Liquids*, McGraw-Hill, New York.
5. Gutmann, V. and Schmidt, R. (1974) *Coord. Chem. Rev.*, **12**, 263.
6. Gutmann, V. (1978) *The Donor–Acceptor Approach to Molecular Interactions*. Plenum, New York.
7. Hughes, E.D. and Ingold, K.C. (1941) *Trans. Faraday Soc.*, **37**, 603.
8. Dimroth, K. *et al.* (1963) *Liebigs Ann. Chem.*, **661**, 1.
9. Reichardt, C. (1978) *Solvent Effects in Organic Chemistry*. Verlag Chemie, Weinheim.
10. Bloeman, H.J. and Burn, J. (1993) *Volatile Organic Compounds in the Environment*. Blackie Academic and Professional, Glasgow.
11. Los Angeles rule 66.
12. Her Majesty's Inspectorate of Pollution (1994) *Environmental and Economic BPEO Assessment Principles*. IPC, London.
13. Cummins, J.E. (1988) *The Ecologist*, **18**, 6.
14. Sax, N.I. (1984) *Dangerous Properties of Industrial Materials*. Van Nostrand Reinhold, New York.
15. Health and Safety Executive (1993) EH40/93 Occupational Exposure Limits.

16. Health and Safety Executive (1988) The Control of Substances Hazardous to Health Regulations 1988, Approved Code of Practice.
17. Hinton, J.F. and Havais, E.S. (1967) *Chemical Rev.*, **67**, 367.
18. Dimroth, K. and Reichardt, C. (1966) *Z. Anal Chem.*, **215**, 344.
19. Popov, A.I. (1976) *Solute–Solvent Interactions.* Vol. 2, Marcel Dekker, New York, p. 271.
20. Show-an Chen. *Chem. Engng Sci.*, **30**, 575 (1975).
21. Blanks, R.F. and Prausnitz, J.M. (1964) *Ind. Engng Chem. Fund.*, **3**, 1.
22. Petsche, I.B. and Debenedetti, P.G. (1989) *J. Chem. Phys.*, **91**, 7075.
23. Hughes, E.D. and Ingold, K.C. (1935) *J. Chem. Soc.*, 244.
24. Cox, B.G. (1994) *Modern Liquid Phase Kinetics.* Oxford University Press, Oxford.
25. Rossky, P.J. and Simon, J.D. (1994) *Nature*, **379**, 263.
26. Hynes, J.T. (1994) *Nature*, **369**, 439.
27. Dubois, J.E. and Ruasse, M.F. (1975) *J. Am. Chem. Soc.*, **97**, 1977.
28. Abraham, M.H. (1971) *J. Chem. Soc. B*, 299.
29. Kornblum, N., Perrigan, P.J. and Noble, W.J. (1963) *J. Am. Chem. Soc.*, **85**, 1141.
30. Kakabadse, G. (1984) *Solvent Problems in Industry.* Elsevier Applied Science, London.
31. Berson, J.A., Hamlet, Z. and Muller, W.A. (1962) *J. Am. Chem. Soc.*, **84**, 297.
32. Atherton, J.H. and Jones, I.K. (1995) Solvent selection. In *The Chemistry of Waste Minimization* (J.H. Clarke, ed). Blackie Academic, London.
33. Freedman, H.H. (1986) *Pure and Appl. Chem.*, **58–6**, 857.
34. Taverner, S.J. and Clarke, J.H. (1995) Phase transfer catalysis. In *The Chemistry of Waste Minimization* (J.H. Clarke, ed). Blackie Academic, London.
35. Wang, M. and Ou, C. (1994) *Ind. Engng. Chem. Res.*, **33**, 2034.
36. Kim, S. and Johnson, K.P. (1988) *Chem. Engng. Commun.*, **63**, 49.
37. Contract Chemicals Ltd Literature (1995) *Envirocats*, Merseyside L34 9HY, England, p. 15.
38. Brown, S. (1993) Tailoring reaction solvents for selectivity and recoverability, *North West Branch Papers*, **3**; *Inst. Chem. Engng*, **8**, 1.
39. Sharratt, P.N. (1993) Waste minimization for batchwise extraction processes. *IChemE Symp. Ser.*, **132**, 81–94.
40. Sharratt, P.N. and Sparshott, M. (eds) (1996) *Case Studies in Environmental Technology.* IChemE, Rugby.
41. McConrey, I. (ed.) (1995) Effluent Treatability and Toxicity, IChemE N.W. Branch Symposium Papers, No. 5, November.
42. Levin, M.A. and Gealt, M.A. (1993) *Biotreatment of Industrial and Hazardous Waste.* McGraw-Hill, New York.

4 Agitation
K.J. CARPENTER

4.1 Agitator selection

The selection of the correct agitator and vessel internal fittings is often extremely important in batch process design. Generally, laboratory reactors will be agitated vigorously unless specifically designed to mimic large-scale operation. Mathematical reactor models also often assume complete homogeneity. However, at full scale the reactor contents will not always be well mixed and may not be even close to homogeneous. There is therefore a real danger that incorrect agitator selection can lead to a vessel's contents separating or a reaction giving much poorer yield, selectivity or rate than expected.

4.1.1 Agitator duties

The most important first step in agitator selection is to define correctly the design duty. The normal classification of agitator duties is presented in Table 4.1. Defining the duty involves determining the key parameter(s) that dominate the selection, bearing in mind that it will be normal for the agitator and vessel to have to achieve several duties, often simultaneously. For example, it would not be unusual for a reactor to have to disperse a gas into a mixture of two immiscible liquids with a solid feed or product. Thus the agitator is required to disperse the liquids, one as drops and one as the continuous phase, disperse the gas as bubbles to achieve mass transfer and prevent the solids from settling out on the vessel base. In practice, the most common cause of an agitator failing to work as it should is not miscalculation of power or rotational speed, but simply incorrect definition of the controlling duty. Close cooperation between the development chemist and the process engineer is required to establish the controlling parameter from laboratory experiments, particularly in complex batch systems.

The controlling duty will normally be selected from those listed Table 4.1. In order to match the available agitator types with the duties, the most common agitator designs will be discussed in turn. The important features of their design and operation will be discussed to show why they are suitable for the duties as listed.

Table 4.1 Agitator duties and impeller selection

Duty	Sub-duty	Impeller design
Difficult rheology	non-Newtonian high viscosity	large diameter close clearance low speed
Miscible liquid blending and reaction	semi-batch, controlled addition pH control fast[*] competing reactions very fast[†] competing reactions	axial flow/mixed flow turbine mixed flow/radial flow
Disperse immiscible liquids	control phase stability control drop size/area design for mass transfer	mixed flow
Gas dispersion (liquid-phase continuous)	control bubble size design for reaction and mass transfer	high gas rate (>0.5 v.v.m.) disc turbine (concave blades) low gas rate (<0.5 v.v.m.) mixed flow (up or down)/disc turbine
Solids in liquid	floating settling deagglomeration	combined turbines/partial baffles axial flow turbine/mixed flow high shear
Heat transfer	low viscosity high viscosity	any turbine close clearance/large diameter

[*] Reactions whose yield is influenced by the extent of macromixing.
[†] Reactions whose yield is influenced by the intensity of micromixing.

4.1.2 Agitator types

For ease of classification the common agitator types are divided into two main groups: close clearance/large diameter and small diameter/high speed impellers.

(a) Close clearance/large diameter agitators

Close clearance agitators are so called because the clearance between the blade tip and the vessel wall is quite small, the agitator diameter will typically be between 70 and 100% of the vessel diameter. They are operated in the laminar to transitional flow regimes, as defined in Section 4.2.2. The rotational speed will be low because of the large diameter and the power and torque limits on most normal designs. They are used without baffles.

The most common designs are listed in Table 4.2, starting with large diameter 'paddles' at the lower end of the viscosity range and increasing in complexity as the viscosity increases. For most batch reactor and blending duties at apparent viscosities up to $20\,\mathrm{Ns\,m^{-2}}$ multiple 'paddles', large diameter turbines and the proprietary Intermig from EKATO are the best designs. In the past the Anchor agitator has been commonly used. It is reasonable for promoting wall heat transfer, as are the other designs, but relatively poor at general mixing. At viscosities above $20\,\mathrm{Ns\,m^{-2}}$ the more complex helical ribbons/helical screw designs are required, followed by dough blenders and specialized proprietary blenders for paste-like materials.

Table 4.2 Close clearance/large diameter agitators

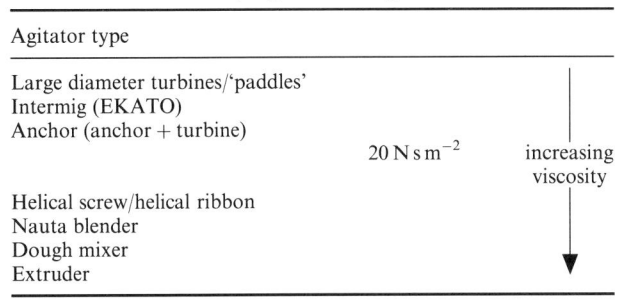

Agitator type		
Large diameter turbines/'paddles' Intermig (EKATO) Anchor (anchor + turbine)		
	$20\,N\,s\,m^{-2}$	increasing viscosity
Helical screw/helical ribbon Nauta blender Dough mixer Extruder		↓

Typical duties for close clearance agitators are general bulk blending for high viscosity and non-Newtonian materials and to promote wall heat transfer. They would be used for reactions where the batch viscosity is high and the material non-Newtonian, for high solids content crystallizations, for high viscosity polymer melts and for distillations from high viscosity materials.

(b) Small diameter/high speed turbines

Turbines are generally used for low to moderate viscosity applications. Their diameters are 0.25–0.5 times the vessel's diameter. They run at higher speeds and normally operate in the turbulent region (Section 4.2.2) at full scale. They will often be used in conjunction with baffles but can be used without. They develop highly complex three-dimensional flow patterns, but a time-averaged picture in a thin vertical plane across a diameter reveals a generally axial or radial flow pattern or a mixture of the two.

Axial flow turbines produce a time-averaged flow pattern as shown in Figure 4.1 for an agitator pumping downwards, the most common mode of operation. The bulk flow generated by the turbine moves centrally down the axis of the vessel, sweeping the base and returning up the wall region. They are designed to produce a high flow to power ratio, achieved by making the blades of an aerodynamic shape with the bulk of the energy used to induce flow, with little turbulent loss. They therefore have a relatively

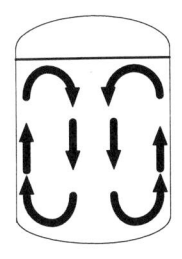

Figure 4.1 Flow pattern for an axial flow impeller.

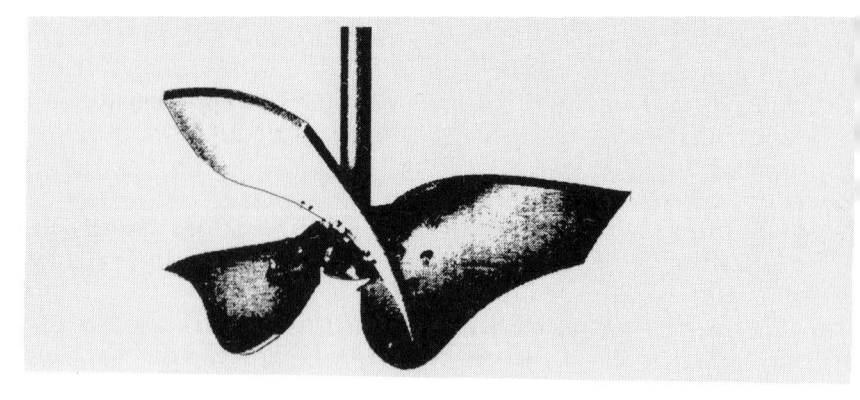

Figure 4.2 Typical axial impeller.

low power number (Section 4.2.2) and are run at relatively high rotational speed. Most designs are derived from three-bladed propellers, with several manufacturers having developed their own patented shape. A typical design is illustrated in Figure 4.2.

Axial flow turbines are good for duties which do not require a high level of turbulent energy, but do require high liquid flow. Thus they are especially good for suspending settling solid particles and for bulk blending of low viscosity liquids. For many simple blending duties they will be a cheap option because they tend to be 'off-the-shelf' items from a number of suppliers.

Radial flow turbines should be operated with baffles for them to provide an effective mixing action, otherwise they will simply produce a circular flow around the vessels with no circulation from top to bottom. The time-averaged flow pattern produced by radial flow turbines with baffles is shown in Figure 4.3. The main liquid flow discharges from the turbine radially outwards and with rotation around the vessel axis. The rotational flow is damped by the baffles and at the wall the radial flow separates to produce upper and lower circulation loops. The upper loop turns over at the surface and returns down the centre to the agitator. The lower loop

Figure 4.3 Flow pattern for a radial flow impeller.

turns along the vessel base and returns to the agitator from underneath. If a radial flow agitator is placed very close to the vessel base, less than $0.1D$, for example, the lower loop cannot exist and only the upper circulation will be seen. In this case the region below the agitator will tend to be relatively poorly mixed and segregated from the bulk, although it will be a small proportion of the total volume.

The blades of radial flow turbines are vertical, either attached directly to the shaft, as in a flat blade turbine, or connected to a disc, as in a Rushton disc turbine, illustrated in Figure 4.4.

A radial flow turbine is characterized by a relatively low flow to power ratio, with much of the energy dissipated by turbulence in the region immediately around the agitator. Calabrese [1] has shown that the maximum rate of energy dissipation occurs in the vortex region that exists behind each blade and can be 120 times the average rate in the vessel.

Radial flow turbines are generally used for duties where their high local energy dissipation rate is an advantage, particularly mass transfer between a gas and a liquid. For this particular duty, curved turbine blades are recommended, as developed by Van't Riet *et al.* [2, 23] and shown in Figure 4.5. Recently, several manufacturers have developed their own profiled versions of the curved blade. Radial turbines are also good for very fast competing reactions. The high local energy dissipation rate can be exploited to maximize the yield of the desired product, provided the reagents are fed directly into the stream entering the impeller. This issue is discussed in greater depth in Chapter 5. Radial flow turbines should not be used for viscous or non-Newtonian fluids.

As the name implies a *mixed flow turbine* has a flow pattern with features of both radial and axial flow. They can be used with either an upward or a downward pumping action. The most common mode is downward but there some relatively recent applications [4] using upward pumping designs for gas dispersion (Section 4.6). The discharge from a downward pumping agitator will be at an angle to the vertical with a large upper circulation. If

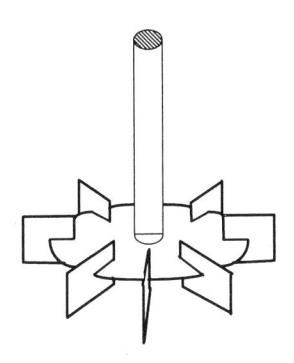

Figure 4.4 Flat blade turbine and Rushton turbine.

Figure 4.5 Curved blade turbine.

a lower circulation loop exists it will be much smaller, as in Figure 4.6. When used at a low clearance from the base, e.g. less than 0.1D, the mixed flow will behave exactly as a radial turbine [3].

The blades will normally be set directly on the shaft at an angle. At a high angle (to the horizontal plane), e.g. 60°, the agitator will be more like a radial turbine with relatively high power compared to flow. At a low angle, e.g. 30°, it will more closely resemble axial flow. The most common angle is 45°.

The 45° downward pumping angled blade turbine is in many ways the workhorse of the fine chemicals industry and can be used reasonably well for most duties, except gas–liquid mass transfer at high gas rates, e.g. greater than 0.5 v.v.m. (volumes of gas per liquid volume per minute).

(c) Special designs

There are several proprietary designs of so-called *high shear agitators*. They are generally characterized by having a very low power number (Section

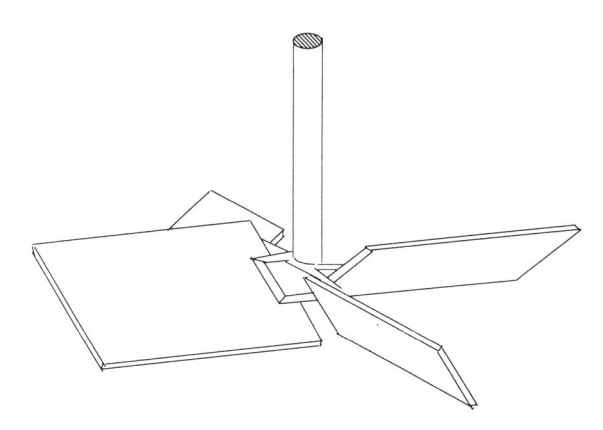

Figure 4.6 Pitched blade turbine for mixed flow.

4.2.2) and therefore can be run at very high speed without excessive power consumption. They also generate very low flow, hence produce an intense region of energy dissipation local to the blade. They can be open or have the blades within a holed or slotted shroud. A shroud makes the intense region even more localized. An open type commonly will be a sawtooth design. This looks like a circular saw blade with the teeth bent alternately upwards and downwards. For further details it is recommended that the manufacturers be consulted. Typical designs are manufactured by Ystral, EKATO, Greaves and Silverson. They are usually used for deagglomerating solids, dispersing filter pastes and forming emulsions. It is common to see them used in combination with a larger agitator in a vessel that provides the bulk circulation necessary to ensure that the material to be dispersed is made to flow through the intense region within a reasonable processing time.

It is common to use agitators in *combinations*, either to achieve a mixture of duties or because of the high H/T ratio of the vessel. Table 4.3 shows the number of impellers generally used for commonly encountered H/T ratios. The most frequent applications of mixed designs are duties where flow and dispersion are both required. For example, a disc turbine in the lower part of a vessel to disperse gas or deagglomerate solids with an angled blade turbine above it to increase the flow and circulation beyond that which the disc turbine could achieve alone. Another common combination design is to use an agitator deliberately to cause turbulence at the liquid surface to draw in solids which would tend to float or to reincorporate sparged gas. The batch hydrogenator is commonly designed in this way (Section 4.6.2).

It is common for the corrosive nature of the reaction mass to favour the use of *glass-coated steel* as the material of construction of the vessel, agitator and baffles. This prevents the use of full wall baffles and requires the use of beavertail baffles to ensure good vertical circulation. Until recently, glass-lined steel also restricted the range of agitator designs that could be specified without specialized manufacture. The most mechanically robust and widely available design was the three-blade Pfaudler retreat curve, illustrated in Figure 4.7. Recently, the novel technique of using interference fit agitators attached to the shaft by a cryogenic technique has enabled a range of designs to be used which more closely resemble the ideal metal shapes that would otherwise be preferred.

Table 4.3 Number of impellers required for different H/T ratio vessels

H/T	Number of impellers
up to 1	1
1–1.3	2
1.15–1.5	3
1.3–2	4

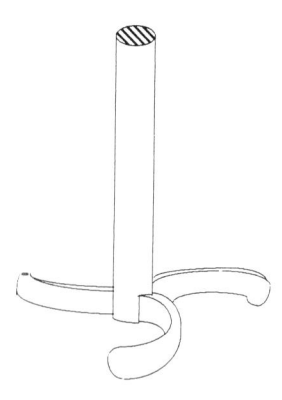

Figure 4.7 Pfaudler impeller.

4.2 Calculation of agitator power, discharge flow and mixing time

4.2.1 Typical power levels

For low viscosity duties, the power input provides a very crude guide to the level of agitation which may be expected. As a general guide, typical values based on general experience are given in Table 4.4.

4.2.2 Calculation of power

The power required to drive an agitator at a given rotational speed is given by:

$$P = \text{Po} N^3 D^5 \rho \qquad (4.1)$$

Po is the agitator power number. It is a function of the agitator design and the flow regime around it, characterized by the Reynolds number (as for the friction factor in pipe flow). In this case Re is the *agitator* Reynolds number:

$$\text{Re} = ND^2 \rho / \mu \qquad (4.2)$$

Table 4.4 Typical power levels

Degree of agitation	Power level (W kg^{-1})
Low	0.16
Moderate	0.5
High	1.6
Very high	3.3

Although every specific agitator with its accompanying vessel and baffles has its own Po, it has been shown that Po is largely independent of scale and can be calculated from a number of standard designs, with correction factors for minor deviations. Agitator manufacturers should be able to provide Po curves for their designs. Figure 4.8 is a typical Po curve for a 45° downward pumping mixed flow agitator with full wall baffles. It can be seen that under fully turbulent conditions, Re > 1000, Po is constant. Under laminar conditions, Re < 100, Po is inversely proportional to Re. Bujalski *et al.* [5] have shown a slight scale dependence for Po for a Rushton disc turbine, but this is generally regarded as insignificant for most normal applications.

4.2.3 Discharge flow

In practice it is not easy to measure nor to define precisely the discharge flow from an agitator. It is anyway not a very useful measure as the flow around the vessel will be significantly greater than the flow emerging directly from the agitator blades. There have been many reported measures of discharge flow over the years and with the advent of laser-Doppler systems, measurement of velocities has become more common and more standardized. The flow direct from the agitator blades (Q) has been correlated as:

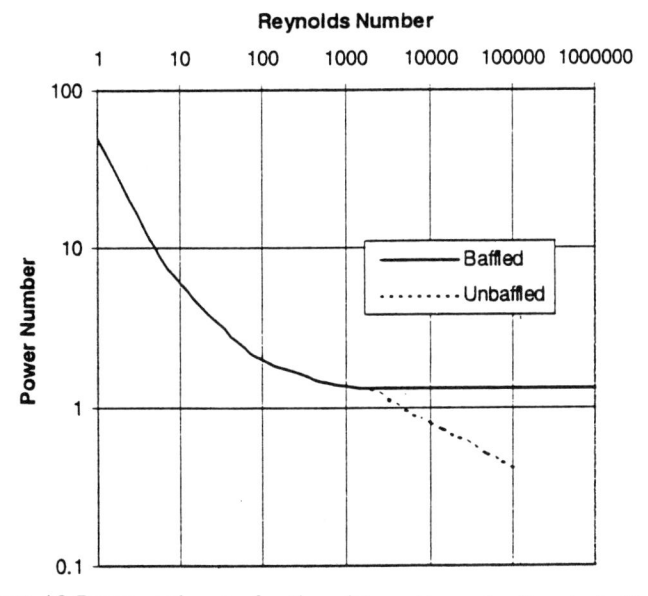

$$Q = K_f ND^3 \tag{4.3}$$

Figure 4.8 Power number as a function of Reynolds number for a typical impeller.

Table 4.5 Agitator flow numbers and power numbers

Agitator	Flow number	Po
Rushton disc turbine	0.7–0.85	5–6
Marine propeller	0.4–0.5	0.33–0.7
45° turbine	0.75	1.2–1.5

where K_f is the agitator flow number. Table 4.5 lists the typical ranges of flow numbers and turbulent Po for the most commonly measured agitators. It can clearly be seen that whilst the range of flow numbers is not great, Po values are vastly different. Hence the ratio of flow to power is vastly different.

4.2.4 Mixing time

Mixing time, or blending time, is a useful measure of how well mixed a vessel is under test conditions, but, like the discharge flow, is not directly useful for calculation purposes in itself. It is normally defined as the time taken to blend a tracer of the same density and viscosity as the bulk into a bulk liquid until the variation in concentration fluctuations measured at a point (or several points) has permanently decreased below a specified value. For example, the 99% blending time means that the fluctuations in the concentration measurement are always below 1% of its initial value. The initial value (the maximum possible measured fluctuation in concentration) is clearly the difference between the injected tracer concentration and the bulk concentration at time zero. Mixing time is dependent on bulk circulation and, as such, it is a very important concept. It is relatively easy to measure and to understand how mixing time varies with changes in agitator design and rotational speed enables an understanding of how other processes which rely on bulk circulation will also vary. The selectivity and yield of many chemical reactions depend on bulk circulation hence mixing time can provide a useful means of comparing or characterizing reactors for reaction yield and selectivity. It also provides a very valuable means of designing large-scale reactors from small-scale laboratory development experiments.

Much of the most accurate and recent measurements of mixing time has been carried out by the UK-based research club Fluid Mixing Processes (BHR Group). They have published a description of their measurement technique [6] and useful correlations to predict mixing time from physical properties and agitator design [7]. Their correlations are separated into above and below a critical Reynolds number Re_c. Above the critical value, in the turbulent region, mixing time is independent of Re; below, in the transition region, it is dependent.

$$Re_c = 6370/Po^{1/3} \quad (4.4)$$

In the turbulent region, $Re > Re_c$

$$Po^{1/3} ReFo = 5.2 \qquad (4.5)$$

where $Fo = \mu\theta/\rho T^2$. In the transition region, $Re < Re_c$

$$Po^{1/3} Re\sqrt{Fo} = 184 \qquad (4.6)$$

4.3 Power and circulation in non-Newtonian fluids

In fine chemicals manufacture the most common forms of non-Newtonian rheology are shear thinning behaviour and yield stress behaviour (Bingham plastics and yield pseudoplastics). Rheological classifications will not be discussed here; for more detail refer to Govier and Aziz [8]. The major difficulties which non-Newtonian behaviour brings are the prediction of a suitable average viscosity to estimate Re hence power demand and to ensure good circulation throughout the reaction mass.

4.3.1 Calculation of power

There are two popular techniques for the calculation of apparent viscosity at a given impeller speed, those of Metzner and Otto [9] and Rieger and Novak [10].

The Metzner–Otto technique is based on laminar flow but has been used in the transitional region. It requires that the relationship between shear stress (τ), hence apparent viscosity (μ_a) and the shear rate (γ) be determined. These data can be obtained using a standard viscometer, usually over the shear rate range 0.1–$500\,s^{-1}$. Increasing and decreasing the shear rate will indicate any time dependency. If there is any, the equilibrium curve should be used in subsequent calculations. The presence of yield stress can be seen by inspection of the data. A pseudo-mean shear rate for the flow around the agitator can be estimated from:

$$\gamma_{mean} = K_{mo} N \qquad (4.7)$$

Metzner–Otto constants (K_{mo}) for a range of the most common agitators are presented in Table 4.6.

The Rieger and Novak method, which is generally regarded to give a better measure of the rheology the agitator will experience, is to use a model agitator in a viscometer. By measuring the torque, at a range of rotational speeds in the laminar region, the power can be obtained from:

$$P = \pi N \Gamma \qquad (4.8)$$

Table 4.6 Metzner–Otto shear rate constants

Agitator	K_{mo}
Propeller	10
Disc or flat blade turbine	11.5
Angled blade turbine	13
Anchor	25*
Helical ribbon	30
EKATO Intermig ($D = 0.8T$)	40

*The value for an anchor depends on the rheology of the material, see Nienow and Elson [11].

and Po obtained from the power (using equation 4.1). From a predetermined Po–Re relationship for that agitator type, the relevant value of Re, and hence μ_a, can be obtained directly as a function of rotational speed.

4.3.2 Circulation

The major problem in handling non-Newtonian fluids is to ensure good bulk movement throughout the mass. This is particularly true of yield stress fluids. In this case the analysis of Nienow and Elson [11] as further developed by Etchells [12] provides a good design correlation. It has been observed that the fluidized region around the agitator approximately adopts the shape of a right circular cylinder, of height H_c and diameter D_c, with the remaining fluid remaining stagnant. With increasing agitator speed the mobile volume increases up to the point at which it reaches the vessel walls. The fluid is then moving right across the vessel and $D_c = T$. The ratio H_c/D_c is a function of the agitator design. Observed values are given in Table 4.7 for common agitator types. The correlation used to determine D_c is:

$$\{D_c/D\}^3 = \text{PoRe}_y/[(H_c/D_c + \tfrac{1}{3})\pi^2] \tag{4.9}$$

Here Re_y is the *yield* Reynolds number.

After the mobile volume has reached the vessel wall ($D_c = T$), the cylinder height increases in proportion to the speed:

$$H_c \propto N^E \tag{4.10}$$

Table 4.7 also shows the relevant values of E. However, it can be shown that the most appropriate design for a system that is not fully mobile at the point

Table 4.7 H_c/D_c and E values for common agitator types

	Disc turbine	45° turbine	Two-flat bladed paddle
H_c/D_c	0.4	0.55	0.45
E	0.88	0.86	0.88

when $D_c = T$ (i.e. the fluid depth is greater than H_c at $D_c = T$) is to use multiple impellers. These should be placed so that each generates its own fluidized region overlapping with adjacent regions and thereby ensuring good bulk movement throughout the fluid depth.

4.4 Design to disperse solid particles

Solid particles can either sink or float and the agitator selection and design are different for each of these duties.

4.4.1 Disperse sinking particles

There are two states of particle suspension used as criteria for design. The more common is the 'just suspended' condition, where no particles remain stationary on the vessel base for any time. This is the minimum condition for a successful design. It should be noted that the local concentration of solid particles will not be constant throughout the fluid, although in most applications this has little consequence. The second design case is complete homogeneity where the solids concentration is at the mean value throughout the vessel.

The preferred agitator design is the axial or mixed flow turbine and the vessel should have full wall baffles. There are a number of specific designs of vessel base, sometimes with a draught tube to give good suspension with minimum power input, which have been used widely for very large crystallizers, but will not be considered in any further detail here. Note that it is virtually impossible to prevent solid particles from settling into the bottom of a 90° conical bottomed vessel no matter what agitator design is used and these vessels must be avoided if good solid distribution is required.

(a) Just suspended
The correlation developed by Zweitering [13] has been successfully used for many years:

$$N_{js} = S\nu^{0.1} d_p^{0.2} [g(\Delta\rho/\rho_L)]^{0.45} X^{0.13} / D^{0.85} \qquad (4.11)$$

Although it appears quite empirical, it is based on dimensional analysis and an analogy to particles in pipe flow. The constant S depends on the system geometry. A number of standard S values are presented in Table 4.8. For unusual systems where S is not available, a simple experiment can be carried out with model solid particles and fluid with the correct physical properties, preferably in a vessel around 0.3 m diameter. Simple observation indicates when no solid rests permanently on the base. It is then possible to measure or calculate the power input and use the same specific power input

Table 4.8 Zwietering constants for typical systems

	D/T	S
Three-bladed marine propeller	0.5	6.8
45° mixed flow downward pumping	0.5	5.8
	0.25	7.1
Rushton six-blade disc turbine	0.5	4.25
	0.33	7.1
	0.25	12.2
45° mixed flow upward pumping	0.5	7.4

(power/batch mass $= \varepsilon$) at full scale as at the just suspended point in the test. As the Zwietering correlation suggest that reduces as scale increases, this procedure would result in overdesign, which will be safe in that it will not fail to suspend the solid particles. In most fine chemical systems the overdesign will not incur significant cost.

(b) Complete homogeneity

It would be unusual in batch operation to design for complete homogeneity, although it would be more common in continuous systems, where it may be required to feed forward a slurry of constant composition. The power input will be much greater than the just suspended condition, generally greater than $2\,W\,kg^{-1}$ [14]. The design speed and power are best estimated from small-scale experiments in such cases, with the full scale design being at the same specific power (ε) as the successful small-scale operation. Nienow [15] provides further details. It should also be noted that for complete homogeneity in a vessel where $H/T > 0.5$ multiple impellers will be required and D/T should be 0.5.

4.4.2 Draw down floating particles

There are several circumstances where design requires solids which would otherwise float to be drawn through the liquid surface. This may be because the solids have an inherently lower density than the liquid or because they become partially wetted and trap air into an agglomerate. Joosten [16] and Smith [17] have both devised designs and correlations for specific duties. However, general experience in fine chemical manufacture suggests that in order to cope with all cases a two tier partially baffled design as in Figure 4.9(a and b) should be used. The design method is rather empirical; it is recommended that a small-scale test be carried out and the full scale design be based on the same specific power input ε. Again, this will be an overdesign, i.e. there will be more power than required. Generally this will be safe. However, in the case where excess gas can be drawn in with the solid and where this may be a problem, it is recommended that a drive system be used whose speed can easily be altered on start up. This can be achieved, e.g. using pulleys, so that the speed can be reduced slightly on

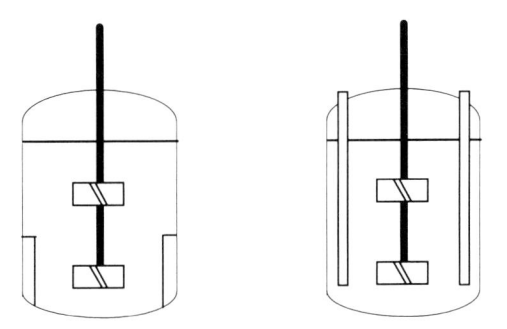

Figure 4.9 Configurations for the incorporation of floating solids. Systems to incorporate floating solids with (a) part baffles and (b) beavertail baffles.

start up by changing the size of one or more pulleys. Such problems can arise where the final mixture is viscous, in which case the gas drawn in will be difficult to disengage.

4.5 Design for two or more liquid phases

There are very few industrial processes outside water treatment or aqueous dyestuff manufacture that do not involve handling mixtures of immiscible liquids. Usually, mass transport across the interface is required either for reaction or simply to extract one component selectively. In either case the requirement for the full scale design will be to match the performance of laboratory experiments either in rate, selectivity, rate of subsequent coalescence for separation, control of drop size or size distribution or any combination of these. The underlying physical processes are the distribution of both phases, break up of one phase into drops, mass transport (possibly with reaction), coalescence and separation.

4.5.1 Miscibility

By far the most common immiscible liquid dispersions contain an organic (o) and an aqueous (w) phase. There are very few process stages in the fine/effect chemical industry which do not involve such mixtures. They occur most frequently as reactions between aqueous soluble materials and organic compounds either neat or in solution in solvent, or extraction stages for separating compounds by selective partition between the phases. There are literally thousands of examples across the range of fine chemicals processing.

Immiscible liquid pairs are not exclusively organic and aqueous, however. There are many organic–organic immiscible systems, e.g. dimethylsulphoxide (DMSO) and hydrocarbons such as heptane. A miscibility chart such as Figure 4.10 (after Godfrey [18]) shows which common materials are

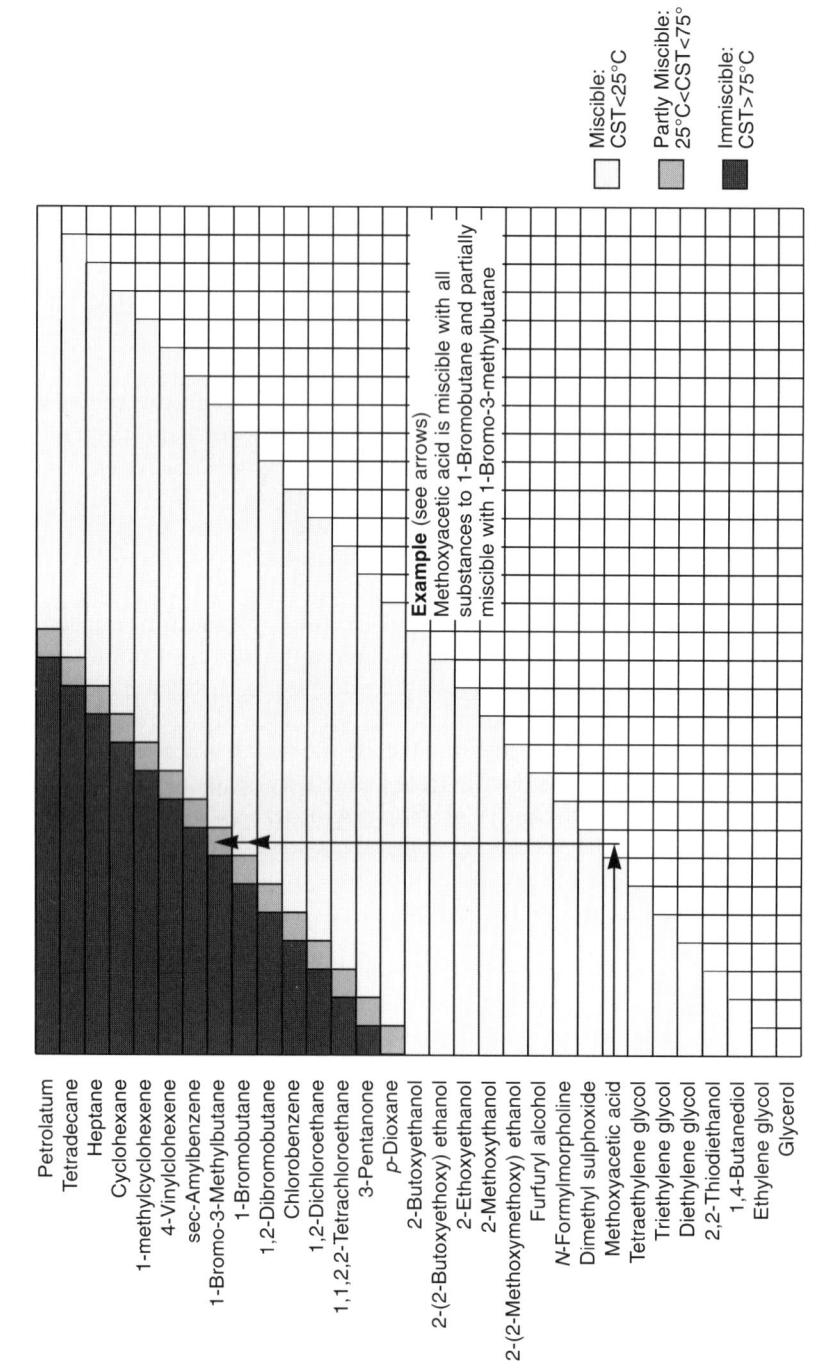

Figure 4.10 Miscibility chart (after Godfrey [18]). Reprinted with permission from Chemtech, Copyright 1972, American Chemical Society.

immiscible. There are also aqueous–aqueous immiscible systems, although less common. The most frequently quoted example is the PEG-surfactant-loaded system used for protein extraction. It is also possible for organics with a high aqueous solubility in conjunction with an hydrophobic organic and water to form a three-liquid-phase system. Silicon-based oils are also capable of forming two- and three-phase systems with organics and water. In each case the basic underlying principles are the same and the following analysis can be applied.

4.5.2 Phase continuity and phase inversion

When designing for any liquid pair, it is usually important to know which phase should be controlled as the dispersed phase – present as discrete drops – and which should be continuous. For example, consider the case illustrated in Table 4.9. Here the aqueous phase is derived from a biotransformation reaction and the objective is to recover selectively one component by liquid extraction with a suitable solvent. The aqueous phase contains products of cell breakdown, which have the effect of stabilizing the aqueous film. The coalescence time for the aqueous continuous system shows this. In effect the drops do not coalesce (coalescence time 15 h) and certainly not quickly enough to run a practicable process. The organic solvent phase is essentially clean; hence, for organic continuous operation, the organic film drains very easily and the coalescence time is even rapid enough (150 s) for continuous operation. It is therefore clear that this system must be controlled organic continuous. However, as is common in biologically derived systems, the feed is very dilute and in order to achieve realistic concentrations to improve capital productivity, it is desirable to increase the concentration in the exiting organic phase which requires the phase volume to be minimized. The result is that the minimum organic must be used which still allows the organic phase to be continuous. This means that the system is run very close to the phase inversion boundary – the point at which the organic phase would disperse and the aqueous phase become continuous. (Phase inversion is discussed further below.) In this example, a phase ratio of 0.6:0.4 (a:o) can be operated robustly, with the absolute boundary at

Table 4.9 Example of the effect of continuity on coalescence rate

Parameter	Value
Aqueous density	1060 kg m^{-3}
Organic density	800 kg m^{-3}
Separation time (o in a)	15 h
Separation time (a in o)	150 s
Phase ratio for minimum organic (a:o)	1.37:1
Volume fraction for minimum organic (a:o)	0.6:0.4

0.7:0.3. Note that in clean systems of solvents typical of chemical manufacture it is common for the aqueous continuous system to coalesce more slowly than the organic. Surface active materials can have a dramatic effect on coalescence.

An example of the effect of differences in phase continuity on reaction selectivity is illustrated in Figure 4.11. Here, the reaction requires the aqueous feed to mix rapidly with the aqueous phase in the reactor to achieve the maximum yield. For the aqueous dispersed case, if coalescence is relatively slow, the feed has to disperse throughout the reactor and coalesce with the existing dispersed drops, before the necessary mixing can take place. The result is that the drop population contains a range of reagent concentration from fresh feed to totally depleted. Operation with aqueous continuous gives a much more robust system as the aqueous concentration is almost homogeneous at the bulk average. Aqueous continuous will give the better yield.

Figure 4.12 is a simplified phase boundary diagram. It illustrates ranges of volume fraction where it is guaranteed that the heavy phase will be continuous, where the light phase will always be continuous, and where either can be continuous or dispersed. This is known as the ambivalent region. Consider moving up line A–B at constant agitator speed. Initially there is little light phase there and the system must be heavy continuous, region HC. As light phase is added, eventually the point B will be reached where the volume fraction of the heavy phase is so small that it cannot remain continuous. Here the system will phase invert and become light continuous. This is the phase boundary. Consider the opposite case of starting with light phase there and adding heavy, i.e. moving down the line C–D. Initially the light phase is continuous, region LC and the volume fraction of the heavy phase is so small that it remains dispersed. At point D, however, there is so much heavy phase there that it must be continuous and the system

Aqueous Dispersed   Organic Dispersed

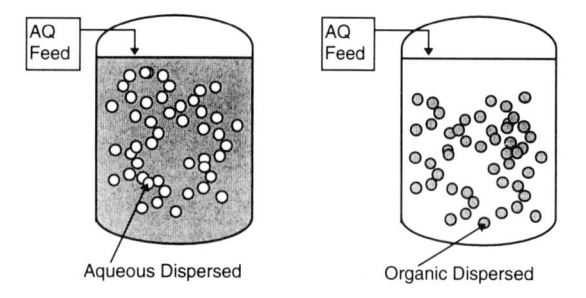

Figure 4.11 Selectivity for liquid–liquid reaction. (a) Aqueous phase dispersed. Aqueous drops deplete rapidly. Their concentrations range from C_{feed} to 0. Selectivity depends on addition rate and droplet break-up/coalescence. (b) Organic phase dispersed. All aqueous phase is at C_{bulk}. Selectivity depends on addition rate and bulk circulation. This is easier to model in the laboratory.

AGITATION 125

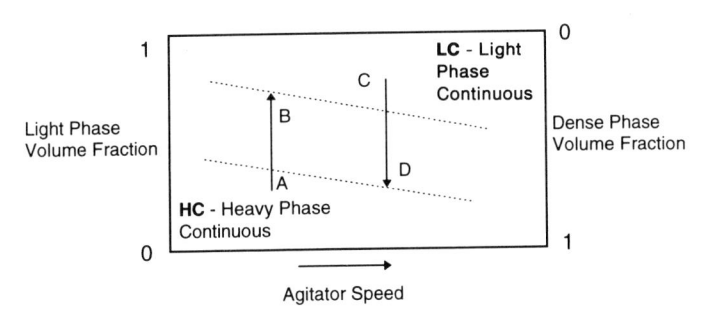

Figure 4.12 Phase boundary diagram.

again phase inverts. This is the other phase boundary. Thus region HC is always heavy continuous and LC always light continuous. In between the boundaries, which phase is continuous depends on how the point was reached, i.e. either phase can be continuous. This is the ambivalent region. The boundaries are largely dependent on the physical properties of the two liquids, with a minor effect of agitator type. Note that the heavy continuous (lower) boundary has no significance for a heavy continuous dispersion, similarly the light continuous (upper) boundary has no significance for a light continuous dispersion. Within the ambivalent region, many system properties can affect stability.

As a general rule in normal chemical manufacture, phase inversion should be avoided. It can lead to a fine dispersion and can give a step change in mixture properties. For example, the dispersion's viscosity can be affected dramatically. The dispersion's viscosity depends mainly on the viscosity of the continuous phase and the volume fraction of the dispersed phase. A typical industrial example involved a high viscosity organic oil to be dispersed in an aqueous phase to undergo reaction. However, as the reaction proceeded, with more oil phase added, the system phase inverted. The result was a stable, high viscosity emulsion and very poor heat transfer. With a high interfacial area, hence potentially high reaction rate and poor heat transfer, the system could only be operated with a very restricted addition rate after inversion had occurred. In this example, the key to achieving reproducible reaction time was to maintain the organic phase dispersed. In some applications, however, a phase inversion process may be used to generate a stable fine dispersion. The most common examples are the manufacture of low fat spreads and butter.

4.5.3 Phase distribution – the just dispersed condition

Having established in the laboratory which phase should be dispersed and an acceptable volume ratio for stable operation, the next step in the design of an agitated vessel for the duty is to select the required agitator and baffle shape and size and the necessary speed and power input. In practice for low viscosity dispersions any turbine-type impeller can be used, provided the vessel is baffled. Without baffles, the two liquids will simply rotate with a vortex at the surface and at the interface. As speed is increased the vortex at the interface approaches the impeller and when it reaches it the light phase will start to disperse in the heavy, but not before. In glass-lined equipment even a Pfaudler three-bladed retreat-curve can be used provided a high mass transfer rate is not required.

The minimum condition for each phase to take part fully in the process is when there is no free separated phase at the top or bottom of the dispersion. This is often referred to as the 'just dispersed condition'. There is little scientific treatment of the parameters which determine the just dispersed condition and design is largely empirical. The physical processes involved are bulk circulation of the two liquids and break-up of the dispersed phase. An empirical, but generally 'safe' approach is to use constant specific power input (ε) as a criterion based on small-scale tests. The recommended procedure is:

(i) Run a test at least at $T = 0.2$ m.
(ii) Observe the point where the phases are dispersed.
(iii) Calculate the specific power input (W kg^{-1}).
(iv) From the large-scale mass calculate the large-scale power input and hence the impeller speed.

A number of workers have produced correlations based largely on small-scale model experiments, which can be used in the place of running tests. For turbine-type impellers, the correlation of van Heuven and Beek [19] was derived from observations up to $T = 1.2$ m using a Rushton disc turbine, $D/T = 0.3$ with full wall baffles (Po = 5.8):

$$N_{vHB} = 7.9 D^{-0.77} \sigma^{0.08} (\Delta\rho/\rho_c)^{0.38} \mu^{0.08} \rho_c^{-0.16} (1 + 2.5 V f_d)^{0.9} \quad (4.12)$$

Lines [20] showed that the 'light-in-heavy' dispersion should be correlated separately from the 'heavy-in-light', using Pfaudler retreat curve impellers and two beaver tail baffles. His experiments were at $T = 0.3$ m scale with $D/T = 0.6$:

$$N_{\text{L-in-H}} = 0.039 \rho_c^{0.22} \Delta\rho^{0.33} \mu_c^{-0.133} V f_d^{0.008} \quad (4.13)$$

$$N_{\text{H-in-L}} = 15.14 \rho_c^{-0.38} \Delta\rho^{0.35} \mu_c^{0.04} \mu_d^{0.06} \sigma^{0.08} V f_d^{0.14} \quad (4.14)$$

Lines' correlations are wholly empirical, based on a very wide range of physical properties with some of the solvent mixtures chosen by statistical design to separate the effects of the variables. Hence, they can be used to estimate the effect of a change in physical properties. For design for a specific liquid pair, Lines' correlations can be simplified and when used to calculate a specific power input (*in W kg^{-1}*), also cover the data of van Heuven and

Beek. To engineering accuracy, the following correlations can be used for all turbine-type impellers.

$$\varepsilon_{\text{L-in-H}} = 6.138 \times 10^{-7} \rho_c^{0.66} \Delta\rho^{0.96} \mu_c^{-0.4} V f_d^{0.24} \quad (4.15)$$

$$\varepsilon_{\text{H-in-L}} = 1.89 \times 10^{-3} \Delta\rho^{1.1} \mu_d^{0.21} V f_d^{0.36} \quad (4.16)$$

This procedure should provide more than the minimum power required, which is generally a safe design approach. The only potential danger is that putting in too much power in very sensitive systems, particularly near the phase boundaries, may cause emulsification, or for a light phase-dispersed system, premature inversion. Such sensitive systems are rare in practice.

4.5.4 Drop sizes

The size of drops generated during the agitation of a two-liquid-phase system is of interest for estimation of interfacial area, mass transfer rates and for the ultimate rate of phase separation. In some systems, the drop size is determined only by the mechanisms tending to break up droplets. In others, the drop size distribution depends both on drop break-up and coalescence phenomena.

(a) Drop break-up

For there to be no coalescence in a real system, careful steps must have been taken to exclude it, for example the surfactants and protective colloids used in suspension polymerization. There are several models of drop break-up, differing mainly in their assumption concerning the fluid mechanical forces responsible for the disruption of the interface. Calabrese [21] has reviewed the most common models. The Kolmogoroff–Hinze model as tested by Davies [22] for a range of equipment, is the most generally accepted for turbulent systems at other than very small-scale or high continuous phase viscosity. The basis for this model is that the maximum stable drop size is determined by the balance between the pressure difference across the interface $(2\sigma/r)$ and the turbulent energy dissipation rate $(\rho[v'(d)]^2$, where $v'(d) = c\varepsilon^{1/3} d^{1/3})$ provided that d is greater than the smallest scale of turbulence $(\lambda = [v^3/\varepsilon]^{1/4})$. The resulting correlation is then:

$$d_{\max} = \text{const.}\, \varepsilon^{-0.4} \rho^{-0.6} \sigma^{0.6} \{\mu_d \text{ correction terms}\} \quad (4.17)$$

The constant is expected to be of the order of 1. Different workers have proposed terms to correct the basic theory for the effect of a high dispersed phase viscosity. Thus for a reactor agitated by a turbine generating $5\,\text{W}\,\text{kg}^{-1}$ in a low viscosity water-like system, with an interfacial tension of $0.022\,\text{N}\,\text{m}^{-1}$ and assuming that the maximum energy dissipation rate is around $200\varepsilon_{mean}$ then $d = 100\,\mu$m. In practice this type of correlation can only be used to provide an order of magnitude value, particularly as the effective interfacial tension can be extremely difficult to estimate in a reacting or surfactant-stabilized system. It is more common to use the derivation to justify the use of constant ε to design from small-scale tests, assuming that $\varepsilon_{max} \propto \varepsilon_{mean}$ and $d_{mean} \propto d_{max}$.

When designing a full-scale reactor from small-scale tests, it must also be remembered that the above derivation refers to the steady-state situation. Several researchers using model systems have shown that it can take a long time for the drop size distribution to reach equilibrium. Chang and Calabrese [24] have proposed an estimate of $21\,000\,N^{-1}$ as the time to reach equilibrium. It is therefore important to design the system for the correct processing time (as a function of the impeller actual speed or mixing time) as well as the specific power input. Inevitably, the large-scale system will take longer in real time to achieve the desired drop size. This is because with a fixed geometry and specific power input the mixing time is longer at larger scale.

(b) Drop size for break-up and simultaneous coalescence

In this case, the system has not been deliberately stabilized against coalescence. The correlations for mean drop size are semi-empirical and derived from the break-up analysis outlined in the previous section. By assuming that:

$$\varepsilon_{max} \propto \varepsilon_{mean} \propto P/(V\rho) \tag{4.18}$$

then for a system of fixed geometry:

$$d_{32}/D \propto We^{-0.6} \tag{4.19}$$

Calabrese et al. [26] have extended this analysis and fitted their measured data from experiments with a Rushton disc turbine in a 0.213 m vessel to give:

$$d_{32}/D = 0.054 We^{-0.6}\{1 + 4.42Vi(d_{32}/D)^{1/3}\}^{0.6} \tag{4.20}$$

where Vi is a correction term for the dispersed phase viscosity.

$$Vi = (\mu_d ND/\sigma)\{\rho_c/\rho_d\}^{0.5} \tag{4.21}$$

Again this correlation can only be used to provide an order of magnitude estimate of the mean drop size. At the current state of knowledge, the preferred design method is small-scale testing with full scale design based on constant ε and constant processing time relative to mixing time. When used to design from small-scale tests, for geometrically similar systems, the above argument gives:

$$N_L = N_S\{D_S/D_L\}^{2/3} \tag{4.22}$$

Static mixer manufacturers have also correlated dispersion data in this form, e.g. Kenics [27] give the mean drop size for their mixer as:

$$d_{32}/Dt = 0.41 We_K^{-0.5} \qquad (4.23)$$

This is for the case $\mu_d/\mu_c = 1$, for other values corrections must be applied. Here, Dt is the pipe diameter and We_K is $\rho v^2 Dt/\sigma$.

4.5.5 Mass transfer

The widely used mass transport rate equation can be applied.

$$J_i = K_c\{\Delta C^*\} \qquad (4.24)$$

where J_i is the flux of transferring material $(\text{kmol s}^{-1})\,\text{m}^{-2}$ (total interfacial area). The total interfacial area can be estimated from the mean drop size and dispersed phase volume and the mass transfer coefficient can be estimated from correlations. However, it must be noted that prediction from correlation can only be regarded as a very crude estimate because of the sensitivity to trace impurities, e.g. surfactant species can increase the interfacial area, but can also dramatically block transport across the interface.

The standard two film theory can be used to obtain overall mass transfer coefficients on the basis of the mass transfer coefficients either side of the interface.

$$1/K_c = 1/k_c + CP/k_d \qquad (4.25)$$

Estimates of the relevant mass transfer coefficients can be obtained from:

$$k_d = 2/3\pi^2 \mathscr{D}/d_{32} \qquad (4.26)$$

and

$$k_c = \text{Sh}\mathscr{D}/d_{32} \qquad (4.27)$$

where the Sherwood number, Sh, is given by Rowe et al. [28]

$$\text{Sh} = 2 + 0.72 Re_p^{0.5} Sc^{1/3} \qquad (4.28)$$

and the total area is given by $A = 6V_d/d_{32}$.

In practice mass transfer rates are not usually limiting, except where combined mass transfer and fast competing reactions determine selectivity. In such cases the normal design procedure would be to carry out laboratory experiments in equipment which can be characterized, i.e. an estimate of power consumption and mixing time can be made. The full-scale design would then aim to match the specific power input and mixing time/reagent feed rate.

4.6 Design for dispersing gas

The gas to be dispersed can be driven into the liquid by an external compressor and gas sparger, sparged and reincorporated through the liquid surface or induced through specially designed impeller blades.

4.6.1 Sparged gas

The most significant difference between gas dispersion and other duties is the effect the gas has on the agitator performance. The presence of large amounts of gas reduces the power drawn by the agitator and reduces its pumping capacity. Van't Riet et al. [2] showed that the fall in performance is due to gas collecting within the vortex structures which form behind the blades. Generally the fall in power as gas rate is increased is measured by the ratio of power with gas to ungassed (P_G/P_U). It is normally plotted against the gas flow number $(F_L = Q_G/ND^3)$ either for increasing gas rate at constant speed [29] or sometimes for decreasing speed at constant gas rate. Nienow [29] has reviewed the formation of gas 'cavities' behind the impeller blades. The understanding of the interaction between the gas and the blade shape has led to the development of curved blade disc turbines (Figure 4.5) for which the P_G/P_U ratio remains close to 1 even at high gas rates. For the majority of applications it will be necessary to install enough power to ensure that the agitator can rotate with no gas, hence the full ungassed power will be installed.

As described below, the mass transfer performance depends on the energy input. The best design therefore utilizes as much of the installed power as possible, hence the need to keep P_G/P_U high. In order to achieve this it is generally recommended that a curved blade disc turbine be used with 12 or six blades, with a diameter $D = 0.5T$, a typical turbulent power number for a six curved blade turbine is around 4.

Nienow et al. [29] have also shown how the ability of the agitator to disperse the gas bubbles throughout the vessel can be correlated. In order to ensure that the gas bubbles are dispersed (at least) throughout the liquid above the agitator:

$$\mathrm{Fl} = 30\mathrm{Fr}(T/D)^{-3.5} \qquad (4.29)$$

He has also shown that in order to disperse gas completely above and below the agitator;

$$\mathrm{Fl} = 0.2(T/D)^{-0.5}\mathrm{Fr}^{0.5} \qquad (4.30)$$

although this is not likely to be practicable or really necessary for large-scale vessels.

It has also been shown that upward pumping angled or profiled blade impellers can be used quite efficiently. Nienow demonstrated this for

4.6.2 Sparged with surface incorporation

For many sparged gas–liquid reactors, the off-gas can still contain a significant amount of reactant which will require handling, either by separation, compression and recycle or by absorption. This is especially true if a pure gas feed is used. For hazardous gases it is preferred not to separate and recompress them and for many other gases absorption and disposal causes an unnecessary effluent load. In order to avoid these problems, internal recycle of the off-gas via the liquid surface is used. The batch hydrogenator is the most common example. The design illustrated in Figure 4.9(a) is normally used. It is relatively difficult to predict surface behaviour accurately from first principles and the semi-empirical design rule of constant specific power (W kg^{-1}) based on laboratory-scale experiments is recommended. The smallest reliable scale of experiment is generally regarded as about 5 l.

4.6.3 Specialized gas-inducing impellers

The major uncertainty with a surface incorporating design is the sensitivity to liquid height above the top impeller. This can be avoided to some extent by adopting an impeller with a specialized design whereby the low pressure region behind the blades is connected to the vapour space either by a hollow shaft or via a stator. These designs are sold for hydrogenation duties and are also common in mineral industries as flotation devices. Forrester and Reilly [30] have studied the effects of design parameters on performance and the reader is referred to their work for a more detailed design method.

4.6.4 Mass transfer

Mass transfer can be estimated in a similar way to liquid–liquid dispersions. In the case of dispersing a gas, it is simpler to base the transfer on the continuous liquid-phase concentrations:

$$J_i = K_L a \Delta C_i^* \tag{4.31}$$

$$1/K_L = 1/\text{He}_i k_G + 1/k_L \tag{4.32}$$

For small bubbles, e.g. less than 0.5 mm, it is recommended to use:

$$k_G = 2/3\pi^2 \delta / d_b \tag{4.33}$$

and for larger bubbles, estimate a bubble lifetime, t, and use:

$$k_G = 2(\delta/\pi t)^{1/2} \tag{4.34}$$

For many systems, $He_i \gg k_L$ and $K_L \sim k_L$. K_L and a are difficult to separate in a general way and often are correlated together. Correlations are presented by Middleton [31] for coalescing and non-coalescing systems (many industrial systems are non-coalescing):

$$K_L a = 1.2 \varepsilon^{0.7} V_s^{0.6}; \quad \text{coalescing (e.g. air–water)} \quad (4.35)$$

$$K_L a = 2.3 \varepsilon^{0.7} V_s^{0.6}; \quad \text{non-coalescing (e.g. air–electrolyte)} \quad (4.36)$$

A similar form of correlation has been presented for the gas hold-up E_G (volume of gas/volume of dispersion). However as many trace impurities can have a dramatic effect on gas hold-up it is recommended that this correlation only be used for extrapolation from small-scale tests:

$$E_G \propto \varepsilon^{0.48} V_s^{0.4} \quad (4.37)$$

In general, gas hold-up will be up to 20% for most operating industrial systems.

The mass transfer rate is very system specific, because of the unpredictable effects of surface active impurities and coalescence behaviour. In practice it is therefore not reliable to attempt to predict K_L nor a from system properties. The recommended procedure is to carry out laboratory experiments at 2:1 scale upwards with the real materials and test for sensitivity to mass transfer rate by increasing and decreasing agitator speed and gas rate, using a scale model turbine agitator, baffles and gas sparger. A gas rate of 1 v.v.m. and a power input of $1 \, W \, kg^{-1}$ are reasonable starting values. It would be normal to maintain specific power input (gassed, P_G) constant. If the gas volumetric rate (v.v.m.) is held constant then the superficial gas velocity, V_s, will increase at large scale, suggesting that an increased mass transfer rate can be achieved. Note that at full scale the power input due to the gas expansion should also be taken into account:

$$P_{Gas} = \rho g H Q_G \quad (4.38)$$

4.7 Design for heat transfer

4.7.1 Heat transfer surfaces

Stirred vessels can be provided with either (or both) an external jacket internal coils for heat transfer. Various types of jacket can be used – a full conventional jacket, a dimpled jacket or a half-pipe jacket, often called a limpet coil. The designs are compared by Markovitz [32]. Internal coils may be full helical coils or a number of smaller ringlet coils. A full helical coil is most commonly used, giving maximum surface area, but requires a two piece vessel. Smaller ringlet coils can be designed to be inserted from the top through large branches. Ringlet coils may cause mixing problems as they can leave quiescent or unmixed regions inside the coils' diameter.

In cases where coils and jackets are insufficient, an external heat exchanger (typically a standard shell and tube unit) with fluid recirculation may also be used. This section will only consider heat transfer to and from coils and jackets – the design of standard heat exchangers will not be considered.

Jackets are favoured for highly corrosive or highly reactive materials as they have no additional materials in contact with the vessel contents. Cleaning of the vessel between batches or campaigns is also easier. There is less risk of contact between the heating or cooling fluid and the vessel contents. Because no vessel internals are required, the use of a jacket does not restrict the choice of agitation systems. This means that processing of rheologically complex materials is easier than with coils.

On the other hand, coils give better heat transfer performance because they have a lower wall resistance and a higher process side coefficient. A coil can provide a large surface area in a relatively small reactor volume. Coils are more versatile for scale-up. While jacket heat transfer area only rises as the 0.67 power of vessel volume, it is much easier to instal coil area in proportion to volume (providing that the coils are not so densely packed as to interfere with flow).

The instantaneous rate of heat transfer can be calculated using the standard expression:

$$Q_h = UA \Delta T_m \qquad (4.39)$$

The time taken to reach a temperature T, with a starting temperature T_0 and a service temperature T_S is given by equation (4.40):

$$t = (Mc_p/UA) \ln[(T_S - T_0)/(T_S - T)] \qquad (4.40)$$

The overall heat transfer coefficient U may be found using the usual expression:

$$1/U = 1/h + \delta_w/k_m + 1/h_s + 1/h_f \qquad (4.41)$$

where h and h_s are the process and service side heat transfer coefficients, and $1/a_f$, the fouling resistance on the service side, can be obtained from Kern [37] or local knowledge. Usually, overall coefficients for a well designed coil are in the range 400–600 W m^{-2} K^{-1}. Overall coefficients typical of agitated jacketed vessels are given in Table 4.10.

4.7.2 Service side heat transfer coefficient

Usually, the overall coefficient is dominated by the process side. To avoid significant service-side resistance, the following features are desirable.

For jackets:

(i) A conventional jacket should be fitted with spiral baffles.

Table 4.10 Typical overall coefficients for jacketed glass lined steel vessels

Duty	U (W m^{-2} K^{-1})
Distillation/Evaporation (glass-lined steel vessel)	350
Heating (stainless steel vessel)	400
Heating (glass-lined steel vessel)	310
Cooling (stainless steel vessel)	350
Cooling (glass-lined steel vessel)	200
Cooling (chilled service, stainless steel vessel)	150
Cooling (chilled service, glass lined steel vessel)	100

(ii) Service injection nozzles should be used to direct the service flow (especially for lined vessels).
(iii) A vent should be fitted to avoid accumulation of inerts.
(iv) For a plain jacket containing liquid the circumferential velocity should ideally be 1–1.5 m s^{-1}.
(v) For a half-pipe jacket containing liquid the minimum velocity should be 2.3 m s^{-1}.
(vi) In a dimpled jacket, pressure drop may limit liquid velocity to 0.6 m s^{-1}.

For internal coils:

With a liquid service, the minimum velocity should be 1.5 m s^{-1}.

Several correlations can be used to estimate the service side coefficient; which is chosen depends on the jacket or coil design and the flow regime of the service fluid. Fletcher [33] discusses this and provides guidance. For commonly encountered situations the following correlations are recommended.

(i) For a conventional, unbaffled jacket with liquid service at high flow [34] use

$$\mathrm{Nu} = \frac{0.03\, \mathrm{Re}^{3/4}\, \mathrm{Pr}}{1 + 1.74\, \mathrm{Re}^{-1/8}(\mathrm{Pr}-1)} \left(\frac{\mu}{\mu_w}\right)^{0.14} \quad (4.42)$$

where

$$\mathrm{Nu} = h_s d_e / k_m \quad (4.43)$$

$$\mathrm{Re} = [d_e \rho (\sqrt{v_i v_A} + v_B)]/\mu \quad (4.44)$$

$$d_e = 0.816(D_j - T) \quad (4.45)$$

$$v_B = 0.5 \sqrt{2 z \beta g \Delta T_s} \quad (4.46)$$

$$v_i = 4Q/\pi d_i^2 \quad (4.47)$$

In these expressions v_i is the inlet nozzle or branch velocity and v_B is the component of the velocity arising from buoyancy, and D_j is the jacket diameter, z is the wetted height of the jacket, d_i the diameter of the inlet and ΔT_s the temperature rise of the service fluid. The value of v_A (the rise velocity in the jacket) depends on whether the inlet is radial:

$$v_A = 4Q/\pi(D_j^2 - T^2) \tag{4.48}$$

or tangential:

$$v_A = 2Q/(D_j - T)z \tag{4.49}$$

(ii) A conventional unbaffled jacket containing liquid at low flow rates, with heat transfer dominated by natural convection [35] is represented by the expression

$$\text{Nu} = K(z^3 \rho^2 \beta g \Delta T_m / \mu^2)^{1/3} \text{Pr}^{1/3} \tag{4.50}$$

In this case, $K = 0.15$ for upward flow, heating and downward flow cooling while $K = 0.128$ for downward flow heating and upward flow cooling, and ΔT_m is the mean temperature difference between the service and the vessel wall

$$\text{Nu} = h_s z/k \tag{4.51}$$

(iii) The service side coefficient for a baffled or dimpled jacket containing liquid should be estimated as for the unbaffled jacket above (case 2). This is conservative as the actual coefficient will be higher.

(iv) For a half-pipe external coil containing liquid the following expression (a slightly modified version of the Sieder–Tate equation) should be used.

$$\text{Nu} = 0.023 \, \text{Re}^{0.8} \, \text{Pr}^{1/3} (\mu/\mu_w)^{0.14} E \tag{4.52}$$

The area should be taken as the whole jacketed area. E is an effectiveness factor that typically takes values in the range 0.8–1 [36]. In this equation, Nu and Re should be based on the hydraulic mean diameter, d_e, given by:

$$d_e = \pi d_i/2 \tag{4.53}$$

(v) Condensing heat transfer (steam service) coefficients in a jacket should be extremely high compared to the process side and any resistance can normally be neglected.

4.7.3 Process side heat transfer coefficient

The process side coefficient will be determined by the agitator type, baffle arrangement and speed. For low viscosity fluids, most turbine-type high speed agitators can give good performance. For high viscosity and non-Newtonian fluids, larger diameter agitators will be required

For non-Newtonian fluids, the apparent mean viscosity is required to calculate Nu from Re. The normal practice is to use an estimate from the Metzner–Otto approach described in Section 4.3.1. It is also essential to ensure that a material with a significant yield stress is fluidized right up to the vessel wall, the method for which is described in Section 4.3.2.

For the vessel wall surface, the process side coefficient can be calculated from:

$$\mathrm{Nu} = A\,\mathrm{Re}^{2/3}\,\mathrm{Pr}^{1/3}(\mu/\mu_w)^{0.14} \qquad (4.54)$$

$$\mathrm{Re} = ND^2\rho/\mu \qquad (4.55)$$

$$\mathrm{Nu} = hT/k \qquad (4.56)$$

while for transfer to/from an internal coil the expression to use is:

$$\mathrm{Nu} = B\,\mathrm{Re}^{0.62}\,\mathrm{Pr}^{1/3}(\mu/\mu_w)^{0.14} \qquad (4.57)$$

The constants A and B are a function of the impeller type, and are given in Table 4.11 for common agitators.

Table 4.11 Heat transfer constants for impellers

Impeller	A	B
Propeller	0.46	1.4
45° turbine	0.61	1.4
Disc turbine	0.87	1.4
Retreat curve	0.33	0.87
Anchor	0.33	
Intermig	0.54	

4.7.4 Wall resistance

The conductivity of the wall material can be found in standard texts (Kern [37]). The resistance may be significant for lined vessels, and in these cases the manufacturer's data should be consulted.

4.8 Nomenclature

A	total interfacial area (m^2)
a	interfacial area per unit liquid volume (m^2 m^{-3})
CP	partition coefficient (concentration in continuous phase/concentration in dispersed at equilibrium)
c_p	heat capacity (J kg^{-1} K^{-1})
D	impeller diameter (m)
$\mathscr{D}$	diffusion coefficient (m s^{-1})
d	maximum stable drop size (m)
d_{32}	mean drop size (volume/area mean) (m)
Fl	gas flow number (Q_G/ND^3)
Fr	Froude number ($N^2 D/g$)
h	heat transfer coefficient (W m^{-2} K^{-1})

H	depth of liquid (m)
He_i	Henry coefficient for component i (gas phase concentration/liquid phase concentration at equilibrium)
J_i	total flux rate (kmol s^{-1}) m^{-2} (total area)
J_L	flux per unit liquid volume (kmol s^{-1}) (m^2 m^{-3})$^{-1}$
K_c	overall continuous side mass transfer coefficient (m s^{-1})
k_c	external (continuous) side film mass transfer coefficient (m s^{-1})
k_D	internal (drop) side film mass transfer coefficient (m s^{-1})
k_m	wall thermal conductivity (W m^{-1} K^{-1})
M	mass of vessel contents (kg)
N	impeller rotational speed (s^{-1})
Nu	Nusselt number (hD/k) (dimensionless)
P	impeller power (W)
Pr	Prandtl number ($c_p\mu/k$) (dimensionless)
Q_G	gas flow (m^3 s^{-1})
Q	liquid flow (m^3 s^{-1})
Q_h	heat flow (W)
Re$_p$	drop Reynolds number ($d_p u_s \rho/\mu$) (dimensionless)
Re$_y$	yield Reynolds number ($\rho N^2 D^2/\tau_y$)
Sh	Sherwood number ($K_c d_p/\mathscr{D}$) (dimensionless)
Sc	drop Schmidt Number ($\nu/\mathscr{D}$) (dimensionless)
u_s	drop settling velocity (m s^{-1})
U	overall heat transfer coefficient (W m^{-2} K^{-1})
t	time (s)
T	vessel diameter (m)
v'	fluctuating velocity (m s^{-1})
v.v.m.	volume of gas per volume of liquid per minute
We	impeller Weber number ($N^2 D^3 \rho/\sigma$) (dimensionless)
β	coefficient of thermal expansion (dimensionless)
δ	diffusion coefficient, gas in liquid (m^2 s^{-1})
δ_w	wall thickness (m)
ε	energy dissipation rate (W kg^{-1})
λ	Kolmogoroff microscale (m)
μ	dynamic viscosity (Ns m^{-2})
ν	kinematic viscosity (m^2 s^{-1})
ρ	density (kg m^{-3})
σ	interfacial tension (N m^{-1})

References

1. Calabrese, R.V. and Stoots, C.M. (1989) *Chem. Eng. Prog.*, **85**, 43.
2. Van't Riet, K. *et al.* (1976) *Trans. I ChemE*, **54**, 124.
3. Kresta, S.M. and Wood, P.E. (1993) *Can. J. Chem. Engng*, **71**, 42.
4. Bujalski, W. *et al.* (1990) *Chem. Engng Sci.*, **45**, 415.

5. Bujalski, W. *et al.* (1987) *Chem. Engng Sci.*, **42**, 317.
6. Ruszkowski, S.W. (1994) *IChemE Symp. Ser.*, **136**, 283.
7. Grenville, R.K., Ruszkowski, S.W. and Garred, E. (1995) *Mixing*, 15, June.
8. Govier, G.W. and Aziz, A.K. (1977) *Flow of Complex Mixtures in Pipes.* Krieger, New York.
9. Metzner, A.B. and Otto, R.E. (1957) *Am. IChemE J.*, **3**.
10. Rieger, F. and Novak, V. (1973) *Trans. IChemE*, **51**, 105.
11. Nienow, A.W. and Elson, T.P. (1988) *Chem. Engng Res. Des.*, **66**, 5.
12. Etchells, A.W. *et al.* (1987) *IChemE Symp. Ser.*, **108**, 1.
13. Zwietering, T.N. (1958) *Chem. Engng Sci.*, **8**, 244.
14. Gates, L.E. *et al.* (1976) *Chem. Eng.*, 24 May, 144.
15. Nienow, A.W. (1985) In *Mixing in the Process Industries* (N. Harnby *et al.*, eds). Butterworth, London.
16. Joosten, G.E. *et al.* (1977) *Trans. IChemE*, **55**, 220.
17. Smith, D.L. (1989) In *Proc. 6th European Conf. on Mixing*, BHRA, p. 259.
18. Godfrey, N.B. (1972) *Chemtech*, **359**, June.
19. van Heuven, J.W. and Beek, W.J. (1971) In *Proc. ISEC*, paper 51, pp. 70–81.
20. Lines, P.C. (1990) *IChemE Symp Ser.*, **121**, 167.
21. Calabrese, R.V. (1992) *FMP Report 054*. July.
22. Davies, J.T. (1987) *Chem. Engng. Sci.*, **42**, 1671.
23. Warmeoskerken, M.M.C.G. and Smith, J.M. (1989) *Chem. Engng Res. Des.*, **67**, 193.
24. Chang, K-C. and Calabrese, R.V. (1989) *Mixing*, X11, August.
25. Chesters, A.K. (1991) *Trans. IChemE*, **69**, 259–270.
26. Calabrese, R.V. and Wang, C.Y. (1986) *Am. IChemE J.*, **32**, 4.
27. Kenics design manual K-TEK 5 (1988).
28. Rowe, P.N. *et al.* (1965) *Trans. IChemE*, **43**, T14.
29. Nienow, A.W. *et al.* (1985) In *Proc. 5th European Conf. on Mixing*, BHRA, p. 1.
30. Forrester, S.E. and Reilly, C.D. (1994) *Chem. Engng Sci.*, **49**, 5709.
31. Middleton, J.C. (1985) In *Mixing in the Process Industries* (N. Harnby *et al.*, eds). Butterworth, London.
32. Markovitz, R.E. (1971) *Chem. Eng.*, 15 November, 156.
33. Fletcher, P. (1987) *Chem. Eng.*, April, 33.
34. Lehrer, I.H. (1970) *Ind. Engng Chem. Process. Design Dev.*, **9**, 4.
35. Barton, E. and Williams, E.V. (1950) *Trans. IChemE*, **17**, 3.
36. Kneale, M. (1969) *Trans. IChemE*, **47**, T279.
37. Kern, D.Q. (1950) *Process Heat Transfer*, McGraw Hill, New York.

5 Mixing and the selectivity of fast chemical reactions
J.R. BOURNE

5.1 The problem

Among the problems that arise in the scale-up of batch reactions is that of reactions whose outcome is sensitive to mixing. While many reactions do not exhibit such sensitivity, those that do can exhibit substantial yield loss on scale-up.

It is the main purpose of this chapter to describe recent advances in understanding the interactions between the physical operation of mixing reagents and the distribution of products from fast reactions. Whereas much is known about the chemical factors influencing the yield of a particular product, the role of mixing is often less clear. A few examples from the chemical literature soon indicate that observed yields are not always equal to the values expected from well established chemical principles and that the ways in which reagents are mixed can also be important.

Competitive-consecutive reactions, represented by equations (5.1) and (5.2), form the first problem category. The yield of the intermediate R is considered to be the performance measure.

$$A + B \xrightarrow{k_1} R \qquad (5.1)$$

$$R + B \xrightarrow{k_2} S \qquad (5.2)$$

This yield should be high when $k_1 \gg k_2$ and when the stoichiometric ratio (number of moles of B charged to the reaction vessel per mole of A) is around unity. (If this ratio equals or exceeds two, R will be fully converted to S as the reactions run to completion.) Experiments on the single-phase nitration of durene (1,2,4,5-tetramethylbenzene) contradicted these chemically based predictions. Although k_1 is orders of magnitude greater than k_2, measured yields of mononitrodurene (R) were much smaller than those of dinitrodurene (S) [1, 2]. The yields of R and S were increased and decreased, respectively, by increasing the intensity of agitation in the reaction vessel and by measures which retarded the reactions. Increases in the yield of the intermediate (R) as the agitation intensity was raised were also observed in the iodination of 1-tyrosine [3], the bromination of 1,3,5-trimethoxybenzene [4] and in some diazo couplings to be discussed in detail later. Some competitive-consecutive reactions are characterized by $k_1 = 2k_2$ and the maximum yield of R will be 50% when the stoichiometric ratio of B to

A is unity. (The corresponding conversion of A is 75% and the yield of S is 25%.) This expected product distribution has been observed for diamines reacting with acids to produce amides [5] and with isothiocyanates to give substituted thioureas [6]. More reactive isocyanates and diamines produced, however, yields of disubstituted ureas (S) beyond 25% and correspondingly yields of monosubstituted (R) ureas below 50% [6]. Similar trends, attributed to reactions taking place simultaneously with the mixing of reagents, were observed when acylating diamines [7].

Parallel (or competitive) reactions, represented by equations (5.3) and (5.4), constitute the second problem category where the product distribution can depend upon the reagent mixing.

$$A + B \xrightarrow{k_3} P \tag{5.3}$$

$$A + C \xrightarrow{k_4} Q \tag{5.4}$$

A high yield of Q should be favoured by $k_4 \gg k_3$ and a high ratio of the concentration of reagent C relative to reagent B. When these concentrations are equal, the ratio of the rates of formation of Q to P will be k_4/k_3 according to chemical principles [8]: this is the basis of the so-called competitive method for determining k_4 knowing k_3 and the measured product ratio Q/P. The literature reveals also in this category significant deviations from these classical predictions. For the single-phase nitration of an equimolar mixture of benzene (B) and toluene (C) with a nitronium salt (A) the measured ratios of nitrotoluene (Q) to nitrobenzene (P) increased from 1.3 to 27 as the agitation intensity was raised [9]. Since $k_4/k_3 \approx 25$ it seems that only at high agitation levels were the reactions free from a mixing limitation. Further work, involving the nitration of dibenzyl [10], confirmed this interpretation. A second example refers to the need to improve the rate of mixing reagents when applying the competitive method to measure kinetics [11]. A third example describes a base-catalysed hydrolysis occurring during pH adjustment as concentrated base was added to an acidic solution [12]. Because of incomplete mixing local, transitory pH gradients promoted unexpected hydrolysis of a valuable product. During the bromination of resorcin their existence explains the mixing-dependent ratio of 2,4- to 4,6-dibromoresorcin [13]. During a particular diazo coupling the simultaneous formation of two monoazo dyes in proportions depending upon the mixing has been interpreted in terms of local pH gradients [14].

There is thus clear experimental evidence that the selectivity of some chemical reactions is determined not only by classical kinetic principles, but also by the way in which the reagents are mixed. To gain a deeper understanding of such phenomena, it is necessary to analyse mixing mechanisms and in particular how mixing can influence transient concentration distributions and hence chemical reactions.

5.2 Mixing mechanisms and modelling

5.2.1 Semi-batch reactor: micromixing

This section concentrates on the results of detailed studies of the coupling between mixing and reaction during semi-batch reaction – the addition over some period of one reactant to a batch of another reactant already in the vessel. Reference should be made to the literature cited for more complete information, e.g. the basis and derivation of the results.

Consider, by way of example, competitive-consecutive reactions, described by equations (5.1) and (5.2), with $k_1 = 20k_2$ which are to be carried out in a stirred semi-batch reactor, whereby 1 mol of B will be fed at a constant rate for every 1.1 mol of A in the tank. The product distribution is to be found when reaction is complete.

When mixing of the reagents in the tank is perfect, classical kinetic principles (e.g. a Jungers diagram) give 85.7% conversion of A, 100% conversion of B and 80.5% yield of R referred to the A initially charged to the tank. An alternative way to express product distribution is the fraction of the limiting reagent (B) present in the secondary product (S). Thus

$$X_S = 2c_S/(c_R + 2c_S) \qquad (5.5)$$

and calculation based on the yield gives $X_S = 0.115$.

When, however, mixing limitation totally dominates, the primary product of competitive-consecutive reactions does not survive and only A and S are found in the final mixture. Thus, the yield of R is zero and $X_S = 1.0$.

Between these two extremes both mixing and reaction kinetics determine yields. Some modelling is needed to predict the product distribution and how it depends on the operating conditions. This requires consideration of two mixing mechanisms – macromixing and micromixing. Figure 5.1

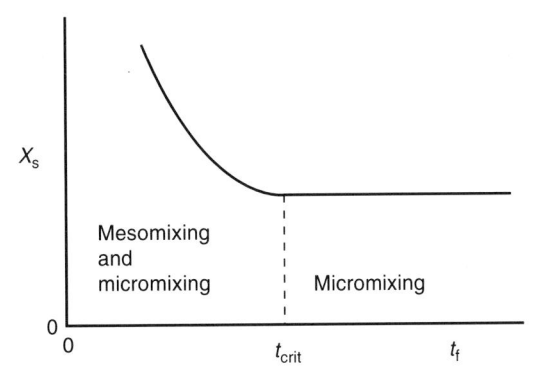

Figure 5.1 Variation of the yield of secondary product (S) from competitive-consecutive reactions as feed time of B-solution to a semi-batch reactor varies and all other parameters are unchanged.

shows how the yield of secondary product (X_S) has been observed to vary with the feed time (t_f) for the reagent B, when all other systems parameters and operating conditions are held constant. Product distribution is independent of feed time, when this is sufficiently long, whereas, for short feed times, the higher the feed rate, the more secondary product is formed [15–18].

The mechanism controlling product distribution at long feed times is micromixing, which in principle depends upon three steps: (i) molecular diffusion, (ii) fluid deformation, and (iii) mutual engulfment of A-rich and B-rich regions [19]. Under two frequently encountered conditions only step (iii) engulfment is important: these are (i) $Sc \ll 4000$, where the Schmidt number is ν/D (ν is the kinematic viscosity and D is the diffusion coefficient) and (ii) $c_{B_0} \gg c_{A_0}$, where c_{B_0} is the concentration of B in the feed and c_{A_0} is the initial concentration of A in the tank [20, 21].

The action of engulfment is due to the formation of fine-scale vortices in the fluid. Figure 5.2 illustrates a growing vortex, whose engulfed volume V_e is growing at a rate of EV_e by inflow from the surroundings

$$dV_e/dt = EV_e \qquad (5.6)$$

The engulfment rate coefficient, E, is related by equation (5.7) to the two properties determining most of the fine-scale properties of a turbulent flow, i.e. ν and ε. The symbol ε represents the rate of dissipation (to heat) of the kinetic energy of the turbulence per unit mass of fluid. (The units of ν and ε are $m^2 s^{-1}$ and $W kg^{-1} \equiv m^2 s^{-3}$, respectively, so that E is a frequency).

$$E = 0.058(\varepsilon/\nu)^{0.5} \qquad (5.7)$$

Because diffusion within the deforming engulfed structure is usually much faster than engulfment, the concentration of any substance i is uniform and is represented by c_i. The corresponding concentration in the immediate

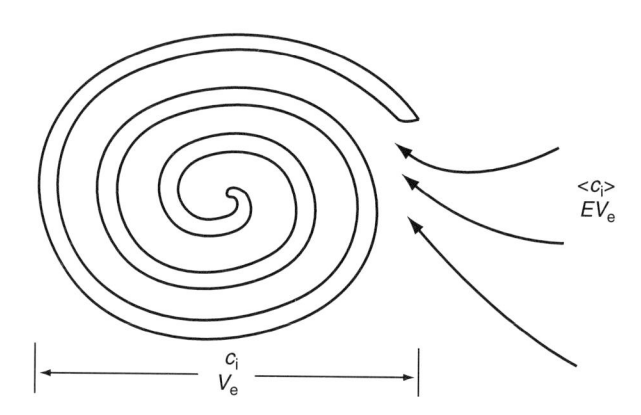

Figure 5.2 Main characteristics of the engulfment model of micromixing.

surroundings of the growing vortex is $\langle c_i \rangle$. For semi-batch reactor operating with a low feed rate, adequate time is available for the flow and mixing to homogenize the vessel, so that $\langle c_i \rangle$ is the average concentration of i in the vessel. Any inhomogeneity exists then only in the reaction zone and is due to inadequate micromixing (Figure 5.2). Moreover, at long feed times, the concentration of not yet engulfed B in the surroundings is negligible. The mass balance for i in the reaction zone is therefore

$$d(V_e c_i)/dt = E V_e \langle c_i \rangle + r_i V_e \tag{5.8}$$

where r_i is the rate of formation of i by reaction (mol m^{-3} s^{-1}) as given by chemical kinetics. Simplifying equation (5.8) by substituting from equation (5.6), the mass balance becomes

$$dc_i/dt = E(\langle c_i \rangle - c_i) + r_i \tag{5.9}$$

Micromixing and reaction in semi-batch reactor can be followed by first discretizing the volume of B-rich solution (the feed), V_B, into σ equal parts. The first part, having a volume V_B/σ and a concentration c_{B_0}, enters the vessel, which initially contains a volume V_A with concentration c_{A_0} (R and S are initially not present). With known values of ε and ν, equation (5.7) determines E and with second-order kinetics for equations (5.1) and (5.2) the source terms become $r_A = -k_1 c_A c_B$, $r_R = k_1 c_A c_B - k_2 c_R c_B$, etc. Numerical integration of equation (5.9) allows the concentrations of the four substances in the growing reaction zone to be expressed as functions of time. Integration stops when the limiting reagent B has been consumed. The necessary time t' for micromixing and reaction is noted and the volume to which the reaction zone has grown is evaluated as $(V_B/\sigma)\exp(Et')$ from equation (5.6). When this final mixed volume is smaller than the total liquid volume $(V_A + V_B/\sigma)$, the mean concentrations of all substances are found by averaging the contents of the mixed volume and the surroundings. These mean values form the composition of the new surroundings when the second part of the feed enters the vessel. If, however, the engulfment rate is high, the whole volume might be engulfed and hence mixed before the limiting reagent has been completely consumed. The reactions are then completed in a perfectly mixed solution. In this limiting case, as E increased, the engulfment formalism reduces to classical kinetics in a uniform system.

By progressively updating the products and the unreacted reagent A, the evolution of R, S, the yield and X_S during semi-batch operating can be followed until all the limiting reagent B has been added and consumed. Figure 5.3 presents the results of many such calculations for the reactions of equations (5.1) and (5.2), when $k_1 = 20k_2$, $V_A c_{A_0} = 1.1 V_B c_{B_0}$ and one part by volume of B is added slowly to twenty parts by volume of A in the tank ($V_A/V_B = 20$). The dimensionless group on the abscissa is a form of Damkoehler number, defined as

$$\overline{Da} = V_B k_2 c_{B_0} / E(V_A + V_B) \tag{5.10}$$

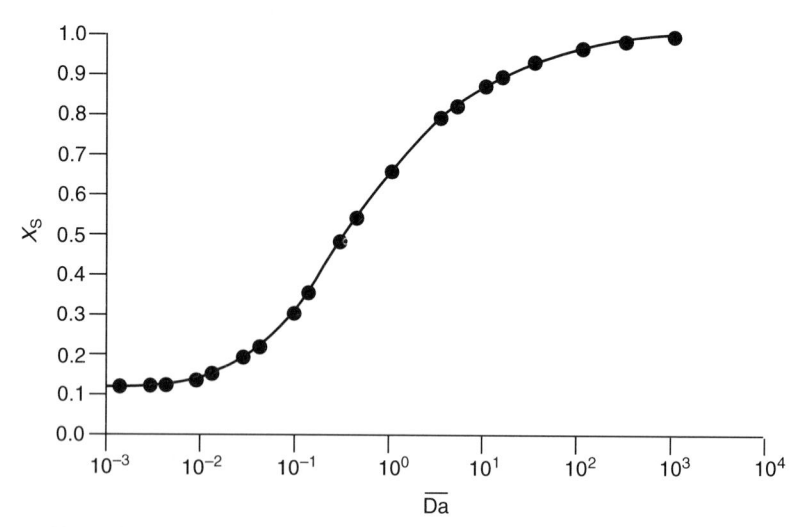

Figure 5.3 Product distribution (X_S) as a function of Damkoehler number $\overline{Da}$ for a semi-batch reactor controlled by micromixing.

By simple re-arrangement $\overline{Da} = E^{-1}/(k_2\overline{c_B})^{-1}$, where $\overline{c_B} = V_B C_{B_0}/(V_A + V_B)$, so that $\overline{Da}$ is the ratio of an engulfment time (τ_E) to a reaction half-life (τ_R). Three regimes can be identified in Figure 5.3.

(a) $$\overline{Da} < 0.001 \qquad (\tau_E < 0.1\%\tau_R) \tag{5.11}$$

X_S tends to 0.115, the value found by classical reaction kinetics. In this slow reaction regime, the time needed for reaction is not influenced by mixing.

(b) $$\overline{Da} > 1000 \qquad (\tau_E > 1000\tau_R) \tag{5.12}$$

X_S tends to 1, so that the yield of R becomes 0. This is the instantaneous regime, where the reaction time is totally determined by micromixing.

(c) $$0.001 < \overline{Da} < 1000 \qquad (\tau_E \approx \tau_R) \tag{5.13}$$

X_S has an intermediate value in the fast regime. The reaction time and the product distribution depend upon both chemical kinetics and micromixing. Modelling of the interaction between these two processes is essential in understanding the factors determining X_S. The E-model (engulfment) consists basically of equations (5.6), (5.7) and (5.9) for the semi-batch reactor.

5.2.2 Extensions

The modelling of micromixing and reaction for semi-batch operation is one of the simpler cases. Other topics already studied include

(i) Two and three parallel reactions – as in equations (5.3) and (5.4) – whose produce distributions are influenced by micromixing, have been modelled using very similar equations to those given above [22, 23].
(ii) pH gradients, controlled by micromixing, require more differential equations for their complete description, but no new physical principles [20, 24, 25].
(iii) Other operating modes have been modelled, e.g. continuous operation of a tank which is macroscopically well mixed, but partially segregated on the molecular scale. Another example is plug flow as in a turbulent tubular reactor [20].
(iv) Temperature segregation denotes a different temperature in the reaction zone to that in its immediate surroundings. This can arise through the reaction enthalpy or through a difference in temperature between a feed stream and the reactor's contents.
(v) Concentration and temperature differences can be present during reaction not only on the microscale [actually near the Kolmogorov scale $(\nu^3/\varepsilon)^{1/4}$], but segregation at larger scales is also possible. For example, feed solution has usually to mix with the whole contents of a tank reactor. This is macromixing which proceeds primarily by convection and so depends upon the bulk flow pattern. Although widely studied, bulk blending is probably not important in semi-batch operation, where the feed time is usually much longer than the blending time. Mesomixing signifies a scale of segregation between those of micro- and macro-mixing. In Figure 5.1 short feed times cause X_S to increase, so that the reactions signal a higher level of inhomogeneity relative to that at molecular scale. What now happens is that the dispersion of the fresh feed out into the A-rich surroundings needs more time than when the feed rate of B is very small. This dispersion or spreading transverse to the direction of flow also relies on turbulence but is, however, a coarser scale phenomenon than micromixing. Such mesomixing has been treated in detail [26, 27].
(vi) In many reactors ε varies spatially by about three orders of magnitude when employing a standard Rushton turbine and the flow field has been modelled in two dimensions (radially and axially) [23, 28, 29].

5.3 Applications

5.3.1 Model reactions

The model can be applied to predict the influence of reaction constants, feed concentrations and mixing intensity on the yield of competing reactions. Expressing the results of the engulfment model in dimensionless form, the final product distribution, expressed as X_S, depends upon the following terms:

$$X_S = f(k_1/k_2;\ V_A c_{A_0}/V_B c_{B_0};\ V_A/V_B;\ \overline{Da}) \quad (5.14)$$

The ratio of rate constants and the stoichiometric ratio are classical independent variables: the volume ratio of the reagents solutions and the Damkoehler number refer to the influence of mixing, in particular micro-mixing by engulfment. For a given reaction and a fixed stoichiometric ratio the following mixing effects can be anticipated.

(i) A modest increase in solution viscosity has no effect on the power consumption of an impeller, provided that the flow remains fully turbulent. However, the mixing rate E will be reduced (see equation 5.7) so that $\overline{Da}$ increases. As Figure 5.3 shows, X_S will increase.

(ii) An increase in power input, e.g. through raising the stirrer speed or moving the feed point to a region of more intense turbulence, raises ε in the reaction zone. $\overline{Da}$ therefore decreases and less secondary product (S) and more primary product (R) are formed. X_S thus decreases.

(iii) Lowering the concentration level of all reagents, whilst maintaining the stoichiometric and volumetric ratios constant, also reduces $\overline{Da}$ and X_S.

Through application of the E-model these qualitative predictions become quantitative and have been fully confirmed for various sets of fast reactions. Two examples will be briefly described here.

(a) Diazo coupling between 1-naphthol and diazotized sulphanilic acid
The primary reaction consists mainly of coupling in the *para* position, although a second monoazo dye is also formed by some limited coupling in the *ortho* position. The secondary reaction consists of each monoazo dye coupling in the free position (*ortho* or *para*) to give a single bisazo dye. This system of four reactions approximates equations (5.1) and (5.2), although the full reaction scheme and its kinetics should be employed in computations [30, 31]. Turbulence levels in most stirred tanks are modest with ε often in the range 10^{-2} to 10^2 W kg^{-1} and the primary coupling is then an instantaneous reaction at room temperature and optimal pH. This pH is close to 10 and can be obtained with a sodium carbonate/sodium bicarbonate buffer. Solutions are sufficiently dilute that there is no heat effect.

Suppose, as example, that in semi-batch operation one part by volume of diazo salt solution (25 mol m^{-3}) is added slowly to 50 parts by volume of 1-naphthol solution (0.6 mol m^{-3}) at pH 10 and 298 K. The mixing intensity, expressed as the dissipation rate ε, is to be found so that X_S, defined in equation (5.5), shall not exceed 0.01 or 1%.

The relevant parameters (refer to equation 5.14) are $k_1/k_2 = \infty$, stoichiometric ratio $(A/B) = 1.2$, volume ratio $= 50$, $k_2 \approx 2 \text{ m}^3 \text{ mol}^{-1} \text{ s}^{-1}$ and $\overline{c_B} \approx 0.5 \text{ mol m}^{-3}$. From equation (5.10), $\overline{Da} \approx 1/E$ and from the E-model

$\overline{\text{Da}} = 10^{-3}$ when $X_S = 0.01$. Solving for E it is found that $E = 10^3 \text{ s}^{-1}$ and when $\nu = 0.89 \times 10^{-6} \text{ m}^2 \text{ s}^{-1}$, equation (5.7) gives $\varepsilon = 280 \text{ W kg}^{-1}$.

This turbulence intensity is probably too great to be achieved in a stirred tank and might call for another reactor type, e.g. a static mixer. When $\varepsilon = 1 \text{ W kg}^{-1}$, then $E = 60 \text{ s}^{-1}$ and $\overline{\text{Da}} = 0.0167$, in which case the E-model predicts $X_S = 0.1$. This is an order of magnitude more byproduct (S) than required.

These estimates indicate some applications of the E-model to a particular diazo coupling. Experimentally obtained product distributions can be analysed spectophotometrically to give X_S within about 0.004. Wide ranging comparisons with the E-model showed good agreement for the effects of stirrer speed, feed location, viscosity, vessel size, mixer type, buffer capacity and pH gradients, concentration, etc.

(b) Reactions between sodium hydroxide, hydrochloric acid and ethyl monochloroacetate

Consider the gradual addition of a solution containing 1 mol NaOH to a stirred tank in which 1 mol HCl and 1 mol $CH_2ClOOC_2H_5$ are dissolved [23]. As in equations (5.3) and (5.4), neutralization and hydrolysis compete, whereby perfect mixing would allow only the neutralization to occur, since it is instantaneous. Because it is practically impossible to achieve fluid mixing which is faster than the acid–base reaction, alkaline hydrolysis will produce some ethanol provided that this reaction is 'fast'. Ethanol is conveniently analysed by GC/FID and product distribution can be expressed as

$$X_Q = c_Q/(c_P + c_Q) \tag{5.15}$$

Like X_S, increasing segregation of reactants and rising Damkoehler number causes X_Q to increase from 0 to 0.5 as the hydrolysis moves from the slow, through the fast to the instantaneous regime.

Because of the modest hydrolysis rate ($k_4 \approx 0.03 \text{ m}^3 \text{ mol}^{-1} \text{ s}^{-1}$) and the limited solubility of the ester (around 2 wt%) this system can be used as a tracer to resolve weaker turbulence fields than the diazo coupling. Whereas the analytical error in measuring X_S for the coupling allows ε to be resolved up to around 400 W kg^{-1}, this limit for the neutralization–hydrolysis parallel reactions is about 4 W kg^{-1}.

By extending the coupling reactions to include the competitive diazo coupling of diazotized sulphanilic acid with 2-naphthol a much faster system of five reactions has been created, whose product distribution is sensitive up to $\varepsilon \approx 10^5 \text{ W kg}^{-1}$ [32].

5.3.2 *Characterization of mixers*

A second group of applications refers to characterizing reactors with respect to their suitability for rapidly mixing reagents. There is a wide range of mixers on the market, but in most cases objective information to facilitate their choice for a particular application is incomplete or unavailable. Whereas the energy input to a laboratory device like a stopped-flow instrument is not the primary consideration in its choice, it is necessary with production-scale reactors to pay attention to their power consumption. It would, for example, be uneconomic to provide an intensively turbulent flow pattern throughout a large vessel when fast reactions need such turbulence only near the exit of a reagent feed pipe for sufficient time and over a relatively small volume to complete mixing and reaction. Local information on the flow can be obtained in various ways, e.g. through computational fluid dynamics (CFD) and/or by turbulence measurements especially with laser Doppler velocimetry (LDV). It can also be helpful to run some test reactions with the object, for example, of finding a favourable addition point for reagent or of deducing the energy dissipation rate ε via a mixing model and so assessing whether a mixer efficiently exploits its overall power consumption. Energy dissipation rates in typical mixers usually fall in the range 10^{-3} to 10^5 W kg^{-1} and so engulfment rates, which are proportional to $\varepsilon^{0.5}$, vary over some four orders of magnitude.

Some mixers, whose suitability for obtaining high selectivities from fast reactions has been studied, include the centrifugal pump [33], pipeline [34–36], rotor-stator mixers of various sizes [37, 38], thin liquid sheets [39, 40], static mixers [41, 42], 'T' junctions [43], free jets [44–46], grid turbulence [47], and agitated tanks [15–18, 23, 26, 27, 29, 48–56], as well as a tubular membrane model [57] and a simple injector in a tube [58]. In all these cases the energy dissipation rate, ε, and hence the mixing rate and other fine-scale turbulence properties vary significantly from place to place. Thus, in such inhomogeneous turbulence, the reagents experience locally variable mixing rates along their trajectories. In addition to knowing the ε-field in a mixer, the velocity field is also needed so that the progress of a mixing-controlled reaction may be determined by integrating along its trajectory.

5.3.3 Scale-up principles

A common industrial requirement is to predict the outcome (e.g. product distribution) of some reactions at a large scale based upon usually meagre information about the same reactions conducted on a smaller scale. The rate of micromixing depends upon the kinematic viscosity and the energy dissipation rate. Thus, if the dissipation rate in the reaction zone is held constant, the product distribution under given operating conditions should be independent of scale. There is no reason to doubt this conclusion, but an additional factor has to be considered. For a fast reaction the fraction of the whole reactor volume within which mixing and reaction are proceeding is usually very small: the analogy with a flame in a combustion chamber is helpful in recognizing that reaction zones are small. This fraction is, however, itself scale-dependent in many practical cases and reaction zones spread out to different extents at various scales. This phenomenon introduces complications in applying the constant dissipation rate scale-up rule when turbulence is inhomogeneous. Much, however, is known today about this problem and robust scale-up methods have been developed [15, 17, 23, 26, 29, 52, 53]. Scale criteria for the mesomixing regime, where the feed rate is also important, are emerging [46, 54].

5.4 Extensions

To gain a proper understanding of micromixing, it has been necessary to collect information, mainly from the field of fluid mechanics, about the fine-scale structure of turbulence. Several other processes of importance in chemistry and chemical engineering also depend upon knowledge of the flow structure, e.g. drop break-up in the formation of dispersions and emulsions [59] as well as the kinetics of coagulation and the growth of fine particles.

Rheological complexities often accompany rising viscosity. When the viscosity increases slightly above that of water, the flow remains fully turbulent and the engulfment rate decreases according to equation (5.7) [41, 45, 48, 51, 56]. At still higher viscosity levels, the flow regime becomes transitional or laminar: here theory of reactive mixing and experimental methods [60] are still developing. As soon as non-Newtonian properties are present, there is no adequate theory. The need to distribute reagents rapidly and uniformly through paper pulp reveals many problems [49, 61]. For example, when using a stirred tank the presence of a yield stress causes cavern formation (i.e. a limited well mixed zone) around the impeller, otherwise all motion is damped. There is great scope for theoretical and experimental studies of mixing and reaction in rheologically complex fluids.

5.5 Concluding remarks

Some multistep chemical reactions are sufficiently fast relative to the rate of mixing their reagents that their product distributions depend not only upon their kinetics and mechanisms but also upon the details of mixing. Much has been learned about this interaction within the last decade by interfacing chemical knowledge, particularly physical organic chemistry, with information about the structure of turbulent flows. Today a theoretical framework as well as specific models are available, whose ranges of application have been established by model-experiment comparisons. The main field of application of these recent developments is initially likely to be in the chemical and pharmaceutical industries. By raising the yield and improving the selectivity of chemical reactions it is possible to make more effective use of raw materials, to improve the process economics and perhaps to simplify work-up and product recovery, so saving energy. Tighter control of by-products contributes to respecting environmental constraints. The influence of concentration gradients on the path taken by a chemical reaction is however relevant in a wide range of disciplines, e.g. the environmental sciences.

5.6 Nomenclature

c	concentration (mol m^{-3})
D	diffusion coefficient (m^2 s^{-1})
$\overline{Da}$	Damkoehler number (τ_E/τ_R)
E	engulfment rate (s^{-1})
k	rate constant (s^{-1} mol^{-1} m^3)
r_i	rate of formation of i by reaction (mol m^{-3} s^{-1})
Sc	Schmidt number (ν/D)
t	time (s)
X_S, X_Q	product distribution index $[2c_S/(c_R + 2c_S), c_Q/(c_P + c_Q)]$
ε	energy dissipation rate (W kg^{-1})
ν	kinematic viscosity (m^2 s^{-1})
τ_E	engulfment time constant (s)
τ_R	reaction half-life (s)

References

1. Hanna, S.B., Hunziker, E., Saito, T. and Zollinger, H. (1969) Das Problem der Mononitrierung von Durol. *Helv. Chim. Acta*, **52**, 1537.
2. Hunziker, E., Penton, J.R. and Zollinger, H. (1971) Nitration of durene and pentamethylbenzene with nitronium salts in nitromethane and acetonitrile. *Helv. Chim. Acta*, **54**, 2043.
3. Paul, E.L. and Treybal, R.E. (1971) Mixing and product distribution for a liquid-phase, second order, competitive, consecutive reaction. *Am. IChemE J.*, **17**, 718.
4. Bourne, J.R. and Kozicki, F. (1977) Mixing effects during the bromination of 1,3,5-trimethoxybenzene. *Chem. Engng Sci.*, **32**, 1538.
5. Agre, C.L., Dinga, G. and Pflaum, R. (1956) Reactions of acids with diamines. *J. Org. Chem.*, **21**, 561.
6. Stoutland, O., Helgen, L. and Agre, C.L. (1959) Reactions of diamines with isocyanates and isothiocyanates. *J. Org. Chem.*, **24**, 818.
7. Jacobson, A.R., Makris, A.N. and Sayre, L.M. (1987) Monoacylation of symmetrical diamines. *J. Org. Chem.*, **52**, 2592.
8. Hammett, L.P. (1970) *Physical Organic Chemistry*. McGraw-Hill, New York.
9. Tolgyesi, W.S. (1965) Relative reactivity of toluene–benzene in nitronium tetrafluoroborate nitration. *Can. J. Chem.*, **43**, 343.
10. Ridd, J.H. (1971) Mechanism of aromatic nitration. *Acc. Chem. Res.*, **4**, 248.
11. Wood, P.B. and Higginson, W.C.E. (1966) Kinetic studies of oxidation–reduction of cobalt–ethylenediaminetetra–acetic acid complexes. *J. Chem. Soc. A*, 1645.
12. Paul, E.L., Mahadevan, H., Foster, J., Kennedy, M. and Midler, M. (1992) The effect of mixing on scaleup of a parallel reaction system. *Chem. Engng Sci.*, **47**, 2837.

13. Bourne, J.R., Rys, P. and Suter, K. (1977) Mixing effects in the bromination of resorcin. *Chem. Engng Sci.*, **32**, 711.
14. Kaminski, R., Lauk, U., Skrabal, P. and Zollinger, H. (1983) pH dependence and micromixing effects on the product distribution of couplings with 6-amino-4-hydroxy-2-naphthalenesulfonic acid. *Helv. Chim. Acta*, **66**, 2002.
15. Bourne, J.R., and dell'Ava, P. (1987) Micro- and macromixing in stirred tank reactors of different sizes. *Chem. Engng Res. Des.*, **65**, 180.
16. Bourne, J.R., and Hilber, C.P. (1990) The productivity of micromixing-controlled reactions: effect of feed distribution in stirred tanks. *Chem. Engng Res. Des.*, **68**, 51.
17. Bourne, J.R., and Thoma, S. (1991) Some factors determining the critical feed time of a semi-batch reactor. *Chem. Engng Res. Des.*, **69**, 321.
18. Thoma, S., Ranade, V.V. and Bourne, J.R. (1991) Interaction between micro- and macromixing during reactions in agitated tanks. *Can. J. Chem. Engng*, **69**, 1135.
19. Baldyga, J. and Bourne, J.R. (1984) A fluid mechanical approach to turbulent mixing and chemical reaction. *Chem. Eng. Commun.*, **28**, 231, 243, 259.
20. Bourne, J.R. and Baldyga, J. (1989) Simplification of micromixing calculations. *Chem. Engng J.*, **42**, 83 and 93.
21. Bourne, J.R., and Baldyga, J. (1990) Comparison of engulfment and interaction by exchange with the mean (IEM) micromixing models. *Chem. Engng J.*, **45**, 25.
22. Baldyga, J. and Bourne, J.R. (1990) The effect of micromixing on parallel reactions. *Chem. Engng Sci.*, **45**, 907.
23. Bourne, J.R. and Yu, S. (1994) Investigation of micromixing in stirred tank reactors using parallel reactions. *Ind. Engng Chem. Res.*, **33**, 41.
24. Bourne, J.R., Gablinger, H. and Ravindranath, K. (1988) Local pH gradients and the selectivity of fast reactions. I: Mathematical model of micromixing, *Chem. Engng Sci.*, **43**, 1941.
25. Bourne, J.R. and Gablinger, H. (1989) Local pH gradients and the selectivity of fast reactions. II: Comparison between model and experiments, *Chem. Engng Sci.*, **44**, 1347.
26. Baldyga, J. and Bourne, J.R. (1992) Interactions between mixing on various scales in stirred tank reactors. *Chem. Engng Sci.*, **47**, 1839.
27. Baldyga, J., Bourne, J.R. and Yang Yang. (1993) Influence of feed pipe diameter one mesomixing in stirred tank reactors. *Chem. Engng Sci.*, **48**, 3383.
28. Baldyga, J. and Bourne, J.R. (1988) Micromixing in inhomogeneous turbulence. *Chem. Engng Sci.*, **43**, 107.
29. Baldyga, J. and Bourne, J.R. (1988) Calculation micromixing in inhomogeneous stirred tank reactors. *Chem. Engng Res. Des.*, **66**, 33.
30. Bourne, J.R., Kut, O.M., Lenzner, J. and Maire, H. (1990) Kinetics of the diazo coupling between 1-naphthol and diazotized sulphanilic acid. *Ind. Engng Chem. Res.*, **29** 1761.
31. Bourne, J.R. and Maire, H. (1991) Influence of the kinetic model on simulating the micromixing of 1-naphthol and diazotized sulphanilic acid. *Ind. Engng Chem. Res.*, **30**, 1385.
32. Bourne, J.R., Kut, O.M. and Lenzner, J. (1992) An improved reaction system to investigate micromixing in high intensity mixers. *Ind. Engng Chem. Res.*, **31**, 949.
33. Bolzern, O. and Bourne, J.R. (1985) Rapid chemical reactions in a centrifugal pump. *Chem. Engng Res. Des.*, **63**, 275.
34. Bourne, J.R. and Tovstiga, G. (1988) Micromixing and fast chemical reactions in a turbulent tubular reactor. *Chem. Engng Res. Des.*, **66**, 26.
35. Bourne, J.R. and Maire, H. (1992) Simulation of micromixing in a turbulent tubular reactor. *Chem. Engng Commun.*, **112**, 105.
36. Li, K.T. and Toor, H.L. (1986) Turbulent reactive mixing with a series–parallel reaction. *Am. IChemE J.*, **32**, 1312.
37. Bourne, J.R. and Garcia-Rosas, J. (1986) Rotor–stator mixers for rapid micromixing. *Chem. Engng Res. Des.*, **64**, 11.
38. Bourne, J.R. and Studer, M. (1992) Fast reactions in rotor–stator mixers of different size. *Chem. Engng Process.*, **31**, 285.
39. Bourne, J.R. and Demyanovich, R.J. (1989) Rapid micromixing by the impingement of thin liquid sheets. *Ind. Engng Chem.*, **28**, 825 and 830.
40. Demyanovich, R.J. and Bourne, J.R. (1992) Impingement-sheet mixing of liquids at unequal flow rates. *Chem. Engng Process.*, **31**, 229.

41. Bourne, J.R. and Maire, H. (1991) Micromixing and fast chemical reactions in static mixers. *Chem. Engng Process.*, **30**, 23.
42. Bourne, J.R., Lenzner, J. and Petrozzi, S. (1992) Micromixing in static mixers; an experimental study. *Ind. Engng Chem. Res.*, **31** 1216.
43. Tosun, G. (1987) A study of micromixing in Tee-mixers. *Ind. Engng Chem. Res.*, **26**, 1184.
44. Baldyga, J., Bourne, J.R. and Zimmermann, B. (1994) Investigation of mixing in jet reactors using fast competitive-consecutive reactions. *Chem. Engng Sci.*, **49**, 1937.
45. Baldyga, J., Bourne, J.R. and Gholap, R.V. (1995) The influence of viscosity on mixing in jet reactors. *Chem. Engng Sci.*, **50**, 1877.
46. Baldyga, J., Bourne, J.R., Dubuis, B., Etchells, A.W., Gholap, R.V. and Zimmermann, B. (1995) Jet reactor scale-up for mixing-controlled reactions. *Chem. Engng Res. Des.*, **73**, 497.
47. Bourne, J.R. and Lips, M. (1991) Micromixing in grid-generated turbulence: theoretical analysis and experimental study. *Chem. Engng J.*, **47**, 155.
48. Bourne, J.R., Hilber, C.P. and Petrozzi, S. (1989) The influence of viscosity on micromixing in turbulent flows. *Chem. Engng Process*, **25**, 133.
49. Bennington, C.P.J. and Bourne, J.R. (1990) Effect of suspended fibres on macro-mixing and micro-mixing in stirred tank reactor. *Chem. Engng Commun.*, **92** 183.
50. Ranade, V.V. and Bourne, J.R. (1991) Reactive mixing in agitated tanks. *Chem. Engng Commun.*, **99**, 33.
51. Gholap, R.V., Petrozzi, S. and Bourne, J.R. (1994) Influence of viscosity on product distribution of fast competitive reactions. *Chem. Engng Technol.*, **17**, 102.
52. Rice, R.W. and Baud, R.E. (1990) The role of micromixing in the scale-up of geometrically similar batch reactors. *Am. IChemE J.*, **36**, 293.
53. Paul, E.L. (1988) Design of reaction systems for speciality organic chemicals. *Chem. Engng Sci.*, **43**, 1773.
54. Tipnis, S.K., Penney, W.R. and Fasano, J.B. (1994) An experimental investigation to determine a scale-up method for fast competitive parallel reactions in agitated vessels. *Am. IChemE Symp.*, **299**, **90**, 78.
55. Bourne, J.R., Gholap, R.V. and Rewatkar, V.B. (1995) The influence of viscosity on the product distribution of fast parallel reactions. *Chem. Engng J.*, **58**, 15.
56. Bourne, J.R. and Gholap, R.V. (1995) An approximate method for predicting the product distribution of fast reactions in stirred tank reactors. *Chem. Engng J.*, **59**, 293.
57. Bourne, J.R. and Zimmermann, B. (1994) Micromixing in a tubular membrane-module reactor. *IChemE Symp. Ser.*, **136**, 203.
58. Armand, J. and Bourne, J.R. (1994) Characteristics of a simple injector for mixing and fast chemical reactions in a tubular reactor. *IChemE Symp. Ser.*, **136**, 211.
59. Baldyga, J. and Bourne, J.R. (1993) Drop breakup and intermittent turbulence. *J. Chem. Engng Japan*, **26**, 738.
60. Frey, J.H. and Denson, C.D. (1988) Imidization reaction parameters in inert molten polymers for micromixing tracer studies. *Chem. Engng Sci.*, **43**, 1967.
61. Bennington, C.P.J. and Thangavel, V.K. (1993) The use of a mixing-sensitive chemical reaction for the study of pulp fibre suspension mixing. *Can. J. Chem. Engng*, **71**, 667.

6 Batch filtration of solid–liquid suspensions
A. RUSHTON

6.1 Introduction

Batch filtration involves the separation of suspended solids from a slurry containing a quantity of associated liquid. The concentration of the solids in the system may vary from batch to batch, as may the size distribution of the particles. Both particle size and concentration have important effects on the process time required for the separation. The slurry is directed towards the filter by means of a pumping, by gas pressure (blown transfer), gravitational head created by differences in level between the feed vessel and the filter, or by centrifugal force in rotating units. Solids are retained by a filter 'medium' (screen, woven cloth, paper, etc.) which simultaneously allows passage of the filtrate. Flow of the filtrate is created by the pressure differential generated over the system.

The solids may be captured on the surface in so-called 'cake filtration'. On the other hand, in the clarification of liquids containing a small amount of suspended material, separation may occur predominantly in the internal interstices of the medium. In some systems the particles have value whilst in others the liquid filtrate is the required product. A general case will specify the recovery of both phases at a certain degree of purity. All separations will be required to be completed in a reasonable process time which will include loading the filter, particle–fluid separation, discharge of the deposited solids and cleaning of the filter in preparation for the next batch.

In all cases of cake filtration the surface deposit will retain some of the feed liquor. The latter may have to be removed by 'deliquoring' prior to discharge of the deposit. In other circumstances, deliquoring may be followed (or preceded) by the administration of a quantity of wash liquor, which is to remove residual traces of solutes entering with the feed liquor. In some filtration systems, filter cake washing can constitute a controlling stage in the separation.

The overall process cycle time t_c required per batch includes periods for:

(i) Loading (filling) the filter.
(ii) Filtration time t_f.
(iii) Deliquoring time t_d.
(iv) Wash time t_w.
(v) Discharge and cleaning.

For convenience steps (i) and (v) are represented by the symbol t_{dw} in the equation below. Thus,

$$t_c = t_f + t_d + t_w + t_{dw} \tag{6.1}$$

If the volume of filtrate produced per batch is V, the productivity of the unit P_r can be calculated from:

$$P_r = V/(t_f + t_d + t_w + t_{dw}) \tag{6.2}$$

Each of the steps above must be considered in process productivity calculations. Laboratory tests are usually necessary when dealing with new separation problems, in order to quantify steps (ii)–(iv). These tests can reveal the presence of a possibly 'rate-limiting' step in the overall sequence, as mentioned above.

Again, when dealing with new separations, the type of filter to be used and an appropriate filter medium must be considered carefully. The problem of equipment selection, in both instances, is made difficult by the enormous number of choices available. Table 6.1 below contains a list of some of the equipment used in batch separations. The range of filters and media available reflects the wide differences which exist in the filterability of suspensions.

In view of this situation, it is fortunate that the subject of filtration has received much attention in the technical and academic literature [1, 2]. Process data are available which facilitates the estimation of t_f, t_d and t_w, and guidelines have been published which lead to the 'best' solution to the equipment selection problem [3, 4].

The best solution referred to above does not necessarily mean an optimum filter selection. The latter would involve a large amount of effort and time; resulting from the wide range of variables in any particular system [5]. However, these selection methods will avoid the severe process difficulties which follow the purchase of a completely unsuitable system.

Table 6.1 Classification of batch filtration equipment

Class	Typical machinery
Gravity	Strainer or Nutsche
Vacuum	Nutsche
	Candle and Cartridge
Pressure	Nutsche
	Plate and frame; recessed plate
	Tube, candle and leaf
	Variable chamber
	Cartridge
Centrifugal	Basket
	Peeler

Most of the filters listed above will be discussed in the text to follow.

Another type of process problem involves studies of existing installations which are giving on-line separation problems and low productivities.

6.2 Filtration process fundamentals

6.2.1 Flow of fluids in filtration

Several physical mechanisms are involved in solid–liquid filtration. Filtration theory is used to develop useful quantitative relationships between variables such as filtrate flow rate, particle size and concentration, medium 'characteristics', etc. The chosen medium presents an initial resistance to fluid flow; this resistance will change with use and as particles are deposited on and within the porous medium.

In the surface deposition of solids the process can be described by a system containing two resistances in series, in which the medium resistance R_m is superimposed by a deposit (or cake) resistance R_c as shown in Figure 6.1.

Experimental evidence shows that the initial filtrate velocity v_0 through the medium is proportional to the pressure differential over the latter and inversely proportional to the viscosity of the fluid:

$$v_0 = \Delta P/\mu R_m \tag{6.3}$$

If ΔP remains constant over the system, the fluid velocity will decrease as the particles deposit:

$$v = \Delta P/\mu(R_c + R_m) \tag{6.4}$$

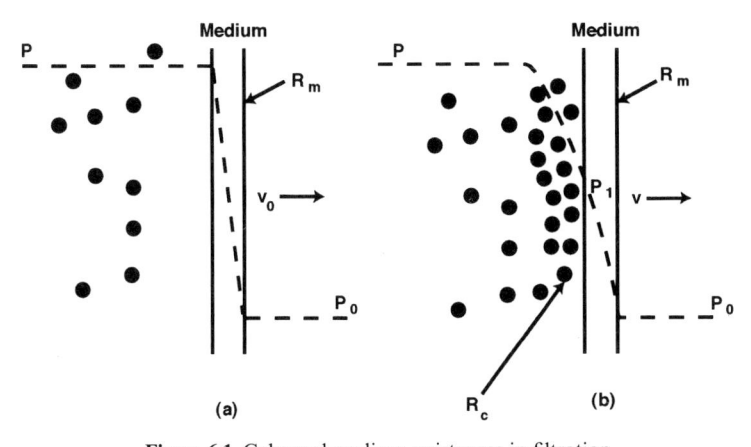

Figure 6.1 Cake and medium resistances in filtration.

where v is the filtrate velocity after deposition. The above relationship can be rewritten as:

$$v = (P - P_1)/\mu R_c = (P_1 - P_0)/\mu R_m \qquad (6.5)$$

where P_1 and P_0 are the fluid pressures at the cake–medium interface and downstream of the medium respectively. This simple development assumes that R_m does not change during the deposition of the first layers of solids.

Liquid flow takes place through the interstices, or pores, of the filter cake and medium. Usually a 'laminar flow' condition is obtained, leading to the above linear relationship between v and ΔP. The state of flow is related to the Reynolds number (Re) of the system [6] where $\text{Re} = (vd_p\rho/\mu)$. A transition region for certain woven filter cloths and perforated plates lies in the range $3 < \text{Re} < 7$ [7]; higher velocities produce turbulent conditions, where v is proportional to $\Delta P^{0.55}$. Laminar conditions are assumed throughout the development of filtration theory, as discussed here.

Filter cake resistances vary widely, with an industrial range including free-filtering sands to high resistance sludges; the latter may be 10 000 times more resistant to flow than sand-like filter cakes. Filter media also exist in a wide range of resistances from open screens to membranes for the processing of coarse solids to dissolved solutes, respectively. The resistance of the medium depends on the diameter and number of the pores present in its construction. Values of R_m for clean media may be estimated by observing the flow rate of a liquid, of known viscosity, through a known area A of the medium, at a controlled pressure differential. The ensuing calculation for R_m is typified by Example 1 below. Filter media in the range of $10^8 < R_m < 10^{14}$ are used industrially.

Example 1. Calculation of R_m. A sample filter of 0.01 m^2 in area is subject to the flow of water (viscosity 1×10^{-3} Pa s) under a pressure differential of 10.0 kN m^{-3}. The measured flow rate is 0.011 s^{-1}. The medium resistance R_m is calculated as follows.

The filtrate velocity in equation (6.3) is given by the quotient (q/A) where q is the volumetric flow of water. If follows that:

$$q/A = \Delta P/\mu R_m$$

Substituting $A = 0.01 \text{ m}^2$, $\Delta P = 10\,000 \text{ N m}^{-2}$, $m = 0.001 \text{ Ns m}^{-2}$ and $q = 10^{-5} \text{ m}^3 \text{ s}^{-1}$ gives:

$$R_m = 10^{10} \text{ m}^{-1}$$

This simple problem serves to record the various units of some of the symbols used in filtration calculations.

It should be noted that the flow characteristics of filter media are often recorded in terms of a permeability K, rather than the resistance R_m. Thus,

$$v_0 = K \Delta P/\mu L \qquad (6.6)$$

where L is the thickness of the medium. Permeabilities are expressed in units of m^2.

Another important characteristic of filter cakes is the porosity, ε, defined by:

$$\varepsilon = \text{volume of pores/volume of wet cake} \qquad (6.7)$$

The permeability can be shown to be related to the porosity and the specific surface S_V of the solids contained in the medium [8]:

$$K = \varepsilon^3/k_0(1-\varepsilon)^2 S_V^2 \qquad (6.8)$$

in which k_0 is the Kozeny 'constant'. This takes a value of 5.0 for sand-like materials, but increases with increased porosity. Thus, in fibrous beds the value may be as high as 30. For sand-like deposits e takes values of 0.45–0.6; beds of fibres or needle-shaped crystals will have much higher porosities where e may be greater than 0.9. Non-woven filter media made from fine polymeric fibres also exhibit high porosities. The importance of the e value lies in the fact that it determines, *inter alia*, the quantity of liquid remaining in the filter cake after its formation. The above remark applies to solid particles and fibres. Sludge-like deposits can contain liquid bound up by clusters of extremely fine particles or colloids; some particulates have internal porosity, where a proportion of the retained liquor may be trapped. This trapped fluid does not take part in the flow processes that are described by equations (6.6) and (6.8).

An increase in pressure differential can cause the particles within the filter cake to be brought closer together or compressed. The solids may also be deformable; soft particles may change shape and adopt lower overall porosities under stress. Most particulate assemblies are compressible to some degree; sand-beds are relatively incompressible so doubling the pressure differential over these assemblies may be expected to double the flow rate of filtrate. On the other hand, increased pressure with compressible material will not have a proportional increase in flow. This effect will be discussed below, in Section 6.2.6.

6.2.2 *Quantitative relationship for cake filtration*

At any stage in the filtration process, the collected volume of filtrate, V, will be associated with a certain mass of solids, M. The latter can be estimated from the known amount of solids in the feed, via a material balance which must account for the liquid left in the cake after filtration. This balance is used to define an 'effective' feed concentration, c^*, defined as the mass of dry solids deposited per unit volume of filtrate:

$$M = c^* V \qquad (6.9)$$

The effective concentration is calculable from information on the wetness of the filter cake immediately after filtration and before any dewatering processes commence. Thus

$$c^* = \rho s/(1 - ms) \quad (6.10)$$

in which s is the mass of solids (dry weight) per unit mass of feed slurry and the wetness ratio m is given by:

$$m = \text{wet cake mass/dry cake mass} \quad (6.11)$$

Experimental values of m can be obtained by weighing a representative quantity of wet filter cake and re-weighing after drying in an oven. Information on the wetness of the filter cake finds application in calculations on the post-treatment processes of cake washing and dewatering, which are discussed later.

Post-filtration measurements of the filter cake thickness, L_c, can also be used to estimate M using equation (6.12).

$$M = \rho_S (1 - e) A L_C \quad (6.12)$$

where ρ_S is the density of the solids in the cake. It seems logical that the overall cake resistance, R_C will depend on the cake thickness, which in turn will be related to the mass deposited per unit area, M/A. This assumed proportionality is used to define a 'specific cake resistance', α (m kg^{-1}), from:

$$R_C = \alpha M/A \quad (6.13)$$

Information is available in the literature on α for many particulates [9, 10]. Values of α usually lie in the range 10^8–10^{13} m kg^{-1}, with low values for easily filtered granular material to the high values exhibited by gelatinous, colloidal and other difficult systems.

Equations (6.4), (6.10) and (6.13) may be combined to produce a relationship for the instantaneous flow rate, q, or filtrate velocity, v, through a filter of area A:

$$q/A = v = A \Delta P / (\alpha \mu c^* V + \mu A R_m) \quad (6.14)$$

Further, the flow rate represents the change in filtrate volume with time (expressed differentially $q = dV/dt$). Equation (6.14) can be rearranged to:

$$\Delta P = (K_1 V + K_2) q \quad (6.15)$$

where $K_1 = \alpha \mu c^*/A^2$ and $K_2 = \mu R_m/A$.

Equation (6.14) represents the basic relationship for surface cake filtration and may be applied to various practical process situations such as:

(i) Vacuum filter calculations $\Delta P = $ constant
(ii) Constant rate pressure filters $q = $ constant
(iii) Filters supplied by centrifugal pump variable ΔP; variable q.

The equations above can be used to determine the area of filter required for a particular separation under the above conditions (i), (ii) and (iii). The performance of actual filters may then be compared with theoretical design predictions. Integrated forms of equation (6.14) are also important in their applications to the analysis of laboratory test data, in acquiring information on α and R_m for new systems. All these applications have been highlighted in the relevant literature [9, 11].

6.2.3 Laboratory tests and filter media in cake filtration

Equation (6.14) presents a basis for laboratory tests aimed at the evaluation of the filter cake and medium resistances. Most preliminary filterability tests are performed using a vacuum system, as shown in Figure 6.2. Other equipment has been described which can be used in elevated pressure tests [12].

Equation (6.14) is rearranged and integrated to yield a theoretical relationship between V and t:

$$t/V = \alpha \mu c^* V / 2A^2 \Delta P + \mu R_m / A \Delta P \tag{6.16}$$

or

$$t/V = K_1 V / 2\Delta P + K_2 / \Delta P \tag{6.17}$$

or

$$\Delta P t/V = K_1 V / 2 + K_2 \tag{6.18}$$

Assuming that α, R_m, μ and c^* remain constant during the test, the equations above point to a graphical method of linearizing the parabolic (V, t) relationship obtaining by integration of Equation (6.14). In Figure 6.3(a), the slope of the graph of t/V plotted against V is related to the specific resistance of the cake deposit, whilst the intercept relates the medium resistance. Repeat of the experiment at different vacuum levels can be used to test for compressibility effects, albeit in a low pressure range. If the cake is compressible, information is required throughout the pressure range expected to be used at large scale, so that an average value may be estimated [5, 13, 14]. In these circumstances, it is generally recommended that test data be obtained on a pilot-scale version of the type of filter which is being considered for production purposes. Plant-scale data are, of course, even more useful, if available.

The t/V test is used in other areas of the overall process problem:

(i) To check the suitability of the filter medium used. This suitability can be estimated by repeated trials, with intermediate cake discharge and cleaning of the cloth (Figure 6.3(b)). Despite cleaning, the increase in the intercept with the ordinate on the various t/V versus V plots indicates that the cloth is retaining fine particulates, or blinding. Generally the medium is required to effect many cycles in the filtration process, otherwise cloth replacement and plant down-time costs may become excessive.

160   HANDBOOK OF BATCH PROCESS DESIGN

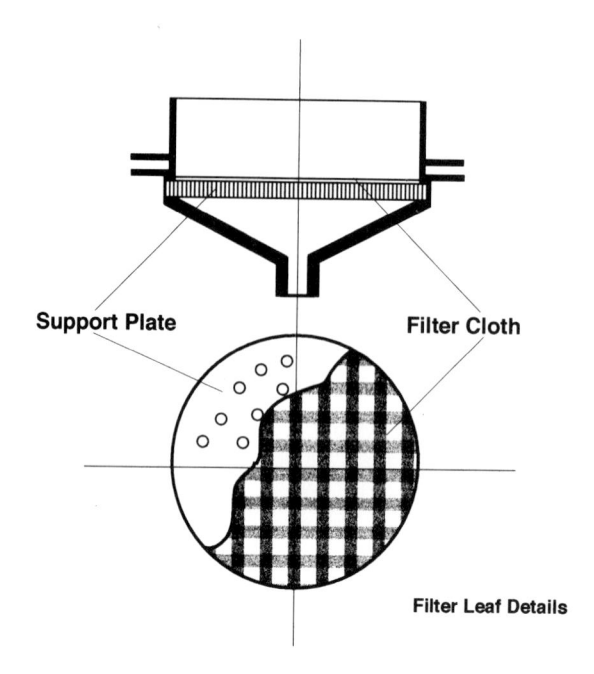

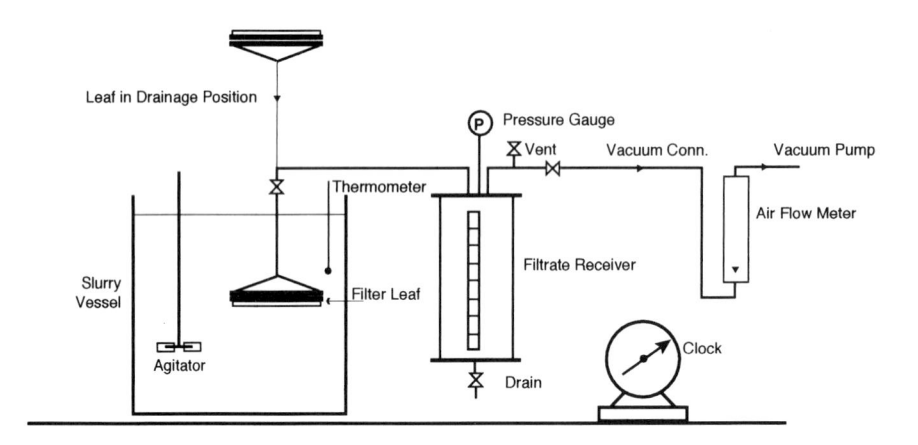

Figure 6.2 Filter leaf test apparatus.

(ii) To test the suitability of methods used for improving the filterability and dewatering of the batch. For example, a high inherent filtration resistance, may be ameliorated by the addition of 'filter-aids' to the system or by increasing the size of the particles involved, by coagulation and/or flocculation. These possibilities are discussed below.

(iii) To check on possible differences between the use of an upward or downward filtration mode. In downward filtration, both the filtrate and

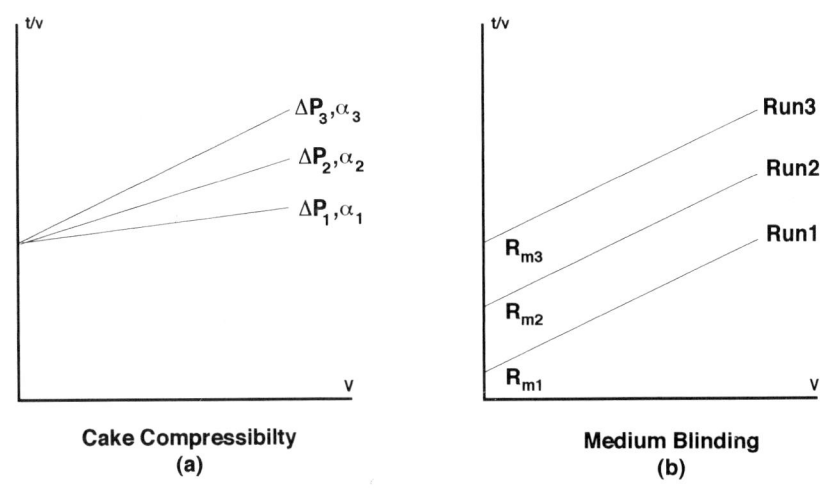

Figure 6.3 Volume versus time graphs for cake filtration.

(a) Cake Compressibility — lines labeled $\Delta P_3, \alpha_3$; $\Delta P_2, \alpha_2$; $\Delta P_1, \alpha_1$ on t/V vs V axes.
(b) Medium Blinding — Run3, Run2, Run1 with intercepts R_{m3}, R_{m2}, R_{m1}.

suspended solids proceed in the same direction. In the upward mode, gravity has an effect on the relative velocities of the liquid and solid phases, so that larger particles may settle away from the filtering zone. If sedimentation effects are pronounced care has to be exercised in plant selection, particularly in those cases where the solids are deposited on a downward facing or vertical surface, where the particles may be classified into fine and coarse layers.

The test described above is sometimes used to check the filterability of batches of the same product, prior to filtration at plant scale. In these cases it is not necessary to evaluate the specific cake resistance; the slope of the t/V versus V graph is a sufficient indicator of the filtration properties of the batch and can be used to decide on the possible need for pre-treatments, perhaps by addition of filter aids, coagulants, etc.

In many tests, non-linearities are exhibited in the t/V versus V graph. It will be realized that the graphical construction is based on an assumption that certain factors remain constant during the separation, but most importantly that the solids do not penetrate the medium and R_m remains constant. Obviously, particle penetration remains a possibility in the case where the pore-size distribution in the medium overlaps the particle size distribution of the solids. The initial interaction between the medium and the particles is all important. Generally it is true that the wider the difference between pore and particle size the better in order to produce surface deposition.

Example 2. Evaluation of cake resistance. To evaluate the specific cake resistance and the medium resistance the laboratory test data shown in Table 6.2 were obtained with a 10 wt% suspension of solids in water. The materials properties and plant data are: solids density $\rho_S = 2500 \text{ kg m}^{-3}$,

Table 6.2 Model laboratory test data

ΔP (N m^2)	107 000		190 000		242 000		323 000	
m	1.50		1.46		1.35		1.30	
V (m$^3 \times 10^3$)	t	t/V	t	t/V	t	t/V	t	t/V
1	19	19 000	14	14 000	12	12 000	9	9000
2	53	26 500	37	18 500	30	15 000	25	12 500
3	102	34 000	69	23 000	57	19 000	46	15 333
4	163	40 750	110	27 500	91	22 750	73	18 250

fluid density $\rho = 1000 \text{ kg m}^{-3}$, fluid viscosity $\mu = 0.001 \text{ Pa s}$ and filter area $A = 0.1 \text{ m}^2$.

Plots of t/V versus V may be prepared and the slope and intercept on the ordinate determined:

$$\text{slope} = \alpha[c^*\mu/A^2 \Delta P]$$

$$\text{intercept} = \mu R_m / A \Delta P$$

Alternatively, the resistances may be estimated by the use of an incremental version of equation (6.15) in setting up a difference table for constant pressure separation (as outlined by Holdich [9]). The effective concentration at each test level is estimated from $c^* = \rho s/(1 - ms)$ for the values of m in the table above and $s = 0.1$.

The results of these calculations are tabulated in Table 6.3. The results indicate a moderate compressibility of the filter cake, that is α increases with ΔP. The compressibility can be quantified by estimating the 'compressibility index' [13], n, from:

$$\alpha = \alpha_0 \Delta P^n \qquad (6.19)$$

The slope of the graph of $\log \alpha$ versus $\log \Delta P$ from the above measurements gives a value of $n = 0.24$. The medium resistance can also be seen to be slightly increased at higher operating pressures.

The above comments on the role of the filter medium suggests that a successful separation depends, to a large extent, on suitable medium selection.

Table 6.3 Calculation of the specific cake resistance for the model problem

ΔP (kN m^2)	Intercept (s^{-1} m^3)	Gradient (s m$^{-6} \times 10^{-6}$)	α (m kg$^{-1} \times 10^{-10}$)	R_m (m$^{-1} \times 10^{-10}$)
107	12 000	7.0	6.4	1.28
190	9400	4.5	7.4	1.78
242	8000	3.5	7.3	1.93
323	6000	3.0	8.4	1.96

Whilst there is probably an optimum filter for every separation, all of the machines listed in Table 6.1 would produce some degree of separation, provided that an adequate filter medium was employed. On the other hand, incorrect specification of the medium can lead to dramatic failure in filtration systems. It follows that the characteristics of filter media and the methods used for their selection have received much attention in recent, relevant filtration literature [1, 15]

A large variety of materials is used in the construction of media for batch cake filtration. These include, *inter alia,* natural substances such as cotton and wool and synthetic polymers, e.g. polyesters, polyamides, etc. Paper in various forms is used extensively throughout industrial filtration processes; metals and ceramic elements are used in high temperature and/or pressure conditions or where corrosive environments exist.

This wide range of materials is coupled with an equivalent diversity in media constructions. Thus filter cloths may be woven or non-woven, as shown in Figure 6.4. The yarns used to prepare woven cloths may be of a 'monofilament' type, composed of solid polymer or metal; the alternative 'multifilament' yarns are made from twists of a selected number of fine filaments. The degree of twist can have important effects on the form of deposit produced when particles impinge on the surface of the medium [15]. Yarns may also be constructed from polymeric 'staple' fibres used to simulate cotton and wool. The filtration characteristics of woven media are also closely related to the weave pattern used in their construction. Thus, factors such as filtrate clarity, cake release, dewatering efficiency, cloth life and resistance to blinding depend on the yarn or fibre type used and the weave pattern, as indicated in Table 6.4 [1].

The filtration characteristics of non-woven materials, which are often used in the construction of cartridge media, depend on the diameter of the fibres used in the preparation of the cloth. The surface properties of these fabrics can be modified by heat treatment of the fibres by hot calendering.

The multiplicity of medium types and constructions makes medium selection a difficult stage in overall process specification. In addition to noting the above general guidelines and in cases where historical evidence is not available on the performance of particular media, tests of the type discussed above are essential. As before, it is generally recommended that pilot scale versions of the filter under consideration be used, on-site, in medium evaluations [16].

Benefits gained from correct medium specification include:

(i) Clean filtrate, with minimal bleeding of solids through the medium.
(ii) Easily discharged filter cake.
(iii) Absence of media deterioration due to blinding.
(iv) Adequate cloth life-time.

Perhaps long-term blinding potential is the most difficult feature to predict using laboratory tests, which are necessarily limited in duration. Interesting

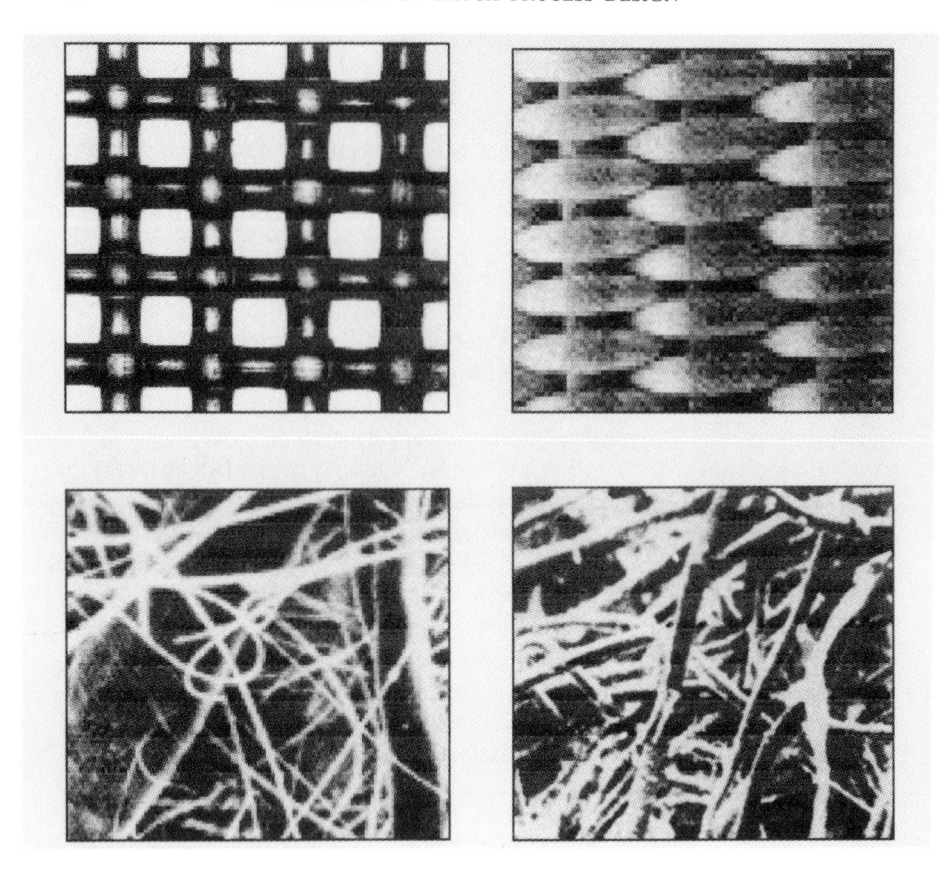

Figure 6.4 Woven and non-woven filter cloths.

Table 6.4 Effect of yarn type and weave pattern in filtration

	Filtrate clarity	Cake moisture	Cake discharge	Medium life	Resistance to blinding
Fibre/yarn	SS	M	M	SS	M
	MF	MF	MF	MF	MF
	M	SS	SS	M	SS
Weave/pattern	PRD	PRD	S	PRD	S
	P	S	PRD	TW	TW
	TW	TW	TW	P	PRD
	S	P	P	S	P

M = monofilament; MF = multifilament; S = sateen; SS = spun staple; PRD = plain reverse dutch. Best performances are to be expected from the media at the top each row of the table.

recent developments include new weave patterns to minimize blinding potential [17] and 'capillary control' media [18]. The latter are in fact, membranes which have sufficient permeability to the filtrate but, because of the small size of the pores in the medium, prevent the access of air. Thus, in dewatering processes by gas displacement of the cake liquor, minimal losses of air are experienced in the dewatering stage. This suggests the possibility for a profitable refit of many existing filter stations. The medium must also prove suitable in terms of gasket performance, strength, abrasion resistance, chemical and/or biochemical stability, abrasion resistance, etc. These factors can generally be settled by contact with suppliers.

6.2.4 Application of basic relationships to centrifugal filters

The equations above may be modified to process information obtained in centrifugal batch filtration (Figure 6.5), where the processes of surface cake deposition on the inside surface of a rotating, perforated bowl centrifuge is depicted. In these machines, often applied in the separation of relatively coarse low materials with low α values such as sugar crystals, particulate deposition and fluid flow occur under conditions of changing filter area. Despite these additional complexities, process equations [13] are available such as:

$$dV/dt = \Delta P/\mu(\alpha c^* V/A_a A_{lm} + R_m/A) \tag{6.20}$$

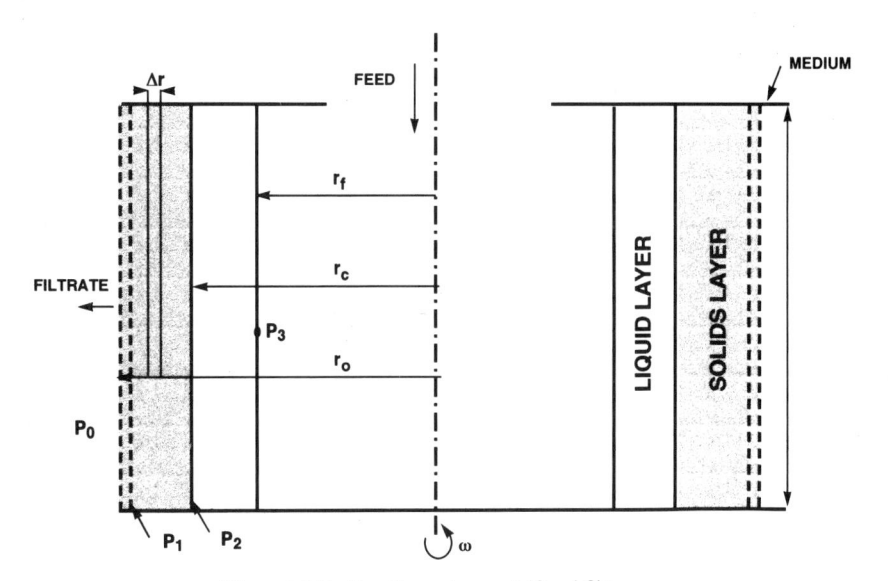

Figure 6.5 Section through a centrifugal filter.

in which A is the centrifuge bowl area for filtration. The average flow area A_a and logarithmic mean area for flow through the cake A_{lm} arise from the decreasing flow area available as the cake thickness increases. The pressure profile developed during centrifuged cake formation is unusual. In Figure 6.5, the fluid pressure in the pool of slurry will rise with increase in radius from P_3 to P_2; flow through the filter cake commences at radius r_c with an associated drop in fluid pressure to P_1 at the cake–medium interface, then down to P_0 at the downstream surface of the medium.

The pressure differential in centrifugal filters is related to the bowl and liquid pool radii, r_0 and r_f by:

$$\Delta P = \rho \varpi^2 (r_0^2 - r_f^2)/2 \qquad (6.21)$$

in which ϖ is the bowl rotational speed. Equations (6.20) and (6.21) have been used in the measurement of α and R_m in dynamic filtration conditions during centrifugation [19]. This work showed that the specific resistance of the cake will increase with time with continued spinning, after filter cake formation. Compression of the cake by fluid drag, manifest in pressure and vacuum filters, is augmented in centrifuges by the presence of a 'body force'. This depends on the density of the particles; the resistance of a high density cake will generally exceed that of a lower density deposit, other factors being the same. These effects may have a bearing on the scale-up of these filters, since the pressures developed are increased at constant speed with increase in diameter.

6.2.5 Filter cake washing

The principal aim in cake washing is the removal of dissolved solutes (and possibly solvents) present in the original feed liquor. As stated above, a proportion of the latter is retained by the filter cake after its formation. Washing is obviously applied in circumstances where the particulates are of value and are required in a certain state of purity.

In process terms, information is required on the quantity of wash liquid required for the removal of solutes; the wash time, t_w, will then depend on the flow rate achieved during washing. Figure 6.6 shows a typical washing curve, where changes in the solute concentration with increased amounts of wash fluid are recorded. The abscissa in this figure may be reported in terms of wash-time, if required. The concentration of the solute obtained at a particular stage, c, is reported, along with the initial concentration c_0; the ordinate has been normalized by using the ratio (c/c_0) in the graph.

The graph refers to so-called 'displacement' washing, in which the wash fluid is caused to flow through the complete thickness of the cake. Washing can, of course, be also effected by dispersing or re-slurrying the wet cake in a quantity of wash fluid, agitating and then re-filtering. This technique is

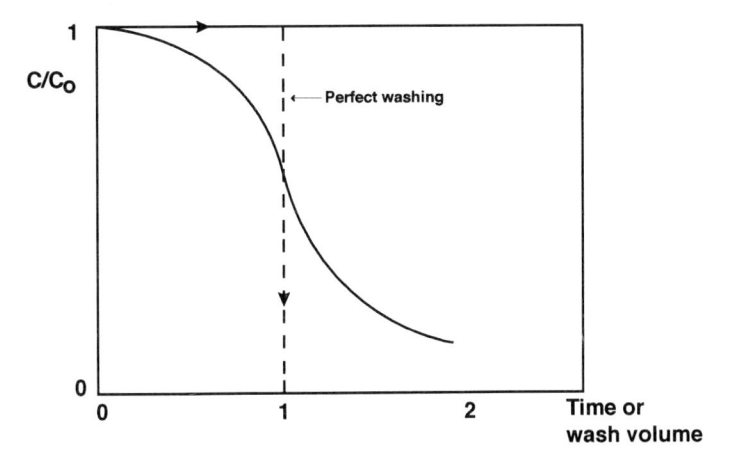

Figure 6.6 Washing curve – solution concentration versus time.

sometimes used in continuous filtration systems. In batch processes, displacement or 'through' washing is more commonly applied.

Washing may be effected immediately after cake formation, in which case the pores within the cake are completely filled with the original liquor at solute concentration c_0. In this case, the volume of liquor is equal to the 'void volume'. Studies of filter cake washing of flooded cakes are sometimes reported in terms of the number of void volumes used. As shown in Figure 6.6, a perfect washing operation would only require the delivery of one void volume of wash in reducing c/c_0 from unity to zero. In practice, the wash liquor tends to proceed through preferential pore pathways, or sometimes cracks in the filter cake, to discharge at a concentration $c < c_0$. The actual value of c attained depends on the processes of mixing and mass transport which occur during the washing process.

In some cases the filter cake may be dewatered before washing. Here, the amount of original liquor retained relates to the cake saturation, S (Section 6.2.6). In this operation, the term 'wash ratio', W_r, is used, i.e. the ratio of the volume of wash used to the saturation level of the cake. For a flooded cake the wash ratio is unity.

The subject of washing has been reported extensively in the relevant literature [20, 21]. Both theoretical and practical approaches have been used to elucidate the factors influencing the washing process. This interest reflects the importance of the washing stage, which can often be the controlling step in a batch separation.

An early model of the washing processing involved the concept of a 'perfectly mixed cell' [22] to describe the changes in c with wash volume or time. It may be shown that, in this case

$$(c/c_0) = e^{-W_r} \tag{6.22}$$

The model ignores the mass transfer effects that are present. Original liquor may be retained in parts of the bed which are not fully irrigated by the wash fluid. The liquor may be contained within clusters of particles or at points of contact between particles. Solute removal from these zones may proceed only by a slow, diffusional process, in which case the equation above will seriously underpredict the wash requirements. The diffusional aspects of cake washing have been the subject of extensive research activity [23].

A generalized approach to washing [24] involves the use of a 'dispersion parameter' (uL_c/D) in which u is the wash velocity, L_c is the cake thickness and D is the 'axial dispersion coefficient'. High values of the dispersion parameter lead to high washing efficiencies whilst low values tend to produce the long washing 'tails' and wash-times. Further details of these methods can be found in the general literature [2, 9].

6.2.6 Filter cake dewatering

Removal of the residual liquor from the filter cake may be effected by:

(i) Application of vacuum.
(ii) Blowing with compressed gas.
(iii) Centrifugation.
(iv) Compression of the cake.

In the last method, the cake pores remain filled with liquor whereas in (i), (ii) and (iii) some of the internal cake space is drained free. In all cases, the mass ratio of liquor to cake solids is reduced. Where the cake is partially drained, use is made of the term 'saturation', S, defined by $S =$ volume of liquor in cake/volume of pores in the cake. At the start of deliquoring by methods (i), (ii) and (iii) the value of S is unity. Complete drainage is not usually achieved; in some cases final drying with hot gas flow through the cake is used. The saturation level prior to drying approaches an equilibrium level, S_e, with continued deliquoring. This process is shown in Figure 6.7.

The kinetics of the dewatering process and the final equilibrium of saturation can be influenced by changes in the applied pressure differential, the surface tension and viscosity of the liquid. In dewatering by methods (i), (ii) and (iii) the applied differential has to overcome the effects of surface tension which creates a 'threshold pressure'. The latter acts against the dewatering effect and has to be exceeded before dewatering can commence; the pressure required to force gas into liquid-filled capillary is inversely proportional to capillary size. It follows that the dewatering of filter cakes composed of small particles, with correspondingly small capillaries, will require high dewatering pressure. In vacuum systems, insufficient pressure may be available to displace retained liquids.

Empirical expressions relating S with t_d are available in the literature for vacuum, pressure and centrifugal dewatering [20, 25]. This information is

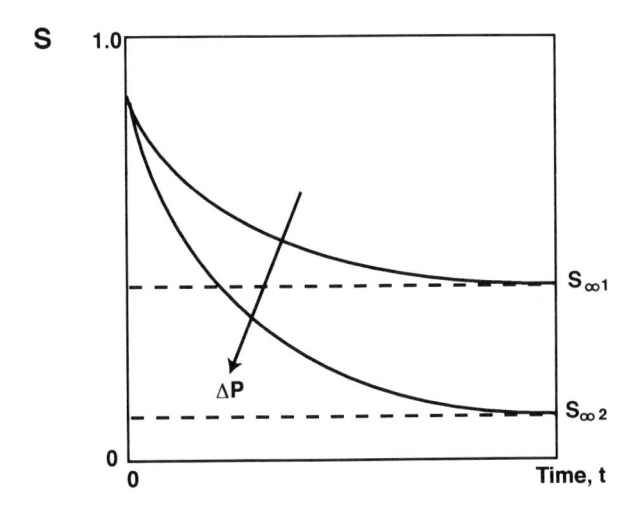

Figure 6.7 Dewatering curve – cake saturation versus time.

based on laboratory investigations into the dewatering of granular solids. Large-scale tests [26] using various slurries of sand, pigments, etc., produced close agreement with these empirical expressions. On the other hand, in studies with centrifugal filters [27, 28], deviations were noted, particularly when processing precipitates exhibited internal porosity, as would be expected. The subject of dewatering has also been analysed theoretically [29, 30] using information on the pore size distribution in the filter cake. These studies lend general support to the empirical approach discussed above and identify the principal process variables influencing dewatering. It is generally agreed that the above approaches should be complemented with practical evaluation of the (S, t) relationship when considering new applications.

6.2.7 Clarification filtration processes

In these processes, the concentration of solids is low and the formation of deposits on the surface of the filter medium is minimal. In fact, effective application of batch clarifying filters follows when internal deposition of solids (within the medium) is the principal separation mechanism. The main removal mechanism is intended to be 'depth' filtration; upstream removal of particles larger than a certain size may be necessary in order to take full advantage of the enormous contact surface available within the filter.

It follows that different process equations are required to describe the change in filtrate volume, or increase in pressure differential with time. In fact, it has been established [13] that the equations reported above for cake or surface deposition are but one solution of a general differential expression applicable to filtration systems. This expression can be applied not only to cake formation, but also to internal depositions of particles, blinding processes at the surface of the medium and combinations of these mechanisms. The general expression takes the form:

$$d^2 t/dV^2 = K'(dt/dV)^g \tag{6.23}$$

for volume–time relationships at constant pressure and

$$d(\Delta P)/dV = K' \Delta P^g \tag{6.24}$$

for operation at constant rate.

It may be shown that the equations produced for cake filtration conditions follow from the assumption that the exponent $g = 0$. Processes of internal deposition in clarifying filtration may be modelled by taking $g = 1.5$, in a so-called 'standard law' model. The corresponding volume, pressure and time relationships [31, 32] are:

$$t/V = K_S t/2 + 1/q_0 \tag{6.25}$$

and

$$\Delta P_0/\Delta P = 1 - (K_S V/2) \tag{6.26}$$

The above expressions can also be developed for non-Newtonian liquids, applicable to the batch processing of molten polymers, etc. through stainless steel clarifying filters [32, 33]. Other relationships can be developed for conditions involving complete surface blinding with $g = 2.0$ and mixed internal/surface depositions ('intermediate law') with $g = 1.0$.

Various mechanisms are involved in depth filtration. The larger particles present may be removed by direct interception with the materials or fibres present in the depths of the filter. The filter is often in the form of cylindrically shaped cartridges fashioned from arrays of metal, polymeric or natural fibres. Interception may occur by straining, sedimentation, inertial impaction, diffusion, and hydrodynamic and electrostatic interactions [9]. Larger particles are more likely to cross the flowlines of the fluid during its tortuous passage through the medium, therefore being collected by inertial effects. Small particles, on the other hand, are more likely to be removed by surface attractive forces present on the filter medium, having first been brought into the vicinity of the surfaces by random bombardments from surrounding fluid molecules. The so-called Brownian motion induced on the general path of the smaller particles enhances their capture possibility. It follows that middle-sized particles which are too small to be influenced by inertial effects and yet too large to adopt the random velocity profile characteristic of the smallest units may exhibit the lowest capture efficiency. The inclusion of membranous materials among such media extends the range of particles down to 0.1 μm in microporous membrane systems, as discussed below.

An alternative theoretical approach, used to describe the process changes taking place during depth filtration, relates the particle concentration changes taking place as the fluid passes through medium to a filtration 'constant' via:

$$(dC/dy) = -\lambda C \tag{6.27}$$

The above 'constant' λ is in fact a time-dependent variable. The initial value of the 'constant', λ_0, which applies at the start of the separation is, however, an intrinsic property of the filter. At short process times, the above expression can be integrated over the complete filter bed depth, L:

$$C = C_0 \exp(-\lambda_0 L) \tag{6.28}$$

These equations and their development to allow for the changes of λ with time have received attention in the literature [34]. Such analyses are usually applied to continuous filtration systems, using deep bed sand filters or the modern dual sand–anthracite combined filter. These units have very large adsorptive capacities for suspended solids and are often used in tertiary stage purification of waste water, produced from municipal and industrial sources. Of course, the filter gradually fills with contaminant and the quality of filtrate (which can be monitored by measurements of the outlet turbidity) starts to deteriorate. At this juncture the feed is interrupted for a period of back-washing involving the fluidization of the filter. Designs are aimed at on stream filtration periods of hours, linked with cleaning times of minutes. In view of the high capacity of these units, reports are starting to appear involving deep bed applications in batch systems.

6.2.8 Laboratory tests and filter media in clarification processes

Filter cartridges are used widely in clarification systems, e.g. in the electronics industry to ensure the clarity of rinse water, in hydraulics to remove abrasive particles above a certain size, in bottling operations in the beverage industry, in off-shore filtration of sea water used for well re-injection, etc. These filters are 'rated' in accordance with their ability to remove particles of declared sizes. The rating can be 'absolute' or 'nominal' In the latter case specifications are of the type: 'removal of 90–95% of particles larger than the nominal pore rating of $x\,\mu$m'. Care has to exercised in the interpretation of these ratings, which can be viewed merely as a guide to performance, rather than a guarantee. In contrast, an absolute rating declares that the medium will be 100% effective in the removal from the filtrate of all particles above $x\,\mu$m. Another important aspect of these specifications relates to the guaranteed absence of media fibre shedding during filtration. In media with unsecured fibres, scouring effects of the filtrate flow may result in the appearance of media in the product.

Filter cartridge efficiency tests are made using dilute suspensions of 'standard' particles. A wide variety of test powders is available, in fine and coarse grades. Industrial users will generally use a test mixture which closely represents their products, for example yeast cell suspensions in the brewing industry. Spherical latex spheres [35] are useful in that they can be produced in narrow size ranges. Glass spheres (ballotini) are applied, *inter alia*, in absolute tests for the largest pore present in the medium. This test is similar in outcome to the (non-destructive) method of measuring the gas pressure required to force gas bubbles to emanate from the surface of a filter, when the latter is immersed below the surface of a 'wetting' fluid. This so-called 'bubble test' can be extended beyond the first bubble pressure, up to the gas pressure level required to cause flow in all the pores present. Simultaneous measurements of these pressures and the concomitant gas flow rate is used in computing the complete distribution of pore sizes in the medium. The wetting fluids used are substances such as isopropyl alcohol which have low surface tensions when compared with water.

Other particulates used in efficiency tests include the 'air cleaner tests dusts', widely used in liquid trials as suspensions of fine ACFTD or coarse ACCTD particles in suitable liquids. The pharmaceutical industry, which has a particular interest in sterile operations and the used of membranes, apply organisms such as *Pseudomonas diminuta* [36], in absolute trials down to $0.2\,\mu$m.

The laboratory equipment used to test these cartridges has been described [37] and is depicted in Figure 6.8. Tests can be arranged to meet the particular application; multi-pass tests are used to simulate the recycling operations commonly used in the hydraulics industry. In other cases single pass tests may be more appropriate. Measurements are made of the concentration of particles, of size $x\,\mu$m, upstream and downstream of the medium. The ratio of these concentrations is used as a measure of the filter efficiency:

$$\beta = \frac{\text{concentration of particles of size } x \text{ upstream}}{\text{concentration of particles size } x \text{ downstream}} \qquad (6.29)$$

High β values are associated with high retention efficiency. Cumulative efficiencies for particles larger than a certain size can be estimated from $E\% = [(\beta - 1)/\beta] \times 100$, which leads to the information shown in Table 6.5.

The use of single-point efficiency to describe the performance of media has been challenged [38] and it is recommended that three particle sizes are reported for factors of 2, 20 and 75 and corresponding removal efficiencies of 50, 95 and 98%, respectively. The highest figure may be viewed as an approach towards an absolute rating and the two other statistics as measures of two values of nominal ratings. It will be realized that these tests are destructive in nature. Other work is in progress [39] which is aimed at extending current non-destructive tests in the validation of expensive cartridge filters which necessarily have to be cleaned and recycled.

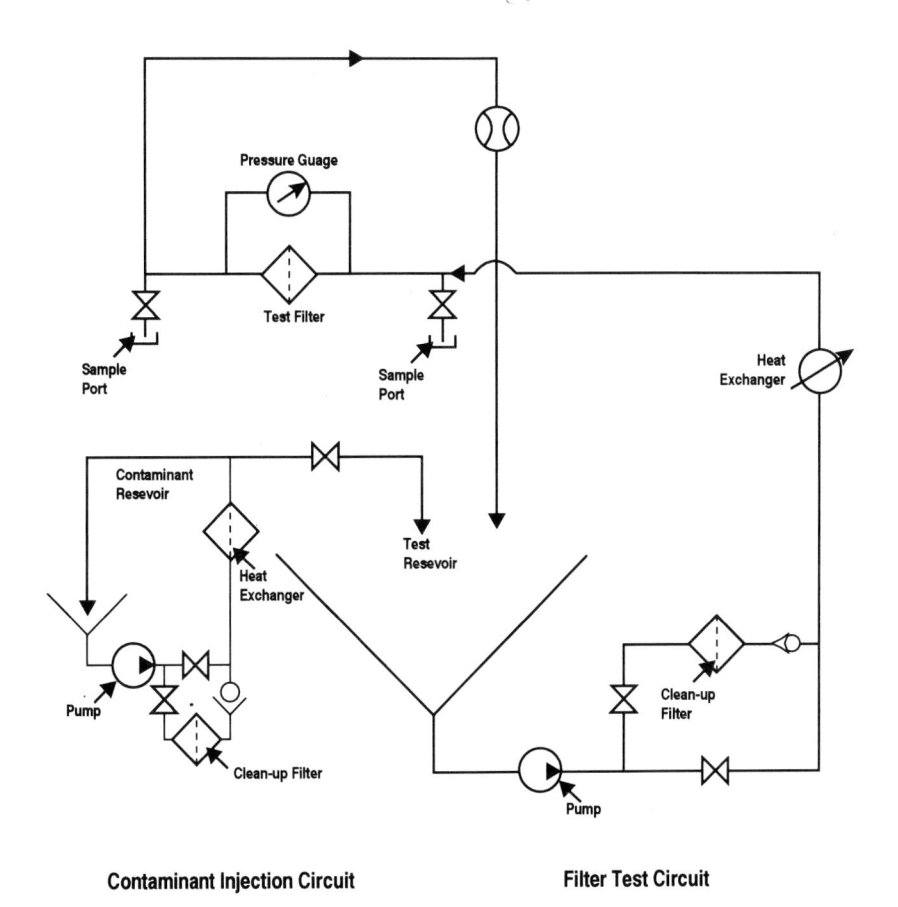

Figure 6.8 Multi-pass test flowsheet [37].

Further information is required on the contaminant holding capacity of the cartridge. The mass of solids producing an 8-fold increase in the clean filter pressure drop characteristics is often quoted as a capacity measure. Details of the multi-pass test system used for dirt holding capacity tests are available in ISO 4572. The frequency of non-cleanable cartridge replacement is controlled to some extent by limiting these applications to low concentrations (below 0.01 wt%).

In view of the cost of replacement, high capacity clarifications may require the application of alternative methods, despite the wide size range of

Table 6.5 Cumulative efficiencies based on point efficiency measurements.

β, for particles of size x	2	10	100	1000	10 000
Cumulative efficiency (E) % for sizes greater than x	50	90	99	99.9	99.99

cartridges available. In this respect, vertical candles and vertical or horizontal leaf filters, discussed below, are often used. The base medium here may be an open monofilament, polymeric cloth or steel mesh on to which a thin layer of filter aid is deposited. This layer is known as a pre-coat; thin layers of 2–4 mm of aid are usually sufficient for clarification purposes. As described elsewhere [9], filter aids are materials of high porosity, low flow resistance and low compressibility. Various grades are available: diatomaceous earth (kieselguhr), perlite (volcanic ash), cellulose, etc. As a pre-coat the filter aid acts as a mini deep-bed filter, which retains suspended contaminants. In some circumstances, where the solids to be separated are of a slimy, highly compressible nature, it may be necessary to add some filter aid into the feed stream, in order to break up the highly resistant layers which may even blind the pre-coat! The correct conditions of applying pre-coats and the dosages required in body feeding have been reported [40]. The amount of body feed varies with the filterability (compressibility) of the contaminant in the range 0.25–8 times the weight of the suspended solids; a low resistance, incompressible material requires the smaller dose of filter aid. The subject of filter aid body feed continues to engage the attention of researchers [41]. Mathematical analysis of constant rate filtration on filter candles, using pre-coats and body feeds has been published [42].

6.2.9 Membrane filtration principles

This section deals with the extension of 'normal' filtration processes, the lower size limit of which is about 10 μm, down to 0.1 μm. This extension is achieved by the use of microporous filtration [or microfiltration (MF)] with membranes which are available in nominal pore sizes of 0.2, 0.45, 0.8, 1.0, 1.2, 2.0, 3.0, 5.0 and 10 μm. The smallest rating of 0.2 μm is popular in biochemical applications, since sterile fluids are produced at this rating. There is, of course, an enormous industrial interest in ultrafiltration (UF) and reverse osmosis (RO) membrane applications. These are directed to the removal of dissolved species and will not be considered. The technologies involved in MF, UF and RO are similar. The trans-membrane pressures used in MF and UF are moderate, up to 10 bar; much higher pressures are required in RO to overcome the osmotic pressures of the dissolved salts. It may also be noted that MF constitutes about 70% of the membrane market, in financial terms.

MF membranes are available in a variety of polymeric materials: polyamides, polyethers, polyesters, polycarbonates, polyvinylidene fluoride (PVDF), polvinyl chloride, polytetrafluorethylene (PTFE), polysulphones, etc. Ceramic membranes are being applied to an increasing number of separation problems; this expansion is principally due to their chemical inertness, which permits vigorous intermittent chemical cleaning of the membranes. An important surface property of membranes when processing aqueous suspension is that of hydrophilicity. Non-wetting polymers requires surface treatment (sulphonation) to counter the hydrophobic nature of these membranes. This is necessary to overcome the capillary pressure which resists pore penetration by non-wetting fluids.

Polymeric membranes vary in construction from sponge-like media with homogeneous pore profiles through the medium to asymmetric membranes consisting of a thin, small-pore layer on the filtering surface supported by a more open-pore, robust underlayer. These constructions are attained by varying the rate of precipitation of the polymer from a solvent by contact with a non-solvent. For example, in the preparation of polyamide membranes from a 15% solution in dimethylacetamide solvent, use of water as a non-solvent will require 50 s for the precipitation; an asymmetric membranes results. An increase in polymer concentration can lead to the sponge type structure, as will a change of non-solvent, e.g. the use of acetone rather than water.

The asymmetric structure is less likely to allow the passage of particles into the internals of the membrane. This desirable feature can also be created by the use of 'composite' membranes, i.e. a combination of a coarse medium superimposed with a fine surface layer.

The 'Nucleopore' membrane is unusual in that the pores are created by the passage of fission fragments of uranium through a polycarbonate film. After bombardment, the film is etched in a warm, caustic batch where the passage sites dissolve to leave almost circular holes in the medium. It follows that these membranes have a tight pore-size distribution, when compared with the sponge and asymmetric media discussed above. The variety of processes used in membrane construction have been described in the literature [43].

A considerable interest developed, with the introduction of MF membranes, in batch separations of difficult materials such as systems with particles below 10 μm in size.

The principal factor influencing filtrate flux is the development of particulate deposits on the surface of the filter. This, of course, is the usual result in cake filtration systems, when the particles are directed normally towards the surface of the medium. Experience with UF and RO systems demonstrated that the detrimental effects on filtrate flux of these deposits and the development of high concentrations of dissolved species at the surface could be ameliorated by cross-flow of the feed over the medium, i.e. flow parallel to the membrane surface.

Cross-flow microfiltration (CFM) has been studied extensively [44, 45]. This method of cake removal by fluid shear has been applied to several filter module designs: tubular, flat sheet and spiral wound. The tubular elements are available in hollow fibre form or as tubes of substantial diameter. All the designs attempt to control the hydrodynamic situation, in providing a relatively clean surface to maintain high filtrate flux and, at the same time, giving as much membrane area as possible, per unit volume of module.

Despite these efforts, particulate separation from a cross-flowing liquid usually involves some deposition of the fine material in the feed. This may be attributed to the flow conditions established at the surface of the membrane where the small particles are influenced by the filtrate flow. This situation of cross-flow, with suction through the fluid boundary layer near the membrane, has been analysed [46] with predictions of boundary layer thicknesses. It may be envisaged that the layer particles of size greater than the boundary layer thickness will remain under the influence of the cross-flow conditions in the main stream.

Figure 6.9 shows a typical result of CFM separation. The decline in flux can be minimized by increases in the cross-flow velocity, v_s, in the range 1–10 m s^{-1}. However, the energy costs associated with this and the possible breakdown of feed particles points to alternative methods of maintaining high flux. It has been suggested [47] than the latter should be in excess of 100 l m^{-2} h^{-1}, in order to make the CFM technique economically attractive. Such fluxes are rarely achieved without some form of *in situ* membrane cleaning, e.g. by periodic, short-time back flushing of the element (see curve B in Figure 6.9). In pipe or conduit flow, maximum fluid velocities are attained at the centre of the duct whilst velocities at the membrane surface, where the shear is required, are at a minimum. This has created the concept of moving the membrane at high speed through the fluid, rather than pumping the latter across the surface. This principle is used in rotary membrane systems which are discussed below.

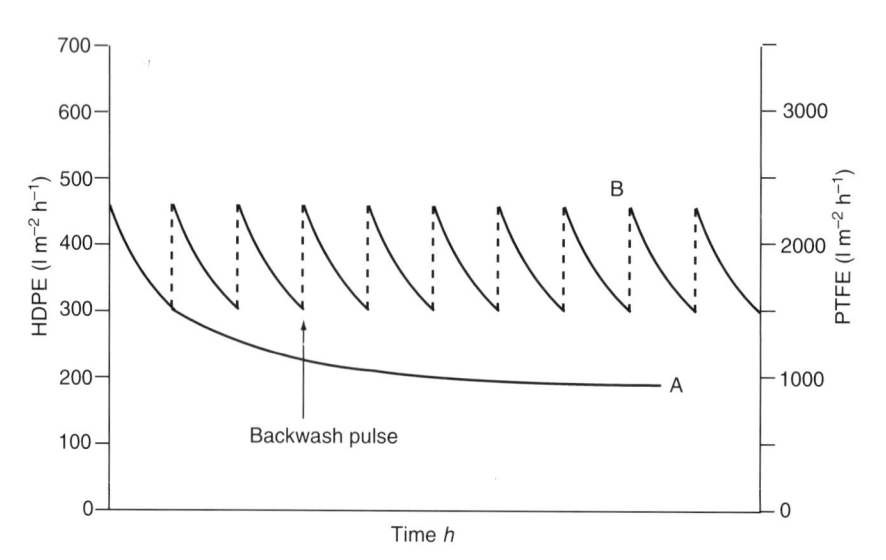

Figure 6.9 Cross-flow microporous filtration: filtrate rates as a function of time with and without back-flow.

6.3 Batch operated filtration machinery

Typical machinery used in batch filtration operations are described below. Only a partial treatment of the filter available can be given, in view of space limitations.

6.3.1 Pressure–vacuum filters

Some of the units listed in Table 6.1 can be operated as vacuum or pressure filters; these include Nutsche filters, horizontal table and leaf filters, and vertical leaf, tube or candle filters.

(a) Nutsche filters
These filters are used throughout industry in small- and large-scale operations. As well as the traditional versions capable only of filtration and cake washing, there are modern, automatic units typified by the Nutrex machine (Figure 6.10). These are particularly applicable where rigorous washing is required, such as in the pharmaceutical industry. The vessel can be used in multiple operations including crystallization/precipitation, extraction, filtration (separation, re-slurrying, washing, dewatering), drying (by vacuum and/or convection). The Nutrex unit can be tilted to facilitate cake discharge. Typical specification ranges are: vessel volume, $0.32–11.0 \, m^3$; filter area, $0.25–6.3 \, m^2$; filter cake volume, $0.08–3.15 \, m^3$. Agitators are designed to provide agitation of the slurry and smoothing of the filter cake prior to washing; the smoothing action tends to eliminate cake cracking, which is a principal cause of porosity. After filtration the agitation system can be used to promote cake discharge.

As with other batch filters, Nutsche units are being continuously developed and extended. Of particular interest in pharmaceutics separations is the use of woven metallic-fibre filter media which, whilst readily sterilizable, offer the double action of filter cake formation and filtrate polishing [48].

(b) Horizontal table and leaf filter
An example of this type of unit is the 'Triune' filter [49] which is described as a stationary bed, horizontal band vacuum filter. The filter consists of a lower filtrate chamber, usually made of stainless steel or polypropylene, fitted with a perforated plate on the upper surface. The filter medium, in the form of a band, rests on the latter plate and is enclosed by an upper chamber during filtration. Application of vacuum produces a tight seal between the upper and lower chambers. The filtration, washing and dewatering stages can be arranged for sequential working, e.g. where different time intervals are required for the three stages. This facilitates optimization of filter cake thickness, wash times, etc. After release of the vacuum, the upper chamber is automatically raised from the cloth, which then moves forward for cake

178   HANDBOOK OF BATCH PROCESS DESIGN

reaction, crystallization

filtration, washing, re-slurry

smoothing, pressing

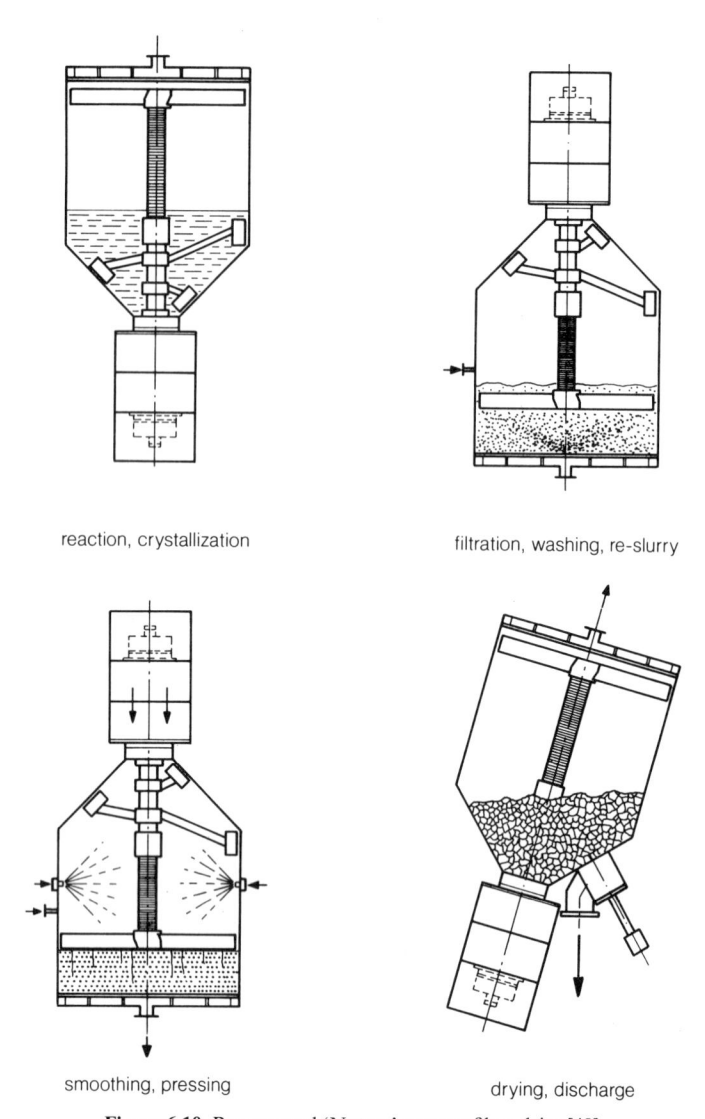

drying, discharge

Figure 6.10 Rosenmund 'Nutrex' reactor filter drier [48].

discharge, cloth washing, etc. Variations in the feed quality, which can produce poor performance in manual units, can be signalled and dealt with automatically. The sealed nature of the machine makes it particularly useful in the processing of hazardous materials. Containment of toxic fumes is possible and the processing of slurries involving solvents such as isopropanol, butanol, acetone, dimethyl formamide, etc., have been reported. Process advantages have been claimed for sticky materials, such as antibiotics. In view of the downward filtration character of the Triune filter, sizing and process development may be effected with a simple Buchner flask system.

Horizontal leaf filters present the advantage of cake stability. In upward and side-ways filtering modes, there is always the possibility that cake-drop-off will be a problem. The cake is stabilized by fluid drag forces generated by filtrate flow. Where low fluid velocities are experienced, the drag effect may be insufficient to stabilize the cake against detachment forces. In downward orientated depositions fluid drag and gravity act together in producing stable cakes; this is of particular importance in systems where efficient cake washing is required. The latter necessitates an even cake development, so that short circuit paths of low resistance are not encountered by the wash fluid.

In systems where a large particle size spread is present in the feed, downward deposition encourages the initial arrival of the faster settling larger particles. These can have a pre-coating effect on the filter medium, which may be an advantage in producing clearer filtrates and easier cake release.

A disadvantage inherent in the horizontal leaf design is that only one side of the leaf is used in the separation process. Further, discharge of the cake may prove to be difficult [1]. This problem has promoted the design of centrifugally discharged filter cakes; in these units, horizontal, circular elements are rotated at high speed, after cake formation and washing, and can be quite effective in reducing the down-time required for cake removal [50].

(c) Vertical leaf and candle filters
Figures 6.11((a) and (b)) and 6.12 show particular types of vertical pressure filters fitted with vertical leaves and candles. Designs of candle or leaf units tend to differ in the manner in which the cake is discharged. Discharge may be effected by:

(i) Reverse backflushing with gas or filtrate [51].
(ii) Sluicing or scraping of the surface of the medium (Figure 6.11(b)).
(iii) Shaking of the medium and its support (the latter may be flexible) [52].

The attention given to solving the discharge problem reflects the importance of this operation in the overall cycle. In some cases adhesion of the particles to the medium is minimized by the use of media made from PTFE. A particularly useful design feature is the facility for isolating individual leaves or candles [53] (Figure 6.11(a)). In this case, failure of one or several elements can be easily accommodated without having to recourse to a complete plant shutdown.

(d) Plate-and-frame and recessed plate filters
These units continue to find widespread use in the chemical and process industries. Modern trends have been directed towards the complete mechanization on the filter. Improved mechanical discharge rates, with the

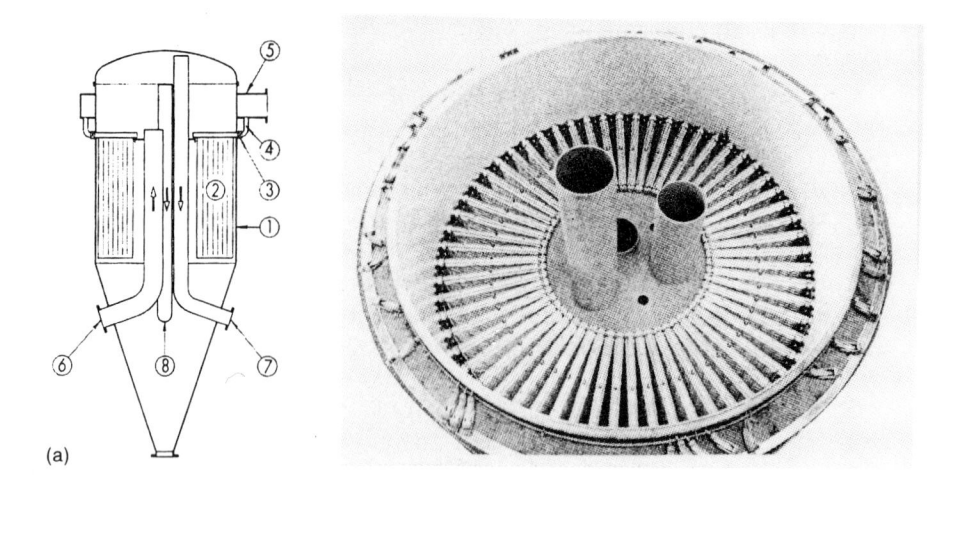

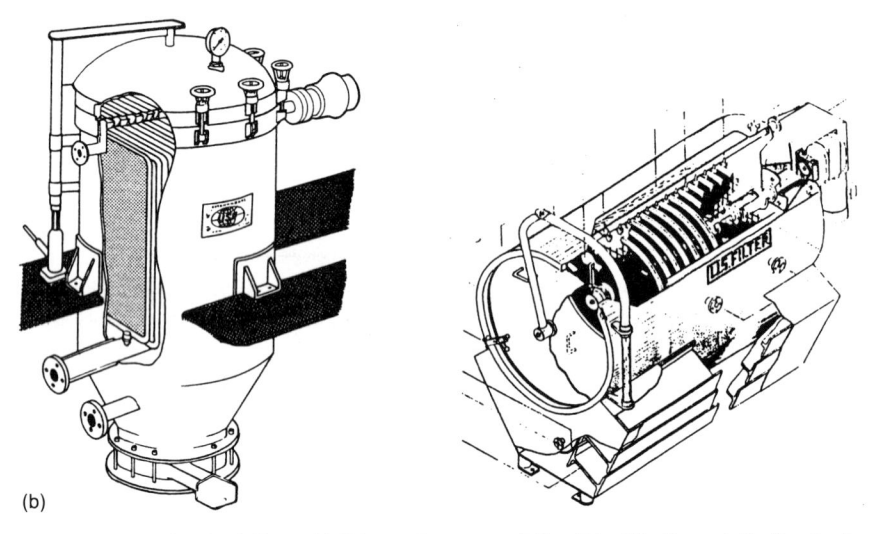

Figure 6.11 Vertical leaf filters: (a) Diastar (courtesy of Gaudfrin SA, France) [1, filter/tank body; 2, drained cloth; 3, collector; 4, isolating valve and inspection hole; 5, general collector; 6, supply/feed pipe; 7, air decompression pipe; 8, levelling pipe] and (b) (courtesy of US Filter/Vanpipe Co. Ltd, UK).

inclusion of the latest methods for automatic cake removal [54], have continued to augment the earlier results of mechanization [55], where cake discharge rates 10 times higher than those obtained manually were claimed.

In plate and frame assemblies, square or rectangular plates are brought together, separated by a frame as shown in Figure 6.13(a). These filters can be arranged to allow for cake washing (Figure 6.13(b)) by the use of two

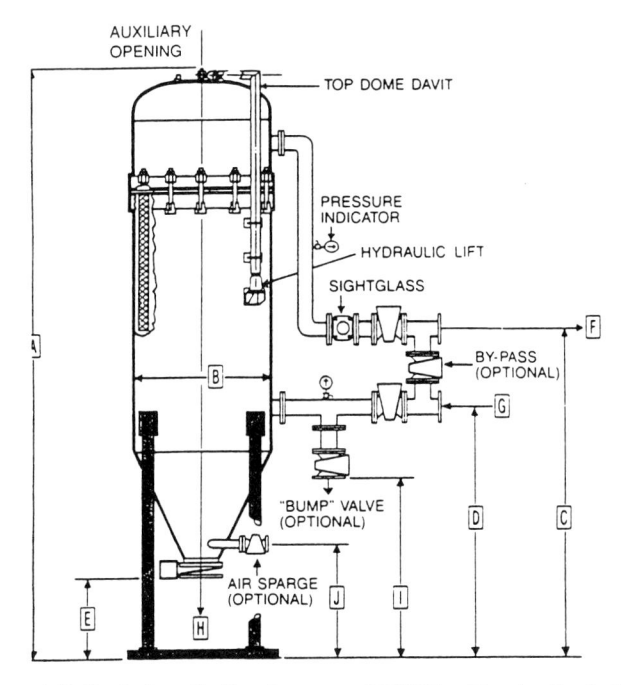

BATCH FILTRATION OF SOLID–LIQUID SUSPENSIONS 181

Figure 6.12 Vertical candle filter (courtesy of US Filter/Vanpipe Co. Ltd, UK).

types of plate. The 'washing plates' are constructed to allow the passage of wash through the back of the filter cloth covering the plate, through the interstices of the cake and out via the cloth covering the 'filtering or non-washing plate'.

An important design feature is the arrangement made for the flow of fluid between the plates and the covering cloth. The surface of the plate may be ribbed or covered with 'pips' to provide channels for fluid flow. In some circumstances an underlay of robust, porous material placed between the cloth and plate can be used to advantage in productivity improvements. A high horizontal resistance to flow can aggravate the tendency for sideways flow of filtrate through the medium. This edge dripping can be minimized by the use of media which has been edge-sealed with rubber or equivalent. Most of the process difficulties which may be encountered in the use of these units can be related to the geometry of the plate assembly. Thus in the cake formation period, maldistribution of the solids may occur [56] if local velocities within the frame or recessed plate are insufficient to maintain the particles in suspension. Any process which leads to a classification of the solids in the filter can lead to differential deposits and cake instabilities. Washing efficiencies can be low in circumstances where there is a preferred flow path through the cake; wash will tend to pass through areas of low

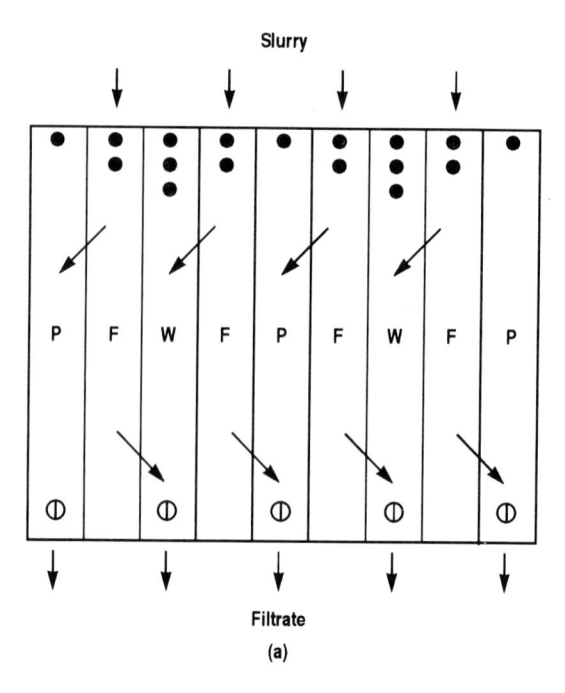

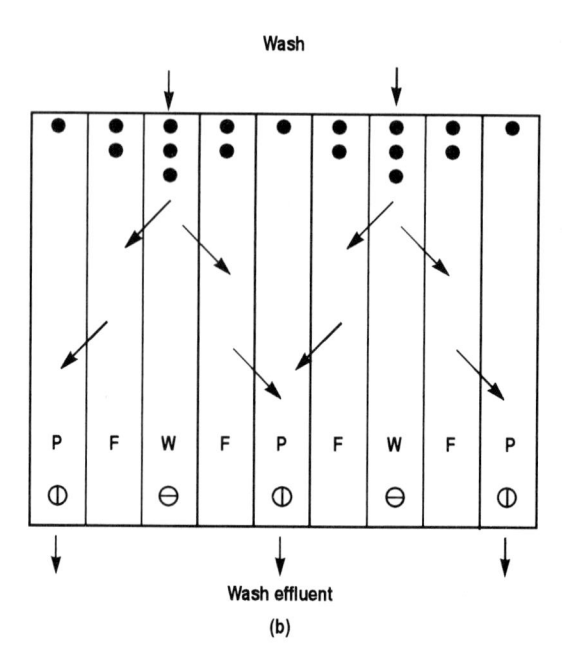

Figure 6.13 Plate and frame arrangements: (a) separation and (b) washing [25].

resistance such as cake cracks or thinly deposited areas within the frame. These malfunctions and others, e.g. cake dewatering, can sometimes be improved by alteration in the process conditions such as higher feed pressures and concentrations. Thus a sewage sludge cake produced at 15 bar may be more readily discharged and handled than one produced at lower pressure, say at 7 bar [57].

Practical experience suggests that difficult-to-filter materials require thin cake conditions, with optimum cake thicknesses in the range 1.5–2.0 cm. Readily filtered materials are optimally produced as thicker cakes. This trend can also be predicted from process models using the basic equations outlined above for filtration and washing [58]; the results of which analysis are as follows. For batch processes involving filtration, washing and discharge stages only, it may be shown theoretically that optimum values for the volume of filtrate processed per batch and the associated cake thicknesses are given by:

$$V = [A^2 \Delta P t_{dw}/\alpha_a c^*(0.5 + K_W)]^{0.5} \quad (6.30)$$

$$L_c = [4\Delta P t_{dw}/\alpha_a \mu (1-\varepsilon)^2 \rho_S^2 (0.5 + K_W)] \quad (6.31)$$

in which $K_W = [4(m-1)Wrc^*]/\rho$.

In the above expressions it will be noted that an average value of the specific cake resistance is used. This represents the effect of pressure in the range of interest and reflects the compressible nature of the filter cakes. As mentioned above, the wetness parameter m relates to the amount of original liquor in the flooded filter cake, immediately prior to washing. The wetness is, of course, related to the porosity of the cake. It will be realized that as α_a increases, the area of filter surface required also increases, for equivalent amounts of feed [9].

From the above discussion, it follows that where batch to batch variations in filterability can be expected, optimal operation at all times is impossible with filters of fixed dimensions. Multiproduct operations using the same filter led to the same situation. Another complication is where the volume of solids to be processed changes from batch to batch. In fixed volume presses this may lead to the necessity of blanking off some plates at low solids concentrations.

Whilst it is generally recognized that drier, more handleable cakes can be produced at higher filtration pressures, these may not be available on a particular station. Where there is no mechanical limitation on the existing fixed volume system, the possibility exists of re-fitting the system with diaphragm covered plates; giving a variable chamber character to the unit. The principles involved in variable chamber pressing are outlined below.

(e) Variable chamber presses
Variable chamber presses contain inflatable diaphragms which are expanded by compressed air or hydraulic fluid at the end of the normal

cake filling stage. The movement of the diaphragm causes a further compression and dewatering of the cake; in some cases this extra squeeze at an elevated pressure can produce a relatively dry cake from a wet, sloppy deposit produced at normal pressures. Generally, drying of the filter cake at the filtration stage is considerably less expensive than thermal drying; the latter involves a considerable amount of energy in creating the phase change inherent in evaporative processes [59]. The overall costs of filtration and drying are reduced if the volume of liquid to be removed thermally is minimized.

A typical batch filtration volume–time curve involving squeezing is shown in Figure 6.14 [60]. The cake formation time is limited to the early, high filtration rate period, after which the wet deposit is squeezed before discharge. This avoids the slow filtration rate filling period, which would be necessary for adequate cake discharge in a fixed-volume press. The squeezing pressures applied depend on the design of the filter. For small diameter cylindrical models, this may be as high as 150 bar. On the other hand, such high pressure are not always necessary since the maximum pressure used will be related to the compressibility characteristics of the cake. Information is available on the equilibrium compression attainable by squeezing. Changes in the solid content of a filter cake, with the application of pressure can be estimated from:

$$(1 - \varepsilon) = (1 - \varepsilon_0) p_s^b \tag{6.32}$$

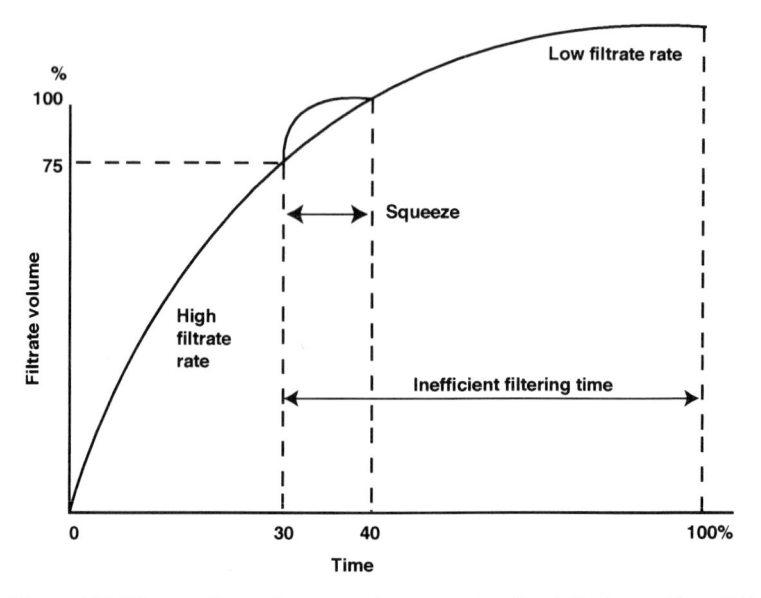

Figure 6.14 Filtrate volume–time curves for conventional and diaphragm filters [60].

Table 6.6 Cake solidity and squeezing exponent for various materials

Material	Pressure range (kPa)	$(1-e_0)$	b
Calcium carbonate	7–550	0.209	0.06
Kaolin	7–700	0.374	0.06
Clay	10–100	0.177	0.10
Titanium dioxide	7–700	0.203	0.13

where p_S is the compressive pressure and $(1-\varepsilon)$ is termed the 'solidity' of the cake. The exponent in the above empirical expression takes values in the range: $0.05 < b < 0.20$, as shown in Table 6.6. At low values of b little effect of pressure is noticed so that other means of dewatering, e.g. gas blowing, may have to be considered.

Other work has been directed towards an understanding of the kinetics of dewatering by squeezing and optimization of the variable volume filter cycle [32]. The pressure level used can have a major effect in some systems, but little effect in others. The influence on pH-sensitive surface forces acting on the particles has been shown to have importance in the expression process [61].

(f) Cartridge filters

Some of the methods used in testing the media used in cartridge filtration have been discussed above, in Section 6.2.8. The complete assembly requires a carefully designed housing into which the filter elements may be secured. The elements range from the tightest membranes to relatively coarse screens. All elements and housings require adequate seals in order to prevent particles bypassing the element. The design of cartridge filtration systems, both from the element and housing point of view, has received extensive attention in the literature [62, 63]. This attention reflects the widespread use of single and multiple element cartridge filters in batch clarification processes.

6.3.2 Centrifugal filters

These machines may be classified by the method used to discharge the filter cake. In manually operated systems, the centrifuge basket is stationary during discharge. This means that a considerable part of the overall filter cycle will have to be repeatedly used for accelerating and decelerating the filter bowl. Consideration of typical literature data [64] gives some indication of the proportion of the cycle which will be occupied in these acceleration/deceleration periods; as shown in Table 6.7 up to 20% of the cycle time is so used.

The 'peeler' centrifuge is a development aimed at reducing the above discharge requirement. Some machines are fitted with a plough-type knife which still necessitates slowing down the bowl, but not stopping it,

Table 6.7 Typical basket centrifuge cycle

Activity	r.p.m.	Percent of cycle
Accelerate to	500	5
Filter at	500	31
Accelerate to	1050	10
Dewater at	1050	13
Wash at	1050	1
Spin dry at	1050	26
Decelerate and unload	–	14

during the automatic removal of the cake. Thin-bladed peeler knives are also available which facilitate cake discharge at full speed, at least with those particles which are not broken down by such a vigorous action. Comprehensive descriptions of these filters are available in the literature [1]. An interesting recent addition to the range of filtering centrifuges available in batch processing is the inverting unit. This filter can be operated simultaneously under pressure, whilst in a centrifuging mode. This combined activity is finding wide-spread applications in the processing of difficult materials such as those handled in the pharmaceutical industries. Some information on typical materials processed by manual basket and peeler units has been published [1, 29] and is summarized in Table 6.8.

6.3.3 Membrane filters

A typical batch system involving CFM is shown in Figure 6.15 for multiple-pass, thickening of a suspension. Here the filtration loop can be considered a well-mixed system [9] and the following mass balance results, for an initial concentration c_0:

$$c_0 J A = V_V (dc/dt) \quad (6.33)$$

where J is the filtrate flux rate ($m^3 \, m^{-2} \, s^{-1}$), A is the membrane area, V_V is the vessel volume and c is the concentration of solids or solutes in the feed to the filter. This applies when the volume of suspension has not fallen greatly, i.e. the slurry volume is close to V_V. If the flux J is constant, integration of

Table 6.8 Centrifugal filter process applications

Machine	Particle size range (μm)	Slurry concentration (wt%)	Filterability α (m kg^{-1})	Production (tonnes h^{-1})
Basket	10–10 000	2–10	$5 \times 10^8 - 8 \times 10^9$	0.1–1.0
Peeler	20–2000	5–50	$8 \times 10^8 - 4 \times 10^{11}$	0.1–5.0

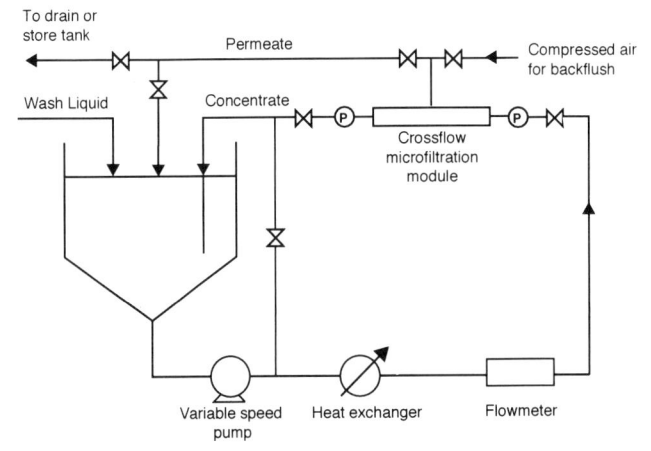

BATCH FILTRATION OF SOLID–LIQUID SUSPENSIONS 187

Figure 6.15 Cross-flow microfiltration flow diagram [44].

the above yields equation (6.34) for calculating the change in concentration with time, for $JAt \ll V_V$.

$$c = c_0 + c_0 AJt/V_V \qquad (6.34)$$

Practical tests can reveal a flux decline with increases in concentration above a certain level, as shown in Figure 6.16. If this decline is linear with concentration, the concentration variation with time t' after the start of the flux decline can be represented by:

$$c = (-k_1/k_2)[1 - \{1 + (k_2 c_t/k_1)\} \exp(k_2 c_0 A t'/V_V)] \qquad (6.35)$$

where k_1 is the initial flux, k_2 is the slope of the flux/concentration graph and c_t is the concentration at the time t (i.e. $t' = 0$) when the permeate flux starts to decline. Note that k_2 will be negative.

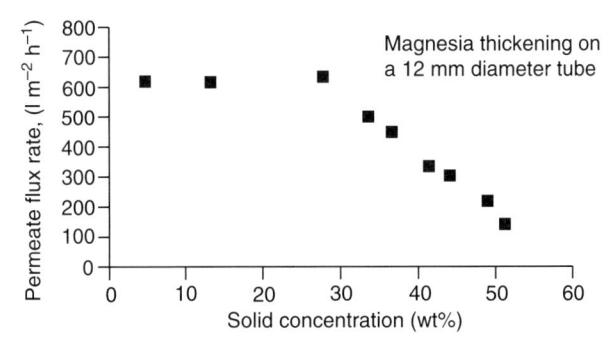

Figure 6.16 Flux decay: magnesia thickening by cross-flow filtration [9].

Extensive investigations into flux decline in CFM have been reported [65]. Means other than back-flushing for reducing the decline have been identified [9] to include: introduction of foam balls to wipe the membrane surface, electrical fields to resist particle deposition, combined ultrasonic and electrical fields, oscillatory or pulsatile flow, surface treatment. Despite these efforts chemical cleaning is often required after prolonged membrane use to remove chemical deposits. Citric acid and or caustic soda are common cleaning agents to be followed by water rinsing.

The CFM technique can be used with effect in the washing of fine suspensions, as shown in the 'diafiltration' equipment in Figure 6.17. Here, the solids are retained in the system and the solute concentration is gradually reduced by replacing filtrate with wash water. If the system is well-mixed, the concentration of solute in the system and the permeate are the same. If the wash is clear of solute, the permeate concentration-time relationship is:

$$c_p = c_0 \exp(-JAt/V_V) \tag{6.36}$$

This applies to displacement wash systems and not where the solute is adsorbed on to the surface of the solids.

Rotary membrane filters [66] have interesting possibilities when separations of small batches of difficult materials are required. The design principle of creating the cross-flow effect by moving the membrane, rather than the

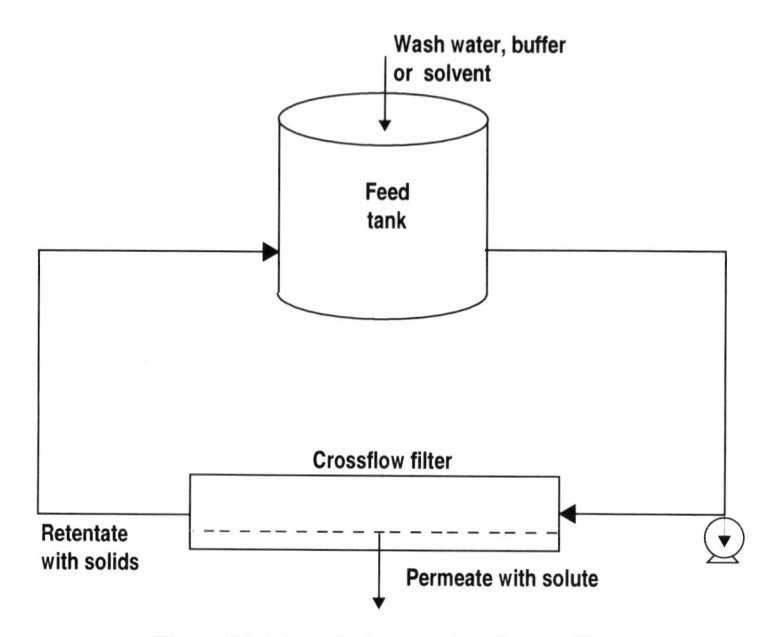

Figure 6.17 Schematic diagram of diafiltration [9].

slurry, finds attractions when processing fragile pharmaceuticals. Again, the operating principles, which involve balancing the opposing effects of inward directed fluid pressure against outward directed centrifugal action, find use in separations demanding (i) gentle treatment of the suspension and (ii) slow filtration conditions, e.g. in biochemical reactions involving antibody production. The alternate use of pressure filtration and centrifugal cleaning has been shown to be effective in maintaining high filtrate flux when separating yeast suspensions [67].

6.4 Nomenclature

A	area (m^2)
A_a	average area (m^2)
A_{lm}	logarithmic mean area (m^2)
b	exponent equation (6.32) (–)
c, c_0	suspended solid or solute concentration (kg m^3)
c_p	permeate concentration (kg m^3)
c^*	dry mass of solids per unit volume of filtrate (kg m^3)
C, C_0	solids concentration by volume fraction and initial concentration, respectively (–)
d_p	pore diameter (m)
D	axial dispersion or diffusion coefficient (m^2 s^{-1})
E	efficiency (–)
g	exponent (–)
J	filtrate flux in membrane system (m^3 s^{-1})
k_1, k_2	constants in equation (6.35)
K	permeability (m^2)
K'	filtration constant in equations (6.23) and (6.24)
K_1, K_2	filtration constants in equations (6.15), (6.17) and (6.18)
K_S	filtration constant in standard law equations (6.25) and (6.26) (m^{-3})
K_w	washing constant in equations (6.30) and (6.31) (–)
L, L_c	depth of medium and cake thickness, respectively (m)
m	wetness ratio: mass dry/mass wet of filter cake (–)
M	mass (kg)
n	compressibilty coefficient
P	pressure (N)
ΔP	pressure differential (N m^{-2})
Pr	productivity (m^3 s^{-1})
q	flow rate (m^3 s^{-1})
r	radius (m)
R_c, R_m	cake and medium resistance, respectively (m^{-1})
s	solids mass (as dry solid) per unit mass of slurry (–)

S	saturation: volume of liquid in filter cake per unit cake pore volume (–)
t	time (s)
u	wash velocity (m s^{-1})
v, v_0	velocity at time t and initial velocity, respectively (m s^{-1})
v_s	cross-flow velocity (m s^{-1})
V	volume (m^3)
y	distance from the surface of depth filter (m)
α, α_a	specific cake resistance at local pressure and average over ΔP (m kg^{-1})
β	concentration ratio of size x solids upstream and downstream of medium (–)
ε	filter cake porosity (–)
λ	depth filtration constant in equation (6.27) (m^{-1})
μ	liquid viscosity (N s m^{-2})
ρ, ρ_s	liquid and solid densities (kg m^{-3})
ϖ	rotational speed (s^{-1})

References

1. Purchas, D.B. (1981) *Solid–Liquid Separation Technology*. Uplands/Elsevier Press, London.
2. Purchas, D.B. and Wakeman, R.J. (eds) (1986) *Solid–Liquid Separation Equipment Scale-Up*. Uplands/Elsevier Press, London.
3. Gaudfrin, G. and Sabatier, E. (1978) *Int. Symp. KVIV* and Belgian Filtration Society.
4. Wakeman, R.J. (1995) *Filtrat. Separat.*, **32**, 337.
5. Tiller, F.M. (1974) *Chem. Engng*, 29 April, 116 and 13 May, 98.
6. Coulson, J.M and Richardson, J.F. (1966) *Chemical Engineering*, 2nd edn. Publisher, Location, Vol. 1, p. 41.
7. Heertjes, P.M. (1957) *Chem. Engng Sci.*, **6**, 269.
8. Kozeny (1927) *Sitz-Ber, Weiner Akad., Abt. Iia*, **136**, 271.
9. Rushton, A., Ward, A.S. and Holdich, R.G. (1996) *Solid–Liquid Filtration and Separation Technology*. VCH, Weinheim.
10. Shirato, M., Tiller, F.M. *et al.* (1987) In *Filtration Principles and Practices*, 2nd edn (M.J. Matteson, ed.). Marcel Dekker, New York.
11. Svarovsky, L. (ed.) (1981) *Solid–Liquid Separation*, 2nd edn. Butterworth, London.
12. Risbud, H.M. (1981) *Filtrat. Separat.*, **18**, 20.
13. Grace, H.P. (1953) *Chem. Eng. Progr.*, **49**, 154, 303.
14. Shirato, M. *et al.* (1969) *Am. IChemE J.*, **15**, 405.
15. Rushton, A. and Griffiths, P. (1987) In *Filtration Principles and Practices*, 2nd edn (M.J. Matteson, ed.), Marcel Dekker, New York.
16. Hardman, E. (1994) *Filtrat. Separat.*, **31**, 813.
17. Maurer, C. (1996) In *Proc. 7th World Filtration Congr.* Hungarian Chemical Society, Budapest, Vol. 1, p. 228.
18. Anlauf, H. and Muller, H.R. (1990) In *Proc. 5th World Filtration Congr.* Société Filtration Française, Nice, Vol. 2, p. 211.
19. Spear, M. and Rushton, A. (1975) *Filtrat. Separat.*, **12**, 254.
20. Wakeman, R.J. and Tarleton, S. (1990) In *Proc. 5th World Filtration Congr.* Société Filtration Française, Nice, Vol. 2, p. 21.

21. Carleton, A.J. and Taylour, J.M. (1991) In *Proc. Filtech Conf.*, Karlsruhe. UK Filtration Society, Horsham, W. Sussex, p. 281.
22. Rhodes, F.H. (1934) *Ind. Engng Chem.*, **26**, 1331.
23. Wakeman, R.J., and Rushton, A. (1974) *Chem. Engng Sci.*, **29**, 1857.
24. Wakeman, R.J. and Attwood, J. (1988) *Filtrat. Separat.*, **25**, 272.
25. Wakeman, R.J. (1981) In *Solid–Liquid Separation*, 2nd edn (L. Svarovsky, ed.). Butterworths, London.
26. Carleton, A.J. and Mehta, K.B. (1983) In *Proc. Filtech Conf.*, London. UK Filtration Society, Horsham, W. Sussex, p. 120.
27. Daneshpoor, S. (1984) MSc Thesis, University of Manchester Institute of Science & Technology.
28. Rushton, A. and Arab, M. (1986) In *Proc. 4th World Filtration Congr.* Ostend, Belgium.
29. Zeitsch, K. (1978) In *Proc. Int. Symp.* Belgium Filtration Society, Louvain la Neuve, Belgium.
30. Dodds, J. and Baluais, G. (1985) *IChemE. Symp. Ser.*, **91**, 161.
31. Hermia, J. (1982) *Trans. IChemE*, **60**, 183.
32. Shirato, M., et al. (1981) In *Proc. 2nd World Congr. of Chemical Engineering*, Montreal, Canada, Vol. 4, p. 107.
33. de Bruyne, R. (1982) In *Proc. 3rd. World Filtration Congr.* American Filtration Society, Northport, AL.
34. Ives, K.J. (1975) *The Scientific Basis of Filtration.* Noordorff, Leyden.
35. Bently, J.M., and Lloyd, P.J. (1992) *Filtrat. Separat.*, **29**, 333.
36. Meltzer, T. (1987) *Filtration in the Pharmaceutical Industry.* Marcel Dekker, New York.
37. Williams, C. and Edyvean, R. (1995) *Filtrat. Separat.*, **32**, 157.
38. Verdegan, B.M. et al. (1992) *Filtrat. Separat.*, **29**, 327.
39. Longworth, B. et al. (1995) *Polymer Filtration Symp.* UK Filtration Society, Horsham, W. Sussex.
40. Smith, G.R.S. (1981) In *Solid–Liquid Separation*, 2nd edn (L. Svarovsky, ed.), Butterworths, London.
41. Wakeman, R.J. (1996) In *Proc. 7th World Filtration Congr.* Hungarian Chemical Society, Budapest, p. 234.
42. Hermia, J. (1993) In *Proc. Filtech Conf.*, Karlsruhe. UK Filtration Society, Horsham, W. Sussex, p. 21.
43. Gutman, R.G. (1987) *Membrane Filtration.* Adam Hilger, Bristol.
44. Bertera, R., Steven, H. and Metcalfe, M. (1984) *Chem. Eng.*, June, 10.
45. Fane, A.G. et al. (1982) *Desalination*, **41**, 263.
46. Schlichting, H. (1968) *Boundary Layer Theory*, 6th edn. McGraw-Hill, New York.
47. Rushton, A., et al. (1979) In *Proc. SLS Practice Symp.* IChemE Yorkshire Branch, p. 149.
48. Evans, A. (1996) In *Proc. Filter Media Symp.* UK Filtration Society, Horsham, W. Sussex, p. 48.
49. Nield, P. (1994) *Filtrat. Separat.*, **31**, 691.
50. Howard, M.G. (1979) *Filtrat. Separat.*, **16**, 150.
51. Mueller, H.K. (1984), *Filtrat. Separat.*, **21**, 259.
52. Rushton, A. and Malamis, J. (1985) *Filtrat. Separat.*, **22**, 368.
53. Simonart, H. and Gaudfrin, G. (1993) In *Proc. Filtech Conf.*, Karlsruhe. UK Filtration Society, Horsham, W. Sussex.
54. Mayer L. (1995) In *Proc. Filtech Conf.*, Karlsruhe. UK Filtration Society, Horsham, W. Sussex, p. 161.
55. Moss, A. (1970) *Chem. Eng.*, **237**, 23.
56. Rushton, A. and Metcalfe, M. (1973) *Filtrat. Separat.*, **10**, 398.
57. Bosley, R. et al. (1986) In *Solid–Liquid Separation Equipment Scale-Up*, 2nd edn (D. Purchas and R. Wakeman, eds). Uplands Press, London.
58. Rushton, A. and Daneshpoor, S. (1983) In *Proc. Filtech Conf.*, London. UK Filtration Society, Horsham, W. Sussex, p. 136.
59. Bailey, P.C. (1979) *Filtrat. Separat.*, **15**, 649.
60. Moon, P.J. (1981) In *Proc. Int. Symp.* Belgium Filtration Society, Louvain la Neuve, Belgium.

61. Wakeman, R.J. and Tarleton, S. (1989) In *Proc. Filtech Conf.*, Karlsruhe. UK Filtration Society, Horsham, W. Sussex, Vol. 2, p. 21.
62. Jaroszczk, T. *et al.* (1987) In *Filtration Principles and Practices*, 2nd edn (M.J. Matteson, ed.). Marcel Dekker, New York.
63. Howard, G.W. and Nicklaus, N. (1986) In *Solid–Liquid Separation Equipment Scale-Up*, 2nd edn (D. Purchas and R. Wakeman, eds). Uplands Press, London.
64. Ambler A.N. (1952) *Chem. Eng. Progr.*, **48**, 150.
65. Rautenbrach, R. and Albrecht, R. (1989) *Membrane Processes*. Wiley, Chichester.
66. Kroner, K. and Nissinen, V. (1988) *J. Membr. Sci.*, **36**, 85.
67. Rushton, A. and Zhang, G.S. (1988) *Desalination*, **70**, 379.

7 Design and engineering of a batch plant
M.J. MAYES

7.1 Introduction

The overall 'product cycle' for design and engineering of a batch chemical plant, measured in days to marketplace, has increasingly been subjected to an innovative approach. The aim has been to reduce design costs and achieve accelerated construction and commissioning periods to enable early beneficial operation. The business driver is securing market share in a competitive environment.

All new products require the establishment of the chemical details and conditions under which the process occurs. This process must be developed to identify the various operations – mixing, reaction filtration, evaporation and drying – to prepare for any pilot trials required to refine the chemical, physical and mechanical conditions necessary for the process. The design of the equipment for the production-scale plant can then commence. It may be that a number of these process design stages have to overlap to some extent in order to shorten the product cycle.

Processes require the selection of the latest technology to suit current and future expectations of productivity requirements and to satisfy a new generation of environmental controls. There is little point in committing a large capital investment to a plant with an operating life expectancy of 15 years, but with a process that is already outdated and will be expensive to modify. The batch process designer today is expected to produce cost-effective operating units, which can be erected and commissioned in the optimum time. The final design of a batch plant is usually expected to have incorporated features to facilitate change in yield requirements and to allow easy integration of future plant expansions. No longer can a process be taken directly 'off the shelf'.

This chapter will discuss the design and engineering of batch processes. The appropriate organization of projects, use of 'fast track' methods to minimize project duration, and management of safety studies and regulatory issues are all important. Case studies will be used to illustrate important points.

7.2 Project definition

Initial project feasibility reports prepared by the client embody the major targets for project execution and provide the start point for the design consultant's involvement in the front-end engineering definition. The project definition for batch processes is now frequently expected to have in-built flexibility, with front-end engineering packages containing the majority of the basic data requirements but often with elements not fully defined and possibly still under development. The detailed engineering phase, consequently, is not only expected to start early but also have to cope with ongoing change.

Projects should have critical success criteria which must be constantly reviewed to ensure that they remain realistic within the programme and budget constraints. These can be weighted according to a priority ranking determined by the business group making the investment. Typical examples of critical success criteria include the following:

(i) The product specification and final formulation.
(ii) Projected capacity of the plant.
(iii) The capital cost of the project.
(iv) Future expansion capability.
(v) The raw material supplies.
(vi) Safety and environmental compliance.
(vii) Disruption to existing plants.
(viii) Timing of the product arriving on the market place.
(ix) Risk factors.

The overall results of an analysis of success criteria can give a typical balance of, say, 70% of the client's priority associated with the ability to deliver and 30% related to timing.

7.3 Project strategy

The client must also determine at an early stage a suitable strategy in relation to the appointment of a main contractor or contractors in order to execute the works. The client may choose to select a managing contractor, to handle the whole project, or become directly involved himself with day-to-day management and overall direction. The strategy adopted must be the most suitable in the particular circumstances to achieve the project deliverables. Different contractors have different expertise. The scope may require several contractors in order to spread the work load to meet an exacting timetable. The design process may need to proceed in defined packages which are reassessed at each stage before the decision is taken to proceed. The project initiator needs to access carefully interface boundaries and co-ordination constraints within the overall project strategy.

The type of contract and methods of payment must also be determined. A process that is not fully defined or subject to ongoing development will be more suitable for a reimbursable engineering contract. The development of the basic process up to the detailed stage will allow the project specifications to be used to obtain lump sum bids for procurement and to allocate the various construction packages. Full lump sum contracts are more appropriate for plants that are of proven design with known variables and established packages of equipment.

7.4 Project organization

The setting up of dedicated task forces for the execution of projects in a 'fast track' environment is essential. The initial team must include the project's founder members, i.e. the chemist(s), the project engineer, the client's consultants and the process engineers. The early selection of the core team who will carry the design through the front-end definition, into the detailed stage and thence into the commissioning and the start-up is essential. The cohesiveness of this blend of expertise, its high visibility to the task force and commitment throughout the project, will directly influence the efficiency of man-hour expenditure and the capital cost of the plant. The project team can include a mixture of the client's and the contractor's engineers. The key is to ensure where possible that the time penalties inherent in engineering data transfer and interpretation between groups are eliminated. Lead engineers will need to be selected on their ability to be front line problem solvers and to handle the requirements of design verification as the work proceeds.

Process design has to be tightly controlled in order to support down stream activities and minimize the effects of disruption. Any potential benefits of changes must be related to costs, with the overall programme targets dictating whether changes due to ongoing developments can be implemented or not. The product must be able reach the market place on time, with the expected quality and quantity, and to sustain sales at the right unit cost.

The team must have a clear organizational structure. Team building activities are an essential element in focusing the individual members on the overall project objective and fostering a spirit of co-operation between different engineering disciplines. The ability of engineers to understand other needs will remove the traditional department barriers and promote the requirements of a fully integrated team who are able to communicate quickly and effectively. These objectives should also be fully supported by senior management and department line managers. These managers also provide resources (often from a matrix organizational structure) to support departments as well as project groups.

A proactive task force also creates the opportunity to look for more efficient ways of organizing work flow; activities can often be shortened or restructured. The capacity to accelerate the project cycle should be owned by the team. Project co-ordination procedures should reflect the work methodology required to do the job, while retaining only the quality checks from the (often cumbersome) company systems. A typical team structure is illustrated in Figure 7.1.

The contractors' staff usually drive the engineering, procurement and construction activities. The client may opt to take direct control of all subcontract activities and on-site construction, with the contractor providing services and resources as required. Clearly, many variations are possible which will reflect the levels and availability of resources/skills available directly to the owner and the additional expertise which is deemed to be required.

An example of a fully integrated task force is shown in Figure 7.2. In this particular instance the management and design team from the client and contractor are intermixed with the lead engineers provided by the client. This arrangement ensures, in principle, that all design developments and information flows are fully captured and communicated throughout the team. This should remove any inherent problems in the transfer of design packages and with their subsequent interpretation by the contractor.

It is essential that in whatever form the final team comprises, the organizational structure and lines of responsibility are clearly defined and integrated with the planning and control systems. All task force/project teams require a strong proactive project management that is able effectively to control the

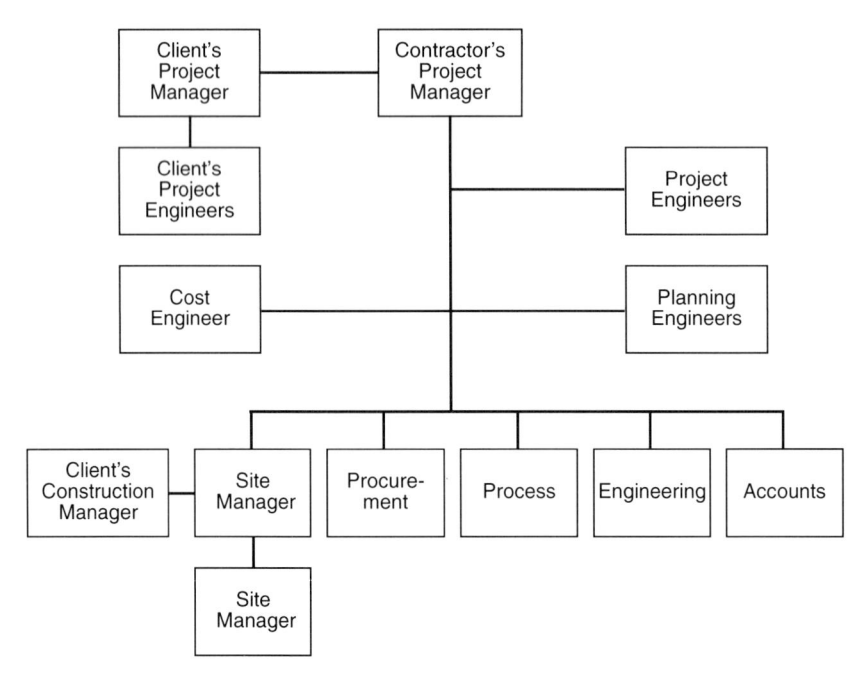

Figure 7.1 Organization chart for an integrated project team.

Figure 7.2 Organization chart for a project task force.

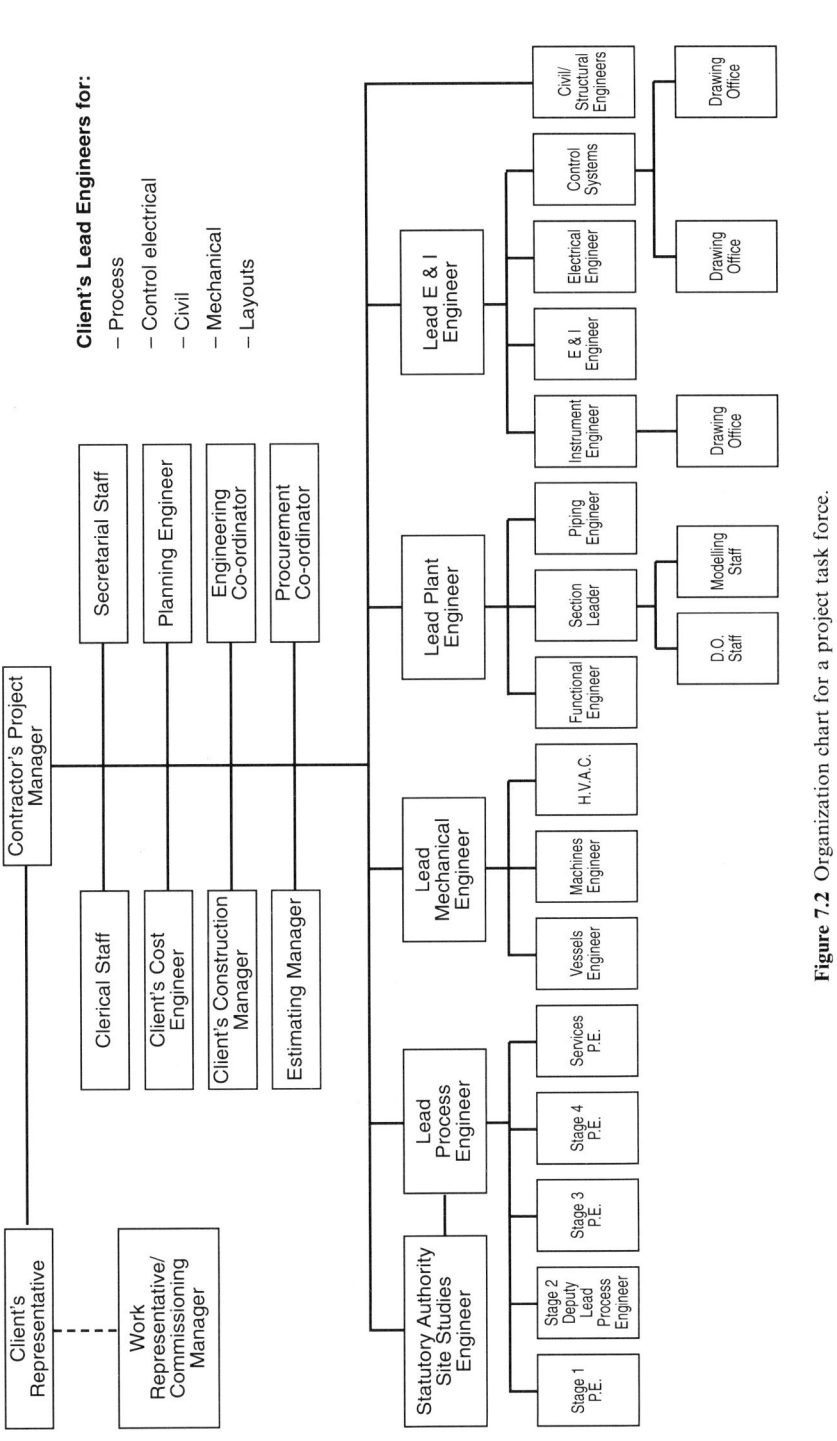

design and construction activities. The ability to communicate the project targets and to reinforce the philosophy of execution in the face of constant change will determine the degree of success of the project. The team must have integrated planning and control systems which are self supporting. The planner needs to attend all key co-ordination and review meetings as a member of the core team, and can often perform a project co-ordination role. Ideally the same planner should extend a role throughout the design phase with secondment to the construction team and fulfil the planning requirements of the commissioning and start-up team. The continuity of staff through any project is essential, especially if a 'fast track' approach is to be maintained.

7.5 'Fast track' projects

All projects in today's climate tend to be described as 'fast track'. However, it is those driven by construction, with a clear focus on the early identification of needs, that enables the engineering and procurement activities to be effectively linked to key target milestones. The advantages of 'fast track' projects are clear – the reduction in time to operation has both financial and marketing benefits.

Figure 7.3 demonstrates the overall relationship of the main project activities and the effect of a 'fast track' approach. The engineering and design activities are accelerated to allow the procurement phase duration from inquiry to final delivery to be shortened. Design may also need to run concurrently with construction activities to gain a substantial advantage. The opportunity then exists to capitalize on maintaining positive float through the remaining construction to mechanical completion using innovative construction techniques. Even one month's additional production can make quite a difference to the total yearly revenue figures for product sold. A project may even be termed 'flash track' with overall savings of up to 10 months possible on a typical 24 month programme.

7.5.1 Techniques

Overlapping activities are a prerequisite to 'fast track' projects, with work needing to proceed on a number of parallel fronts. The utilization of comprehensive interactive information technology/programming techniques is essential. Any system selected must provide constant on-line reporting that is easily intelligible to the project team. The software must have inherent flexibility that can demonstrate any deviations from the original plan and cascade their action down line.

'Fast track' projects utilize work breakdown structures in order to ensure that all activities are broken down into their component parts and

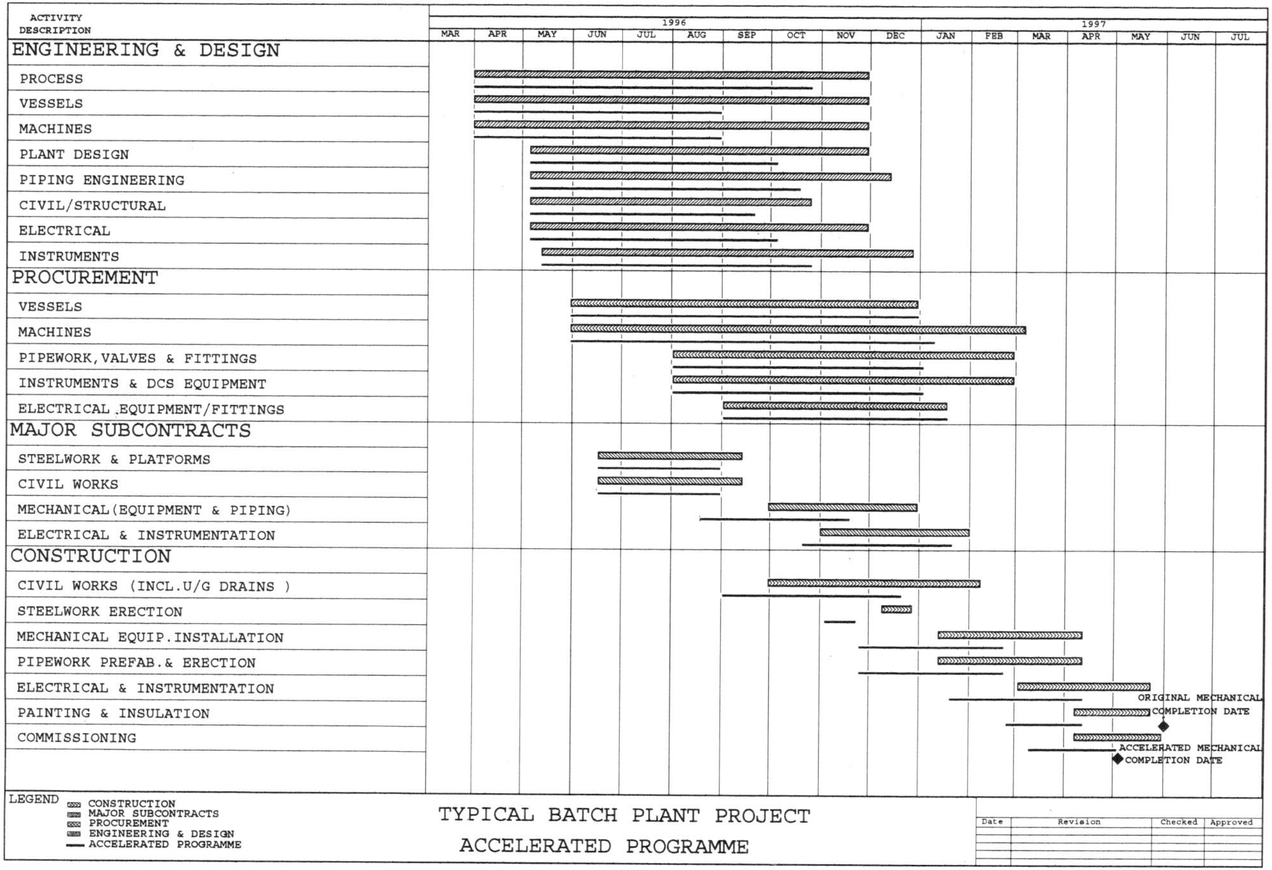

Figure 7.3 Accelerated batch design project.

services. Manageable units of work must be identified for which individual responsibility can be allocated. These individual elements are then integrated into the functional organizational breakdown structure and linked with a cost. Cost control must be dynamic and centralized with the project core team. A 'fast track' approach shortens the decision-making process, so deviations to the cost model must be identified very early in order to ensure programme advantages are not obtained at disproportionate cost to the overall project.

Vendor data flows become critical, with the design and construction data requirements needed even earlier than during a normal project. Preliminary data will have often to be used before final documentation is issued. The advantages in being able to incorporate packages of equipment previously used in the design of similar plants is immediately obvious. Standard equipment will have 'catalogue' data which are relatively secure and can be used for layouts and routeing studies.

A 'fast track' approach requires techniques to manage the higher risks arising from greater uncertainty and to control any changes effectively. Any changes to the original contract scope will cause some 'pain' but these must be restricted to the bare minimum, i.e. only those items that are required to meet the original scope and safety standards. The project will need to utilize effective change control that is fully integrated with the control and information systems. The project team need to be aware of all potential sources of change which must be managed in such a way that they are not allowed to provide excuses for delay or become distractions to ongoing activities. The effect of change must be identified quickly and if a significant cost or schedule implication is apparent, a high profile mechanism for its early rejection or consideration of viable alternatives must operated. Design should be 'fit for purpose' and free of any preference engineering. Any requirement for potential future attributes must not extend original delivery promises and must be part of a defined strategy with a clearly assigned budget.

Procurement policy needs to focus on known single source vendors if possible in order to shorten the inquiry and manufacturing cycle. Critical path equipment will require special monitoring and expediting measures, to ensure any potential delays are identified at commencement of the problem rather than on completion of the delay. Vendor intelligence data need to be current. Take-overs, reorganizations, steeply falling share prices are all indices of potential delay to final delivery of equipment.

The importance of an early input by construction and commissioning representatives should not be underestimated in its effect on reducing commissioning and start-up periods. A sequential handover of the plant in grouped plant systems may be required. This strategy must be fully established at the onset of the detailed engineering phase and not part way through. The design deliverables would need to be presented in work packages, perhaps to suit modularization techniques. These design packages would be delivered to the individual contractors in a priority order so that the sections of the plant that required early completion and commissioning activities were worked on first and completed on the dates required.

7.5.2 Problems

The 'fast track' process, while having much to commend it, does require caution in its application. The reduction of float on the critical path can create the situation where any activities that are not completed in the allocated time frame create a major reprogramming exercise.

Analysis of the plant's constructability can cause changes in the construction sequence. If the design work has been divided into work packages the need to rearrange design deliverables will arise. This may well be handled easily by the planning system but can be very disruptive to the team and can lead to demotivation, especially if the changing targets are perceived as unrealistic or impossible.

In order to achieve early start dates contractors need to be 'short listed' and the bid process commenced well in advance of final completion of design packages. The invitation to bid will embody the state of play at the time of issue which inevitably will require adjustment to reflect design progress achieved up to contract award. This can often protract final negotiations with late adjustments to lump sum prices or schedule of rates. The design package will also contain omissions due to the inevitable slow response from vendors on design data.

The ongoing development and change inherent in a design process always takes a finite time to emerge, despite a 'fast track' philosophy. Any change to scope will attract variations from the contractors and or subcontractors at escalated rates.

The effects of the weather should also never be underestimated. Unusual or exceptional weather (seemingly more common recently) can have a serious impact on projects. Civil works always suffer the most with the site access points and hardstanding often poorly maintained. Steelwork erection is very susceptible to high winds and low temperatures. A 'fast track' programme will have to address the subsequent rescheduling caused by delay and be able to adjust to out of sequence activities. Inevitably acceleration measures will be required to bring target dates back in line; this will attract a barrage of additional claims from site contractors.

Contractors can work extended hours and maintain a high manning profile in order to maintain a schedule. However, this does not automatically lead to increased efficiency especially as only a limited number of workers may be able to work productively in one area. In addition, the amount of supervision and available tools will need to be increased in proportion to the manpower increase, all of which takes time to mobilize. These additional costs may not all be to the contractors' account.

7.6 Regulations and other controls

Interface with external organizations (especially regulatory authorities) is normally allocated to nominated fully qualified individuals with total responsibility for compliance. Check lists for all these requirements should be formally recorded in the project co-ordination procedures to ensure full awareness. Statutory requirements stipulate specific actions that are required by a project team. These approval stages are an inherent requirement of all projects and care must be taken to ensure that none are allowed to cause project delay. They relate to environmental controls, sites plants and buildings, handling and storage of material, health and safety of construction staff, and final notifications before plant start-up. Typical controls include the following:

(i) Planning approval.
(ii) Building warrant (local council).
(iii) Hazard studies.
(iv) Environmental impact statements.
(v) Fire protection philosophy.
(vi) Critical machines/registered pipelines.
(vii) Area classifications.
(viii) Relief stream verifications.
(ix) Registered structures.
(x) Alarm and trip classifications.
(xi) The construction design and management regulations (CDM).

Regional planning approvals can frequently take several years following detailed site studies with local building warrants equally protracted. The project team must ensure that all ongoing applications are monitored. A lack of response does not always ensure the procedure is ongoing: it can mean a lost application with resultant major delays to the civil engineering start date.

7.6.1 Hazard and operability (HAZOP) studies

HAZOP studies carried out in the engineering phase must be controlled with the timing of each phase and implementation of actions geared to suit the overall piping and instrumentation diagram issues. The HAZOP study is usually carried out in six stages commencing at the project feasibility stage and running through the definition and detailed design. The final phases occur just before introducing process materials and 3–6 months after start-up. The typical timing of these studies is shown in Figure 7.4. Any late

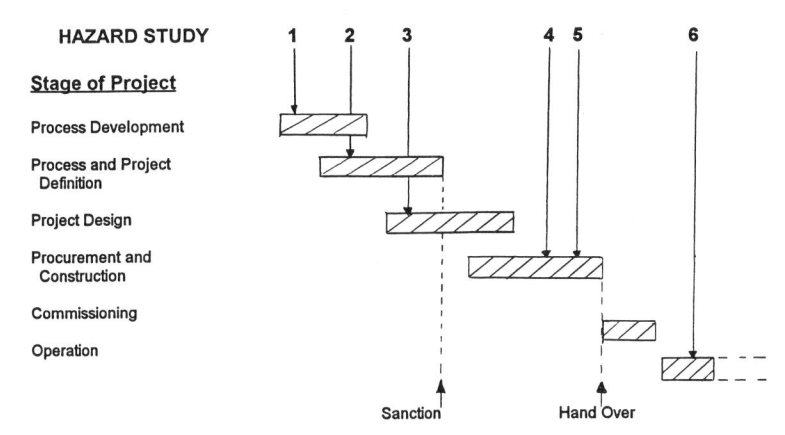

Figure 7.4 Typical project programme showing the timing of hazard studies. *Study 1: Project feasibility stage.* Identification of the basic hazards of the process and materials involved. Establishment of safety, health and environmental (SHE) criteria and contracts required with external bodies. *Study 2: Project definition stage.* Identification of significant hazards with inherent (SHE) requirements and appropriate design features required. *Study 3: HAZOP.* During and at the end of the project design stage. *Studies 4 and 5: Completion of construction stage.* Equipment and procedures as stipulated by Stages 1–3 and as required by company and legislative requirements. Must be completed before introducing process materials. *Stage 6: Post start-up.* Between 3 and 6 months after beneficial production to check all studies completed and operation consistent with design intent.

design development due to poor definition will still need the full hazard study treatment.

The intent of HAZOP studies is to identify hazards and operability problems that may arise during operation, to identify their consequences and to select measures to prevent or reduce such deviations from occurring as far as practicable. The HAZOP team composition need careful selection with an experienced facilitator who needs to be aware of the programme constraints and be prepared to guide the team along at the required pace. This should not be approached as an occasional part-time role. The hazard files, section notes and individual actions need to be constantly expedited to prevent the process drifting. The use of core members of the project team should not be allowed to disrupt the design process. The involvement of representation from operations and the start-up group will provide a practical dimension into the early identification of design changes required. It is important to ensure that the right balance is achieved with the isolation of operating problems and the identification of all significant hazards. The study is usually organized by division of the study into process sections or items of equipment with clearly defined boundaries. The presentation of study results should be flexible in order to ensure that they are understood by others. The execution and application of HAZOP studies within the overall safety process is becoming increasingly regulated with the end results being implemented into the operator's process safety management systems.

7.6.2 Environmental impact studies

Environmental impact studies are likely to be required as part of the process authorization procedures. The detailed requirements will depend on the local regulatory framework. Impact assessments commence with a baseline assessment of the condition of the site before the project is started. This can provide valuable data in avoiding negative impacts on the site and prove useful for any future disputes with local authorities. The study then identifies sources of exposure of people in the vicinity of the site and on the local ecosystems.

7.6.3 Fire and explosion hazards

A fire protection philosophy must be able to comply with the operator's safety policy, the works standards, and local and national regulations. Fire alarms and escape routes must be sufficient for the cumulative possible risks inherent in the process. Pipeline and equipment subject to certain specified temperature and pressure criteria must be individually identified and registered in order to check design and operating compliance.

All processes plants are subject to hazardous area classification. This involves 'zoning' of potential fire and explosion hazards. This activity can be particularly complex with batch process technologies which are placed within a multi-floored building with space constraints. A location where a flammable atmosphere can exist may be divided into three zones, each zone corresponding to a different level of fire risk:

ZONE 0 In which an explosive gas–air mixture is continuously present or present for long periods.
ZONE 1 In which an explosive gas–air mixture is likely to occur in normal operation.
ZONE 2 In which an explosive gas–air mixture is not likely to occur in normal operation and if it occurs will exist for a short time.

Any area not classified as ZONE 0–2 is deemed by implication to be non-hazardous and normal industrial equipment can be used. This classification deals only with risks due to flammable gases/vapours and flammable mists. Dust standards are covered by different zoning classifications, and stipulate size and configuration of protection devices. The development and agreement of the area classification requirements to suit a particular batch process can have a significant influence on the final layout options. It will certainly affect the selection of equipment, which will have to meet the classification for the area in which it is located.

7.6.4 Construction safety

Construction safety is an important issue and is often controlled by regulation. In the UK, the CDM regulations came into force on 30 March 1995, and provide a framework in which to improve the overall management and co-ordination of health, safety and welfare issues throughout all stages of a construction project. Specific duties are placed upon clients, designers and contractors via the appointment of new duty holders, i.e. the planning supervisor and the principal contractor. New documentation is required in the form of health and safety plans and the health and safety file. The impact on design activities is further to raise the profile of health and safety issues. The design process will be independently monitored to ensure that health and safety risks have been adequately identified and analysed. Relevant information must be communicated between disciplines and risk elimination techniques demonstrated. The concept and feasibility stages of a project will provide the greatest opportunity to consider different design options so that potential hazards can be avoided.

7.7 Design techniques

The use of computer systems plays a significant role in the ability of the project team to control the type, quality and timing of design deliverables. An integrated information system that can consolidate project needs and deliver the necessary data and reports for performance measurement will enable the project to be effectively monitored and directed towards the milestone targets required.

Systems need to be selected to suit the specific purpose and not to generate large comprehensive databases which do not directly benefit the project. The most effective databases are those that store basic design information that remains constant throughout the project. Design packages that can be directly imported from previous plants of a similar nature or configuration can enable a 'fast track' approach to be realized by reducing 'on-line' design and engineering time and cost. Computer graphics software such as computer-aided design (CAD) systems or equivalent can provide very visual computer generated drawings in two dimensions and orthographic or three-dimensional with perspective views. They are the nearest equivalent to three-dimensional modelling in the form of the traditional plastic model which is still more frequently used than perhaps acknowledged.

The project strategy plan must decide which engineering systems shall be utilized for the execution of the design packages. Large- to medium-sized projects with the complete design activity centralized in one main office may well justify the use of a full PDMS system, especially if existing databases can be utilized for equipment packages, layouts and pipe routeing studies. Computer-generated drawings have a number of distinct advantages: they can easily be re-scaled, layered or superimposed on others. Drawings can be self-generating using relevant numeric data. The computer can perform self-checking and create bills of materials and take-off quantities from the drawings. There is an in-built ability to check for interference between plant items or to excessively close spacing between piping, steelwork and equipment.

Review packages enable the computer model to be used as a virtual reality tool. The model can then be 'walked through' and interrogated to visualize options. Companies such as Herzog Hart have developed this ability to view the plant as a live three-dimensional model into a distinct methodology. The basis is an extensive database of 'objects' which are derived from an analysis of every physical component assembly on plants of a similar type and origin. The objects also carry many non-physical attributes such as cost estimates, operating instructions, maintenance instructions, safety documentation and purchase orders. Components are also arranged into reusable modules which can be easily modified to suit specific requirements and built into the overall plant. The object is to obtain earliest visualization of a project to ensure both the design team and others less directly involved reach early agreement on concept/layout issues. The model is built utilizing objects incorporated into the object library and includes conceptual layouts for other items such as pipe racks, cable trays, services, HVAC, etc. Typically, within 4 weeks a video can be prepared of the concept model with key principles and areas requiring resolution identified on a commentary. By these means, the project becomes 'live' at this early stage with all parties able to relate to and contribute to developing the model attributes and visualization through the project life.

HAZOP studies can take a new dimension with the monthly or weekly update of the walk-through facility. Construction sequence videos can also be prepared which could assist in avoiding site surprises and enable the fabrication priorities to be clearly understood before placement of sub-contracts. This in turn can also enable the correct scheduling of design packages to suit. Video conferencing facilities can be used as a communication aid with the data potentially transportable into other CAD systems such as PASCE/PDMS and AutoCAD/Microstation.

This type of innovative approach to design creates the opportunity to reduce engineering cost and construction costs by up to 20% while maintaining quality standards. For example, a comparison of the use of manual drafting and material take-offs against the use of a full PDMS package over a range of projects carried out over the past few years revealed a significant benefit for computer-based methods, the manual methods absorbing 16% more man-hours.

There are also inherent disadvantages and drawbacks of PDMS systems, such as the computer connect time costs, the type of hardware utilized and the software update costs. The manipulation of large three-dimensional models takes a lot of memory and processing power, and the response time of the system can slow to a crawl with a consequent loss in productivity. The costs of frequent intermediate hard copies can be high with final issues often produced much later than originally planned. The ability to use electronic transfer to remote locations is currently not always so easily achievable as IT experts would have us believe, but this is of course changing. The implementation of drawing changes can also work out to be time consuming and costly on-line, the manual method being quicker and cheaper. Progress assessments are often more difficult with reliance having to be placed on the system to download all the completed drawings at the touch of a button once the interrelated database fields have all been successfully linked together. Finally, should the system 'crash' data and software need to be backed up and recoverable in a suitable disaster plan.

Medium to small projects with a distribution of design activities may opt for the more limited use of design systems with use of AutoCAD for engineering line diagrams and two-dimensional layouts with a plastic model as the main design tool.

7.7.1 Case study – design for ease of construction

A project undertaken by Zeneca plc at Grangemouth, Scotland, in 1994/1995 for a batch process in the agrochemical business adopted a concept of early involvement of the constructor in order to improve the degree of constructability in the design. The philosophy was to restrict the design deliverables from engineering design to the minimum required by the constructor who would then continue with the design in order to fabricate and construct the plant. This could then result in a more cost-effective plant with resultant improvements to the programme. The responsibility for rework was also placed firmly with the contractor with the added premium for the risk set at a reasonable level. The constructor was entirely responsible for purchase of materials and generation of the construction-related drawings.

The strategy involved the use plastic models of the complete plant in two scales (1:50 and 1:25) for a new batch process to be installed in an existing building. The models were completed in the design contractor's offices with the minimum use of backup drawings. The design team used manually drawn conceptual layout sketches only. These were sufficiently detailed for the modelling engineers to interpret and place directly on to the model. Model reviews were also undertaken by representatives of the start-up and commissioning team and used as an aid in the HAZOP study process.

The models were then transported to the mechanical and piping contractor's offices for completion of manual isometrics and material take-offs. The site building was within a very short distance enabling site verification to take place on routeing and clash studies during the design and fabrication phases. Ultimately the models were placed on site for reference during installation.

7.8 Layout considerations

The starting point for the preliminary layout is the process flow diagram or ELD. The positioning of any equipment such as reactor vessels, wash tanks and storage vessels will be to provide the process function coupled with adequate operational and maintenance access. Batch processes which are inherently labour intensive are usually put into a building to protect the operators and to provide a more controlled environment. These are usually multifloor structures due to the quantity of equipment and tiered arrangements are required to use gravity for material transfers. Clean areas are often required for pharmaceutical and biological material processes. Early visualization using computer modelling from detailed object libraries can provide, for some applications, a more comprehensive aid to determining layout constraints. The initial conceptual three-dimensional model or two-dimensional layout is utilized to identify the basic site and space constraints and provide the basis for the detailed layout studies.

The plant layout will need to allow for the effects of:

(i) Site topography and geology with particular reference to load bearing capabilities.
(ii) Weather, including temperature, wind strength, snow loading, etc.
(iii) Environment covering people, activities and buildings in the vicinity.
(iv) Transport including road, rail and marine access for supply of raw materials, products, etc.
(v) Services such as power, water and effluents.
(vi) Legal constraints include planning and building, effluent pollution limits, fire and safety and other local and regional bylaws and regional regulations.

Site standards will also be established covering the following:

(i) Separation distances.
(ii) Building lines.
(iii) Building construction, finish.
(iv) Road dimensions (width, gradient, radius).
(v) Service corridors.
(vi) Pipe bridges.

Duplicate streams should be laid out in a uniform manner as will apply to any other units placed in series or with grouping of multiple items. Plant items should in general occupy the minimum necessary space, but this principle will need to be modified to prevent and limit accidents and to allow full access for construction, plant operation and maintenance. Equipment that is packed too densely will impede ventilation and allow concentrations of leaked flammable and toxic vapours – this must be dispersed by free air circulation or forced ventilation.

The optimum plant layout will be constrained by the need to observe minimum safe separation distances. Reference should be made to Tables 7.1 and 7.2 which summarize recommended minimum distances between various process units and typical clearances for access. These or similar standards and codes of practice will need to be applied to provide a safe operating plant. If space is a major restriction, equipment will need to be selected

Table 7.1 Typical minimum desirable distance between process units

	Distance (m)		
	From similar unit of equipment	From nearest adjoining process unit	From possible source of ignition*
Pressure storage			
ethylene	15[†]	60	60
C_3	15[†]	45	45
C_4	15[†]	30	30
Flammable liquid storage			
<2500 barrels, flashpoint <66°C	5[‡]	15	30
>2500 barrels, flashpoint <66°C	8[‡]	45	60
>2500 barrels, flashpoint >66°C	5[‡]	45	30
Crude pipe still	7.5	8	30
Light hydrocarbon distillation	7.5	8	60
Catalytic reforming	15	30	30
Alkylation units	30	45	60
Autoclaves >20 bar > 0.3 m^3	7.5	15	30
Hydrogenation >68 bar	15	15	20
Catalytic polymerization	15	15	45
Thermal polymerization	15	15	30
Caustic washers (gasoline)	7.5	7.5	30
Sulphur dioxide extraction	15	15	30
Centrifuge dewaxing	15	45	60
Boiler plant	NL	30	NL
Fire pump house	NL	45	NL
Cooling tower	NL	30	NL
Waste-water separators	NL	20	20
Blowdown stacks	NL	15–30	60
Buildings, general	NL	10–60	NL
Flares	NL	60	NL
Loading racks	NL	15–60	60
Offices and canteens	NL	30	NL

NL, no limitations.
* A common value used in the UK is 15 m regardless of flammability and related properties. This practice should not be followed.
† Use 0.25 × the sum of diameters of adjacent tanks if this is greater. A group of vessels should not exceed 10 000 m^3 unless a single vessel.
‡ 15 m from the bund wall of the flammable liquid tank.

Table 7.2 Typical clearances for initial plant layout

	Distances (m)		
	Distance to similar unit*	Horizontal clearance†	Vertical clearance†
Pumps	0.8–1.5	0.8–1.5	4
Compressors	average width	3	4
Columns	3	1.5	–
Vertical vessels	average 0.5 × diameter‡	1.5	–
Horizontal vessels	average 0.5 × diameter‡	2	1.5
Heat exchangers	1–1.5	1.5	1
Fired heater	3	3	–
Reactor, stirred vessels	7.5§	1	–
Centrifuges, crushers, mills	3	3	–
Main roads to battery limits (BL)	–	9	5.5
Secondary roads to BL	–	7.5	5
Railroads to BL	–	4.5	7
Main piperack	–	4.6	4.9
Secondary piperack	–	3	3.7
Other overhead piperack	–	3	2.1

* Distances to other units depend on hazard and can be much higher (see Table 7.1).
† Allow for items such as longest internal part that must be removed for maintenance and operation, plus construction, erection, walls.
‡ For storage tanks values can vary considerably, according to hazard.
§ Much greater distances may be required, according to hazard; also consider barrier walls.

that is protected against ignition sources in such a way that it can work adjacent to hazardous emissions. There is of course a cost penalty with this sort of compliance. The hazard classification (Section 7.6.3) of an area will define the exact permissible location of types of equipment.

It is sometimes preferable to segregate totally mechanical equipment from the process area, with utility piping placed on the mechanical side. This can maximize access to process equipment and simplify maintenance procedures.

Layouts are usually subject to intensive reviews by the project team. An example of a three stage review is summarized in the checklist shown in Figure 7.5. All the main disciplines should be involved with the reviews which are formally recorded. The results should also be considered alongside hazard study results.

7.8.1 Case study – layout

A project being undertaken by Zeneca plc during 1995/1996 involved the copy of batch process, a Toner Resin plant built about 5 years ago and located in the USA. The plant consists of a three-storey building housing the main process plant with a number of ancillary areas for raw material tankage, intermediate product storage, utilities equipment and final product storage/warehousing. Pipe racks connect the ancillary units.

REVIEW CHECK LIST

Contract ____________________.
Date ____________________.
Lead Engineer

No.	Discipline	STAGE 1 REVIEW								STAGE 2 REVIEW								STAGE 3 REVIEW							
		Construction	Vessels and Machines	Electrical	Instruments	Piping	Process	Structural and Civils	HVAC	Construction	Vessels and Machines	Electrical	Instruments	Piping	Process	Structural and Civils	HVAC	Construction	Vessels and Machines	Electrical	Instruments	Piping	Process	Structural and Civils	HVAC
01	Position/Orientation of Equipment		■																						
02	Equipment Withdrawal and Removal		■								■														
03	Equipment Internal Access for Maintenance		■								■														
04	Equipment Internal Access for Installation		■								■														
05	Maintenance Access and Walkways		■								■														
06	Noise Levels with regard to Equipment Positions		■																						
07	Lifting Equipment for Maintenance		■								■														
08	Construction Access	■																							
09	Temporary Erection Openings	■						■																	
10	Set Down Areas	■																							
11	General Safety, Escape Routes and Operability						■	■							■	■									
12	Position of Safety Showers														■	■									
13	Hazardous Liquids, Escape and Containment						■																		
14	Definition of Hazardous Areas						■																		
15	Commissioning Requirements	■					■																		
16	Cable Routing Requirements			■	■							■	■												
17	Computer Highways				■								■												
18	Location of Instruments, Operability, Access and Maintenance												■		■										
19	Maintenance Positioning of Control Panels and Boxes											■	■												
20	Lighting Design			■								■													
21	Location of Distribution Boards and Socket Outlets											■	■												
22	Location of Telecommunications Equipment												■												
23	Location of Fire Fighting and Detection Equipment														■										
24	HVAC and Building Services								■																
25	Positioning of Major Ductwork								■								■								
26	Preliminary Stress Review of Major Ductwork					■			■								■								
27	Preliminary Stress Review of Critical Pipework													■											
28	Position of Lifting Well / Hoist	■	■					■																	
29	Holes in Walls, Floors and Roofs							■								■									
30	Location of Structural Members, Bracing, Haunches etc.							■								■									
31	Plant Drainage Requirements							■								■									
32	Washdown Areas						■	■							■	■									
33	Client / Licensor Philosophy						■																		
34	Pipeline Functionality Check Against ELD																						■		
35	Location of Valves, Operability, Access and Maintenance													■	■							■	■		
36	Location of In-line Items, Operability, Access and Maintenance													■	■							■	■		
37	Plant Operation and Maintenance Requirements									■					■										
38	Pipeline Venting and Draining Requirements																					■	■		
39	Pipeline Falls													■	■										
40	Commissioning Requirements, Temporary Spools etc													■								■	■		
41	Pipe Support Philosophy													■											
42	Site Testing Requirements									■				■								■			
43	Safety Aspects during Erection	■								■													■		
44	Final Stress Review of Critical Lines																					■			
45	Final Stress Review of Major Ductwork																					■			
46	Pipebridge & Major Pipe Routing Requirements					■	■	■																	

Note : The items to be considered are not limited to those above

Figure 7.5 Project review check list.

The topography of the new location in Grangemouth, Scotland, dictated a re-orientation of the ancillary units, while trying to maintain the concept of the original layout within the various process areas. This was to enable the maximum utilization of the existing design documentation. Some of the range of factors which needed to be considered in implementing this project were:

- The direction of incoming utilities to the new plant.
- The complete re-orientation of all drawings to suit the new location.
- The extent that US imperial measurement systems could be adopted for a UK facility.
- The changes in climate and soil conditions between the different locations.
- The changes in the national design codes and regulatory requirements.
- The changes required to the hazard zoning system.
- The change to IS instrumentation from explosion proof devices.
- A change from electrical conduit systems to cable trays.

The quality of the existing documentation was variable with limited piping arrangement drawings and no original piping isometrics. The decision was made to model the process building on a PDMS system to produce a clash-free pipework design – clashes had been a problem in the erection of the original plant. The site also required a detailed survey to capture the as-built status – a problem that is inherent to many plants. Equipment for the new plant had to be purchased from the UK and Europe which required changes to data sheets and specifications.

This example demonstrates that even where plants exist and they are to only to be copied, the constraints of documentation records and subtle changes to design parameters can have a significant effect on layouts and re-engineering costs.

7.9 Plant relocation/reuse of existing equipment

Many batch process systems use a similar configuration of equipment, often with a number of standard design packages of a comparable size. An extensive market has now matured for the purchase of complete second-hand plants and for the purchase of individual items. Comprehensive inventories of equipment are held by specialist vendors derived from surplus stock to items derived from cancelled projects. Equipment is reclaimed from redundant plant and held as stock or refurbished to order. Lead-times for the manufacturer of new equipment can be eliminated by the procurement of existing items, meeting the required performance criteria, and of adequate quality to suit the application and necessary life expectancy. Equipment that has been totally or partially refurbished will carry new warranties or guarantees for specific actions from date of purchase. The existence of the original manufacturing documentation will provide added value to used equipment or eliminate its potential use in certain environments.

The reuse of equipment, while carrying many advantages such as cost and immediate availability, does carry certain risks. Adequate time must be allowed to carry out complete inspection and to determine what level of refurbishment may be required. The supplier's expectation of condition may not be realized. A detailed examination of internals and exteriors by competent inspectors after shipment from a distant location may lead to 'nasty surprises' and ultimate rejection, often after partial disassembly. The use of such items, perhaps later to be replaced with alternatives from other suppliers, will absorb lead-time for final on-site delivery for construction.

Decontamination certificates, if provided at all, should always be treated with caution. Vessels arriving with a 'cleaned' tag can exhibit subtle signs of residues, such as a thin coating of white powder. The contents may have only been partially removed without neutralizing the original chemicals. This causes ongoing delay while decisions are made whether to steam clean or to apply solvents. These may not be immediately available or difficult to use in the new location.

The use of spare and used equipment serves a vast market in the chemical industry, with the opportunity to reduce costs by reducing down-time of operating plants and lead-times of critical equipment to fit exacting construction schedules. The relocation of complete existing batch process plants is becoming much more attractive in today's market, where the required return on investment has to be within a stipulated time period that is not always viable with the lead-time needed for a completely new plant. The transfer or relocation of an existing plant can result in two very attractive benefits:

(i) Major reduction in total project cost.
(ii) Reduced project duration and earlier start-up.

If a suitable plant can be purchased for a competitive price then overall savings of up to 50% of the cost of a new plant can be achieved. The major cost components of a relocation project are:

- Purchase of existing plant.
- Engineering.
- Dismantling, packing and shipping.
- Refurbishment and design modifications.
- Re-erection.

While the erection costs (including civil works) will generally be similar to the construction of a new plant, the remaining costs are usually significantly less than the design and supply of new equipment for a new plant. Figure 7.6 illustrates the scale of the various component costs from engineering through to final erection that can be expected in comparing the cost of a new plant with one that is relocated.

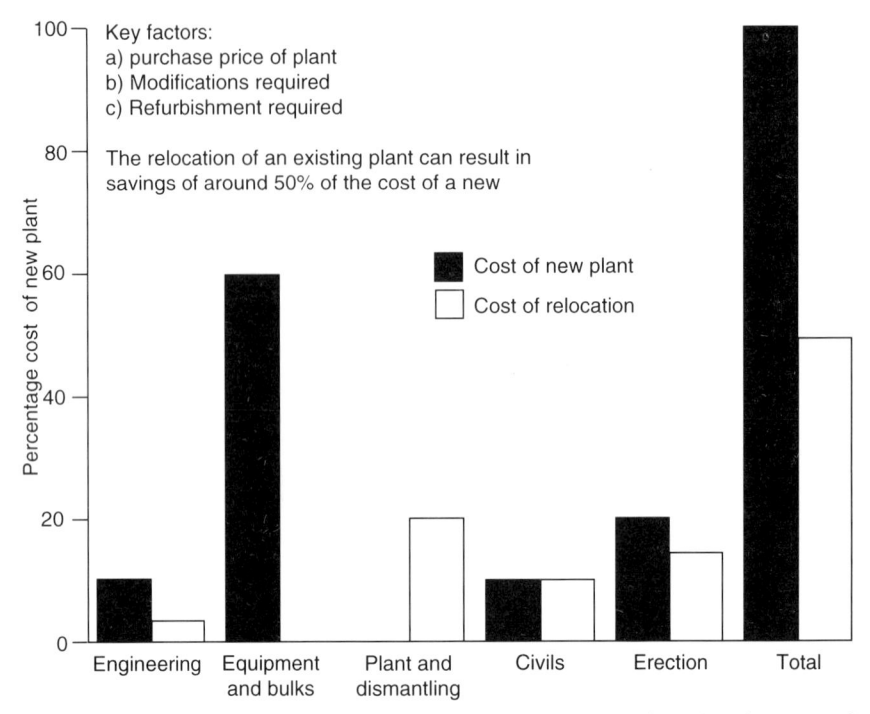

Figure 7.6 Cost savings for a relocation project compared with a new plant. Open bars, cost of new plant; closed bars, cost of relocation. Key factors: (a) purchase price of plant, (b) modification required and (c) refurbishment required. The relocation of an existing plant can result in savings of around 50% of the cost of a new plant.

In most cases the overall project programme for a relocation project will be shorter than that for a new plant. Savings in time are made primarily in the engineering design and procurement stages. While some engineering activities are essential to a relocation project, the duration is far less than that for the complete design of a new plant. If the plant contains complex items of machinery or vessels then the dismantling time will be significantly less than the procurement lead-time required for the equivalent new item, thus resulting in a further shortening of the programme. If careful consideration is given to these two major benefits then the relocation of a complete plant is usually a very attractive proposition.

For a process plant to be suited to a relocation project it should meet one or more of the following requirements:

(i) The plant should contain high value capital equipment and complex items of machinery such as: turbines, compressors, pumps, fans, centrifuges, extruders, spinning machines, etc.
(ii) Some of the vessels and pipework should be manufactured from high value materials such as corrosion resistant stainless steel.

(ii) The plant should generally be well maintained and free from major defects.
(iv) The plant layout can be replicated on the new site.

Ideally, the plant selected for relocation should either still be in operation or should have been mothballed following its shutdown, i.e. systematically decommissioned and preserved, otherwise unnecessary corrosion and other forms of deterioration may have occurred. Another benefit to the buyer in purchasing an operational plant is that actual data for the plant performance and product specification can be obtained first hand.

Once it has been established that the desired type of plant is suitable for relocation, then the next step is to find one which is suitable and, of course, available. A plant should be sought which matches the blueprint for the ideal new plant, taking into consideration the following parameters:

(i) Plant capacity.
(ii) Product grade and quality.
(iii) Age and condition of plant.
(iv) Utility requirements compared to those available at the new site.
(v) Environmental impact and regulations.
(vi) Plant layout and available space at the new site.

The price which has to be paid for the plant to be relocated will depend on a number of factors such as the age of the plant, the cost when new and the number of potential bidders. In most cases a very attractive deal can be reached, as the alternative to selling the complete plant is often demolition and subsequent sale for scrap which rarely generates a significant amount of cash.

The status of the existing documentation again becomes an important factor. This may have to be rejuvenated to represent the as-built plant. The opportunity exists to modify or upgrade the plant. New cables and insulation and other bulk materials will require purchasing, and the scope of supply defined for utilities and off-sites.

One of the most critical activities of the dismantling phase is the clear identification of each item of plant. Computer-based systems have been developed which record the details of each item, together with its unique number. Every single loose item must be allocated its own unique number whether it is a complete pump, motor and bedplate or just one holding-down bolt. This system is used to monitor the progress of each item as it is dismantled, cleaned, refurbished (if required) and packed for transportation. Packing lists for each case or container are automatically produced which both saves time and eliminates errors. Each item is given two labels, one stainless steel disc stamped with the unique number and one multi-layer paper tag. Figure 7.7 shows the flow of items from initial identification through to packing during a relocation project. The system also links each item with the respective area of the plant from which it has been removed.

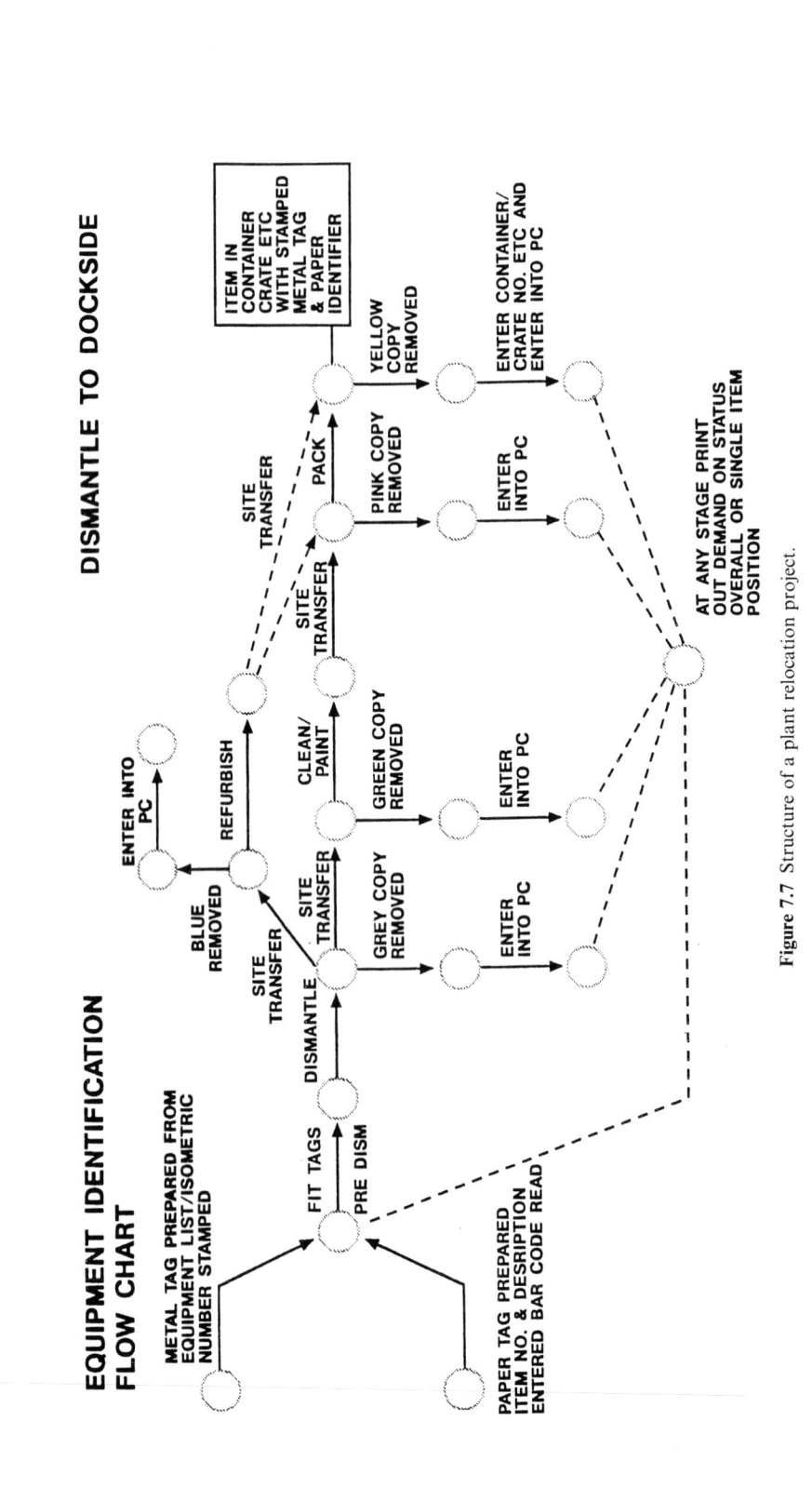

Figure 7.7 Structure of a plant relocation project.

7.9.1 Case study – plant relocation

A project undertaken by Simon Carves in 1991 involved the dismantling of an existing acrylic fibre plant utilizing a dry spinning process for relocation in Hungary. The plant comprised a raw materials tank farm, polymerization reactor, solution preparation, blending and packing equipment, spinning, processing and finishing plant, packaging and baling equipment, monomer and solvent recovery facilities, and laboratories. The plant incorporated 1000 items of equipment and approximately 17 000 m of new pipe and 2000 new valves. The lack of existing documentation proved to be a complication with none of the plant modifications made during the previous operating history adequately recorded. This caused some difficulties in re-assembly as the design adjustments were unique to this particular plant.

Although the plant was progressively shut down with decontamination procedures applied, it proved impossible to remove all process residues. Total draining of liquids, purging of gases, and removal of dust and debris could only be achieved with the plant split for dismantling. Electrical and instrumentation equipment was the most susceptible to damage during dismantling and re-assembly. Replacement and refurbishment strategies were complicated by the fact equipment was outdated and frequently unavailable on a like for like basis. Relatively minor damage could lead to the need for a complete replacement of an item with the consequent re-engineering to suit different attributes.

7.10 Modular plant

In order to achieve the objective of early market penetration, many batch process chemical plant projects need to be executed differently from large continuous plant to reduce the design and construction periods. A modular plant is built in complete units, including steelwork, pumps and instrumentation, before delivery to site. Modularization can be used to reduce project timescales, but it does dictate a change in the traditional engineering sequence. Work can be undertaken on a number of parallel fronts with the site preparation, utility requirements and control buildings progressing unhindered, while the construction plant can be spread among a number of fabrication sites to maximize progress.

Equipment must be specified in greater detail and ordered much earlier than normal. The equipment footprint, nozzle locations, etc., must be defined early in the procurement phase. The simultaneous construction activities will create many critical paths which are entirely driven by equipment deliveries. Vendor relations are key in providing early equipment data and deliveries which must satisfy the required sequence of installation into the module framework. The use of three-dimensional computer technology is widely used to provide accurate interference checking and avoidance of major clash problems. Closure engineering needs to be completed before fabrication has commenced.

Modularization provides the opportunity to reduce construction times and obtain faster plant start-up times as well as reducing labour costs. Its other benefits are tolerance of poor weather, overall increased quality and efficiency, simultaneous production capability, a firm testing base, and fewer interruptions to any existing plant. Modules are restricted in size by the need to satisfy expensive transportation cost and an increased engineering effort. The off loading and setting needs are also more complex and require the appropriate access and cranage.

There are five factors that help to determine the decision to modularize. In order of importance these are:

(i) *Plant location.* Weather constraints, high labour costs or lack of local skills may dictate maximizing off site work. Even where these factors are absent significant cost and schedule savings are evident using modular methods.

(ii) *Labour considerations.* The differential cost of labour performed at a job site and the same work in a shop environment are usually powerful driving forces for using modular construction methods. A reduced job site labour force can provide more stable labour relations and higher field efficiency. The maximization of a local content on international projects may be a positive or negative influence on modular construction depending on the import laws. All projects may require an analysis of socio-economic conditions in the host country to determine the proper balance of conventional and modular construction.

(iii) *Plant characteristics.* These must be evaluated to derive the full benefits of off site work. The modular plan must consider the owner's operation and maintenance requirements and their impact on the extent of modularization, configuration of modules and equipment arrangement within individual modules.

(iv) *Environmental and organizational factors.* The owner must understand the potential benefits of off-site construction

(v) *Project risks.* The additional front-end planning necessary with modularization gives added benefits to the overall schedule with firmer milestone dates and realistic float. The dispersed construction also allows easier absorption of the unforeseen.

8 Control
P.E. SAWYER

8.1 Introduction

Controlling batch processes is part of everyday life. Making a pot of tea or coffee for breakfast, a bowl of soup for lunch or a toasted snack for supper are all examples of batch processes. They involve making a limited quantity of a product according to a recipe. This recipe specifies:

- The product to be made and the ingredients required.
- The equipment needed.
- Operations such as mixing and heating.

To make an acceptable (edible) product, these operations have to be carried out in the order, at the temperature and for the time set down in the recipe. They also involve a combination of automatic control (e.g. boiling water in an automatic kettle) and manual actions (e.g. adding tea to the teapot).

The overall objective in operating any process – continuous or batch – is to convert raw materials into desired products in the most economical way. During production, the process must meet a number of requirements, including the following.

(i) *Safety*. The safe operation of a process is a primary requirement for the well-being of plant personnel and the surrounding community as well as ensuring continuity of production. This means that pressures, temperatures, concentrations, etc., must be kept within allowable limits. These limits may be dictated by the design of the process or by regulations governing operating practices.

(ii) *Production specifications*. The process must achieve target production levels of products that meet specific quality standards.

(iii) *Environmental regulations*. Increasing public concern for the impact of industry on the environment is reflected in a growth in legislation regulating effluent discharges, etc.

(iv) *Operating constraints*. All items of process equipment have constraints inherent to their operation, e.g. pumps must maintain a positive suction head, whilst distillation columns should not be flooded.

(v) *Economics*. The operation of a process must take account of market conditions such as the demand for products and the availability of raw materials. In addition it should be as economical as possible in its use of raw materials, utilities, capital and labour. Only then will operation be at an optimum level of minimum operating cost or maximum profit.

All these requirements indicate the need for continuous monitoring and control of the operation of a process plant. This is possible through a rational arrangement of equipment (instruments, valves, controllers, computers) and human intervention (design and operating personnel) which together make up a control system.

8.2 Control of continuous processes

For a continuous process, the requirements for control will be met by a system that:

- Keeps the process in material and energy balance.
- Maintains a given set of steady-state operating conditions.

Fulfilling these objectives involves holding key process variables – the *controlled variables* – at their *desired values* or *set points*. These set points are determined by the various economic, operational, safety and environmental requirements of the process. Changes in these set points will not normally occur frequently. When they do, however, the control system must be capable of bringing the process to the new steady-state condition quickly and smoothly. In this context, the system acts in the same way as a *servomechanism* that enables a mechanical system to follow a desired trajectory.

The operation of any process will be affected by *disturbances* which cannot be controlled directly. Often, these disturbances alter the material or thermal *load* on a process, so they are termed *load changes*. Their effect on the controlled variables can, however, be minimized by adjusting other process variables – the *manipulated variables*. *Regulation* – compensating for load changes – is the major task of a control system for a continuous process.

8.3 Control of batch processes

8.3.1 A simple example

A typical simple batch plant is shown in Figure 8.1. A single reactor is used to make a single product, using two raw materials. To produce one batch, the reactor is charged with the appropriate quantities of raw materials defined by the recipe for that product. Hot water is circulated through the reactor jacket to raise the temperature and initiate a reaction. Once the reaction is in

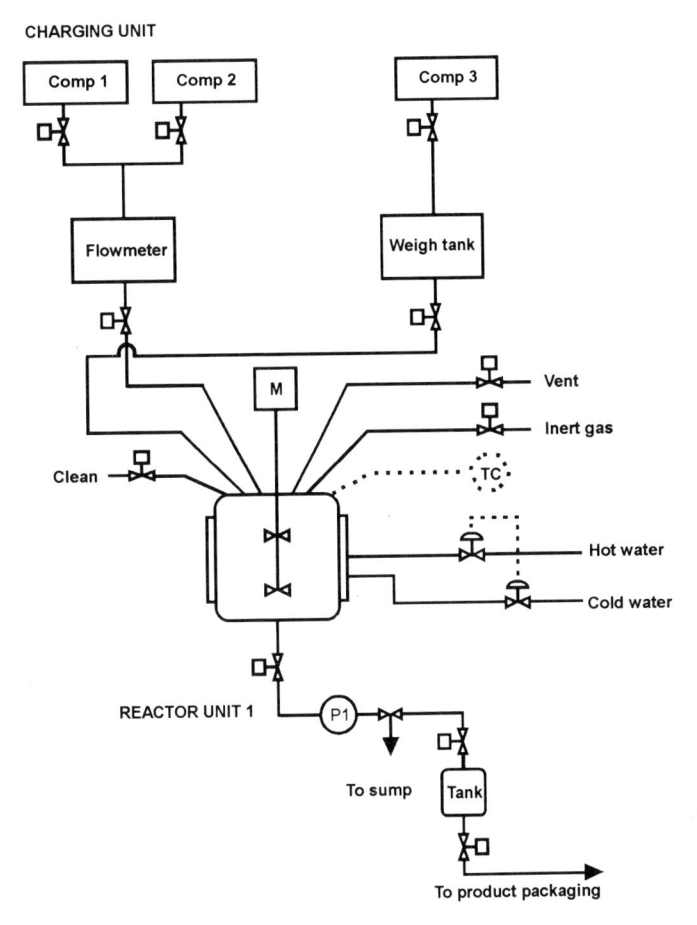

Figure 8.1 A simple batch process. (Redrawn with permission from Sawyer [8], © 1993 IChemE.)

progress, cooling water replaces hot water so as to remove excessive heat and hold the reactor temperature constant for a specified time to allow the reaction to reach completion. Finally, product is cooled and discharged to the storage tank and the reactor is vented and cleaned if necessary. These steps are summarized in Table 8.1.

This example illustrates a number of the basic features of batch processes that affect their control:

(i) Batch processes involve a defined sequence of operations; these operations are a function of the process and the product. The satisfactory progress of a batch through the process units carrying out these operations involves a significant amount of communication between their control systems, e.g. when the system controlling reactor charging has

Table 8.1 Process instructions for the model process

1. Initialize.
2. Weigh charge of component 3.
3. When enough of component 3 has been charged to cover the agitator, turn on the agitator and start charging component 1.
4. When both components are charged, heat to 120°C as rapidly as possible. Minimize overshoot to avoid production of off-specification product.
5. Hold temperature at 120°C for 2 h.
6. Sample for analysis by QC laboratory.
7. When sample is OK, cool and pump out to storage.

completed a charge, it must 'trigger' the reactor control system so as to initiate the reaction operation.

(ii) Time is a key process variable in batch operations. Real-time is one 'trigger' that moves a batch on to the next stage in the sequence of operations. Other 'triggers' correspond to events occurring or conditions being reached, either in the specific items of equipment involved or elsewhere on the plant. Time is still involved, but only indirectly. In the example process, a condition ('reactor full') 'triggers' the transition from the charging stage to the reaction stage. The transition from reaction to product discharge, however, is 'triggered' when the specified reaction time has elapsed.

(iii) Batch processes use large numbers of two-state devices in the process itself and also for sensing and control. In the example process, on/off valves are used to control the various discrete material transfer operations involved. Each inlet or outlet valve will be fitted with limit switches that enable its position to be checked before, during and after a transfer takes place. Other common examples of two-state devices are agitators and pumps.

(iv) Operator intervention plays a significant role in the control of many batch processes. It is often necessary for an operator to initiate critical operations via the control system rather than to allow them to be 'triggered' automatically, e.g. at the end of the reaction stage an operator might intervene to take a sample and then initiate product discharge or further processing depending on the results of an analysis.

(v) Continuous control loops are used in controlling temperatures, pressures, etc., during batch processing. They do not, however, function in the same way as when used to control continuous processes, e.g. the temperature of the reactor charge is controlled by a continuous feedback control loop. This loop need not be active during the charging operation. During the heating phase of the reaction, the loop will be required to manipulate the flow of hot water through the jacket so as to avoid an 'overshoot' in temperature. Once reaction has been initiated, the same loop will manipulate the flow of cooling water so as to regulate the reactor temperature. Control during heating and cooling may also involve 'retuning' the controller. In summary, control loops used in batch processing need to be capable of being easily reconfigured.

(vi) Data often need to be logged on an irregular rather than a continuous basis, e.g. the output from the flowmeter of the charging unit need only be recorded whilst component 1 is being charged.

8.3.2 Multiproduct, multistream and multipurpose operations

Many modern industrial batch processes are more complex than the simple example described above. They can be thought of as 'two-dimensional'. One dimension reflects the number and variety of the products being manufactured. It ranges from the production of successive fixed batches of a single product to production campaigns involving the manufacture of a wide variety of products and batches. The second dimension reflects the way in which the production plant is organized and utilized. It takes account of:

- The provision of a single or of multiple processing streams.
- The dedication of a stream to a single product or its use as a multipurpose facility.
- The ability to reconfigure a processing stream – the 'flexibly-coupled' plant.

These two dimensions can be combined to define the relative complexity of batch processes (Table 8.2).

Figure 8.2 shows how the dedicated, single product plant of Figure 8.1 might be adapted to multistream, multipurpose and multiproduct operation. This example illustrates some additional features of complex batch processes that affect their control:

(i) The same equipment may be used to produce different products or product grades. Each of the reactors could be used to produce a variety of products by modifying the number and quantity of components charged, the reaction temperature and reaction time. These process parameters are specified in the recipe for the batch. The control

Table 8.2 Relative complexity of batch processes (scale 1–5)

Process dimension	Product dimension		
	Single product	Fixed multiproduct	Campaign multiproduct
Dedicated single stream	1	N/A	N/A
Multipurpose single stream	N/A	N/A	3
Dedicated multistream	2	2	3
Multipurpose multistream	N/A	N/A	4
Multipurpose multistream reconfigurable	N/A	N/A	5

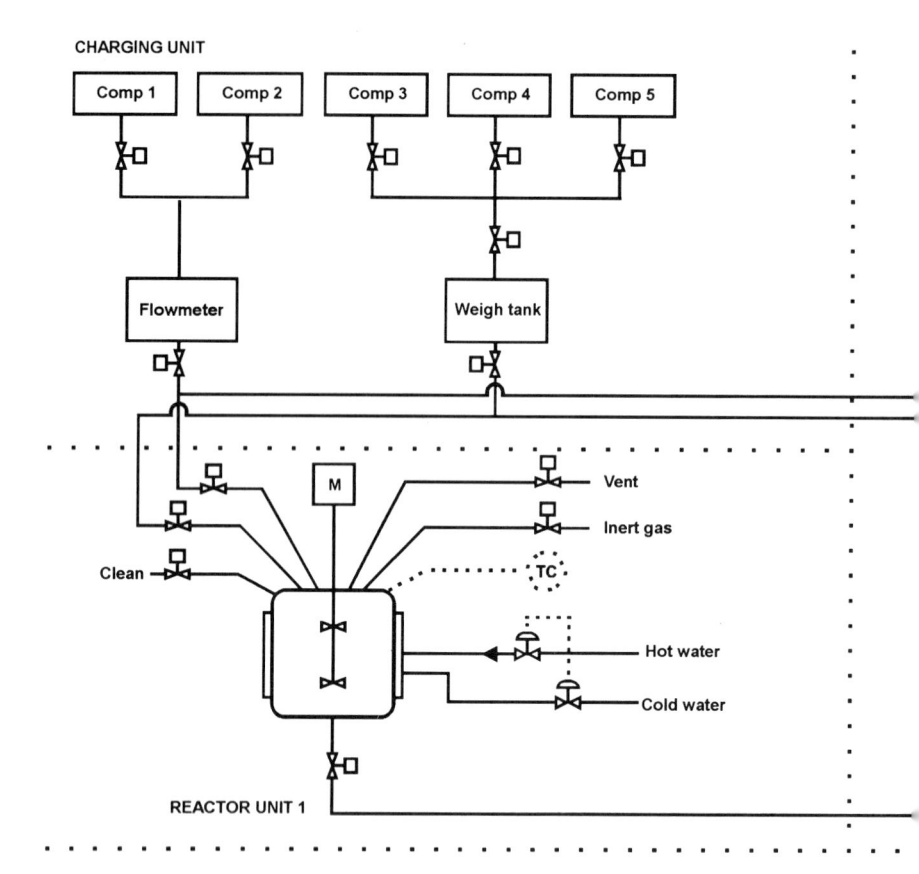

Figure 8.2 A multistream, multiproduct, multipurpose plant. (Redrawn with permission from Rosenof and Ghosh [3], © 1987 Van Nostrand Reinhold.)

system must be able to handle recipe data so as to make the required product.

(ii) Equipment and utilities are shared between processing streams. The two reactors in the example share a single charging unit, product storage facility and hot/cold water services. When both reactors are operating in parallel, the control system must allocate individual batches to them, co-ordinate their operation and arbitrate between any conflicts in the demands they make on these shared resources, including any overdemands on the capacity of utilities.

(iii) At any instant, the plant may contain a number of different batches of different products; each batch may also be at a different stage of processing. In order to preserve batch integrity, the control system must be able to keep track of individual batches so that recipes, process routes and logged data can be uniquely related to the correct batch.

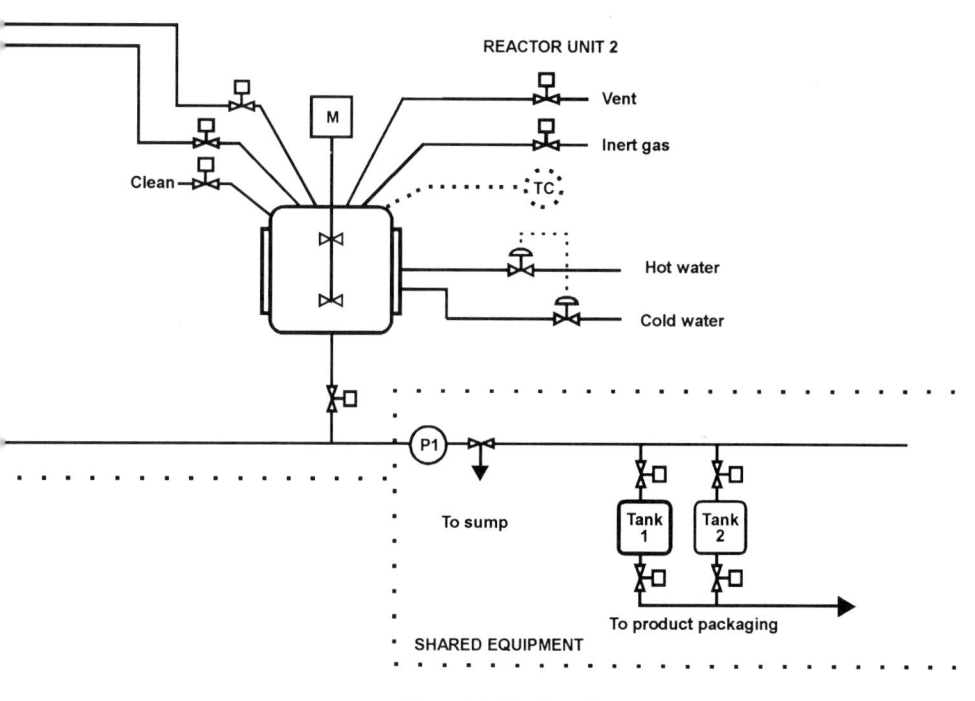

Figure 8.2 Continued

8.4 Batch control systems – structure and functions

The distinctive features of batch operations outlined above have an impact on both the structure of batch control systems and the functions they perform.

8.4.1 Models and terminology – the SP88 standard

Discussing any topic is made easier if you all speak the same language. Anyone who has tried to communicate through an interpreter knows how difficult that can be. Batch control systems are no exception; in the past, terms like 'recipe', 'phase', 'sequence' were often used without being adequately defined. What was required was a standard 'language' – a set of models and associated terminology – in which problems can be described, compared, discussed and solved. This was one of the major needs identified by the Instrument Society of America (ISA) when it set up the SP88 Committee in 1988. Since then, this committee has extended earlier work on batch control standards by a NAMUR working party in Europe and the Purdue TC-4 committee in the USA. Producing the first part of an ISA Standard has also involved a number of user groups, including the European and World Batch Forums.

Part 1 of the ISA Standard was issued in June 1995 [1]. The models and terminology it defines are intended:

- To reduce the 'time to market' for new products.
- To enable vendors to supply appropriate tools for implementing batch control.
- To make recipe development possible by production personnel rather than by control specialists.
- To reduce costs across the process and product life-cycle.

Part 2 of the Standard will focus on issues relating to the implementation of batch control systems in an 'open' multivendor environment. In particular, it will provide:

- Data models and exchange formats that allow information such as recipes to be exchanged between various vendor systems.
- Guidelines for batch control languages that reduce the engineering effort needed to describe control procedures capable of being implemented by various vendor systems.

8.4.2 *Models for batch processing*

ISA SP88 Part 1 is a 95 page document that represents the efforts of 100 people over 6 years. The models and terminology introduced in this chapter are therefore simplified versions that focus on key concepts. The simple process introduced in Figures 8.1 and 8.2 will be used as an example.

(a) A process model

The overall process of making a batch of product can conveniently be subdivided into *stages*, *operations* and *actions* as shown in Figure 8.3.

Each *stage* in a batch process makes several physical or chemical changes in the material being processed. Reaction is a stage in our example process; others might be recovery of unreacted raw material or the purification of the product.

CONTROL 227

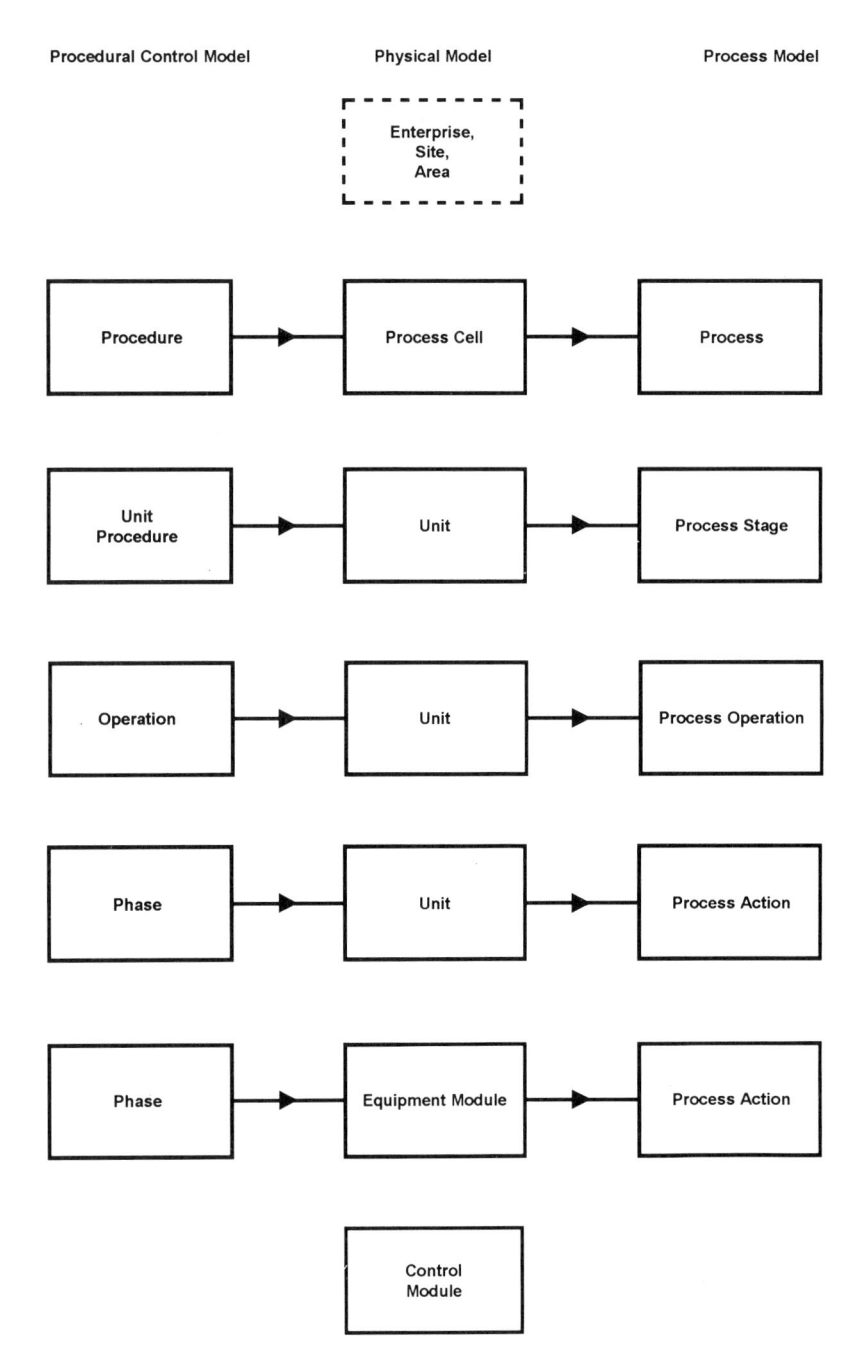

Figure 8.3 Procedural control, physical and control models for batch processes. (Redrawn with permission from ISA-S88.01-1995 [1], © 1995 Instrument Society of America.)

Stages are made up of *operations*. The reaction stage in our example involves operations such as:

- Charge – Raw materials 1, 3.
- React – Heat to required temperature and hold for specified time.

In turn, operations can be broken down into a series of *actions*. In our example, the React operation breaks down naturally into:

- Heat – to required temperature.
- Hold – at required temperature for specified time.

(b) A physical model
This model describes the equipment associated with batch processing. Like the process model it does so at various levels as shown in Figure 8.3. Of these, the top three – *enterprise*, *site* and *area* – are determined mainly by business considerations. Below these are the four levels that deal with the physical equipment required for batch processing.

The first of these is an extension of the traditional idea of a train – the equipment used by an individual batch. A *process cell* contains the equipment needed to make one or more batches. This concept allows activities such as production scheduling and process control to be organized on a broader basis.

The *units* in a process cell carry out major processing tasks. A unit is usually centred on a major item of process equipment such as the reactor in our example process. It also includes – or can acquire the services of – all the processing and control equipment required by its task.

Minor (but important) processing tasks are carried out by *equipment modules*. The weigh tank in our example forms the centre for an equipment module for weighing component 3.

Control modules combine sensors, actuators, etc., into a single entity. Examples from our simple process are:

- The temperature controller consisting of a transmitter, controller, output selector and control valves for hot and cold water.
- The header and automatic on/off block valves directing the flow of component 1.

(c) A procedural model
The features that distinguish the operation of batch processes from that of continuous processes have already been outlined. One result is that three types of control – *basic*, *procedural* and *co-ordination* – go to make up the equipment control needed in batch processing.

Basic control includes regulatory control, interlocking, monitoring, exception handling and repetitive discrete or sequential control. Basic control is common to batch and continuous processes. A distinguishing feature in the batch environment is the requirement to accept commands that modify its behaviour.

Procedural control is the control that enables equipment to perform the sequence of operations involved in batch processing. It consists of a hierarchy of procedures, unit procedures, operations and phases that correspond closely to levels in the physical and process models (Figure 8.3). At the lowest level, phases are often subdivided into steps that correspond to one or more instructions in a high-level control language used to implement sequential control. Table 8.3 illustrates the application of this model to the procedure for making a single batch of product in the simple batch plant of Figure 8.1.

Co-ordination control oversees the other two types of control. It includes:

- Supervising the availability or capacity of equipment.
- Allocating equipment to batches.
- Arbitrating requests for allocation.
- Co-ordinating common resources or equipment.

(d) A recipe model
In order to produce a batch of product it is necessary to know:

- The type and quantity of product required.

Table 8.3 Operations, phases and control steps for the model process

Operations	Control steps	Phases
Initialize	Initialize	Start jacket circulation pump. Put reactor temperature controller in SECONDARY AUTO mode with set point of 120°C.
Weigh	Weigh component 3	Initialize (tare-off weigh tank). Open outlet valve from head tank. When weight of component 3 equals preset, close outlet valve from head tank.
Charge	Add component 3	Open outlet valve from weigh tank. When enough of component 3 has been added, start the agitator. When weigh tank is empty, close outlet valve.
	Add component 1	Initialize (reset flow totalizer to zero). Open outlet valves from head tank to flowmeter and from flow meter to reactor. When volume of component 1 charged equals preset, close outlet valves.
React	Heat	Initialize (put reactor temperature controller in CASCADE mode with set point of 120°C.
	Hold	Initialize (reset timer). Start timer.
	Sample	
Discharge	Cool	Initialize (set reactor temperature set point to 35°C).
	Transfer	Initialize (set reactor outlet valves to correct destination storage tank), i.e. start discharge pump. Set reactor temperature controller to MANUAL mode with output at zero (full cooling). Before agitator blades are uncovered, stop agitator. When reactor is empty, close reactor outlet valves, stop discharge pump, stop jacket circulation pump.

- The process operations involved, the order in which they must be carried out and any associated operating conditions such as control and quality parameters.
- The equipment used in processing.

The use of the description recipe for this information recognizes the analogy between batch processing and cooking food mentioned in the Introduction.

The recipe model shown in Figure 8.4 classifies this information under four headings, as follows.

(i) *Procedure*. This defines, in general terms, the processing required to produce a class of product.
(ii) *Formula*. This is the set of data that links a general procedure to the production of a batch of given size of a specific product. It includes types and quantities of raw materials, operating conditions etc.
(iii) *Equipment requirements*. This section of the recipe specifies the type and size of equipment needed, materials of construction, etc.

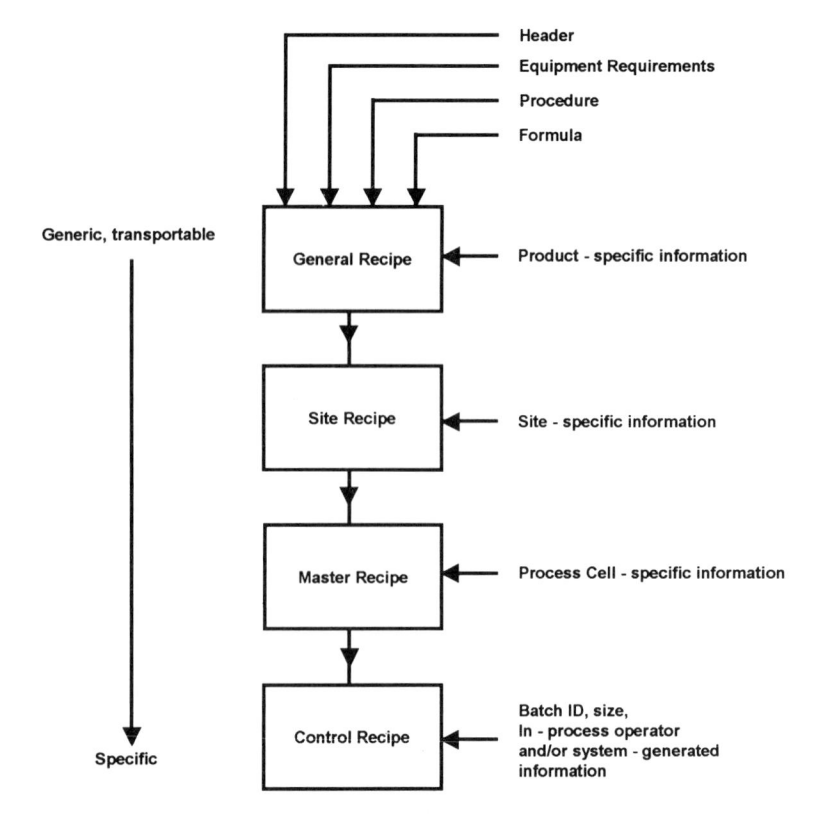

Figure 8.4 Recipe model for batch processes. (Redrawn with permission from ISA-S88.01-1995 [1], © 1995 Instrument Society of America.)

(iv) *Header*. This identifies the batch in terms of the product made, the version, origin and date of the recipe used, etc.

It is worth emphasizing here that the recipe contains information related to the processing needs for a specific product rather than to scheduling or equipment control. This separation of the recipe (that describes how a batch is to be made) from the equipment (that is used to make it) is a key concept. It gives the flexibility to make many different products without having to redefine equipment control for each product. Recipes are also transportable and easier to validate. The concept is also a familiar one. In a cookery book, recipes are not specific to a particular make or model of gas, electric or solid fuel cooker; they also leave you to decide when you actually start cooking.

In practice, more than one type of recipe is needed to accommodate differences in the scope and detail required at various levels within an organization. Four types of recipe make up the hierarchical model shown in Figure 8.4.

At the enterprise level, the *general recipe* provides generic information on raw materials, their relative quantities and the required processing. This can serve as the basis for planning investment and production and for informing customers and regulatory authorities. When information specific to a particular site is added, the resulting *site recipe* typically contains sufficient detail to be used in local long-term production scheduling. The focus of both general and site recipes is on technique – how to do it in principle.

Master and *control recipes* define tasks rather than technique – how to do it with available resources. The master recipe takes account of the characteristics of a specific process cell. It serves as the basis for the control recipe that is specific to each individual batch. This recipe provides detailed information such as actual raw material quantities and equipment to be used.

Where several separate batches are made from a common source of raw materials and using the same recipe, the batches form a *lot*. If further grouping is done, e.g. if the batches in a lot are blended to compensate for changes in feedstock or other processing conditions, the complete production run between recipe changes is traditionally termed a *campaign*.

(e) An activity model

Control systems for batch processes must support a variety of activities at the different levels of the process hierarchy. These activities can also be arranged in a hierarchical structure as shown in Figure 8.5.

Some of these activities are discussed in detail elsewhere in this Handbook. Chapter 2 deals with *production planning and scheduling* whilst *safety and environmental protection* issues implicit in process management and unit supervision are covered in Chapters 9 and 10. The scope of the other activities is discussed below.

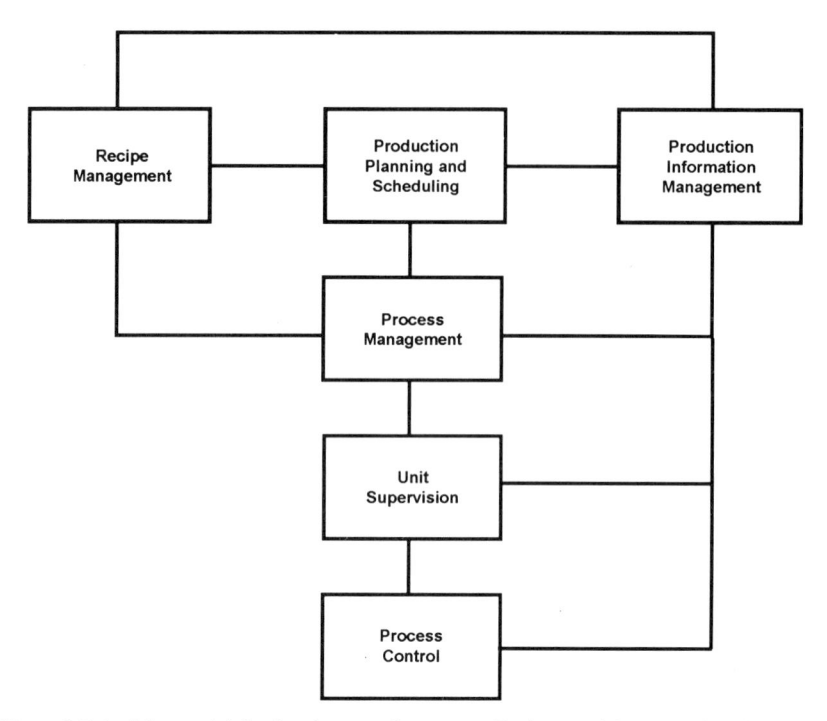

Figure 8.5 Activity model for batch control systems. (Redrawn with permission from ISA-S88.01-1995 [1], © 1995 Instrument Society of America.)

Recipe management. The use of recipes to define batches may not be justified for single product/stream operations. Batch sizes, raw material requirements and operating conditions can be incorporated as fixed parameters in the processing sequences. Any changes will be infrequent and can be made by editing these sequences directly. In multiproduct/stream operations, changes in these parameters and in the allocation of units and resources occur frequently and are best handled by the use of recipes. The control system must provide for:

- Creation, editing, storage and retrieval of recipes.
- Modification of recipes during processing.
- Security of access to these facilities by management and operators as appropriate, and compliance with existing change control procedures.

Production information management. This is a high-level activity whose scope includes the generation of production reports to management and other functions that are not batch specific. The main batch-related function is the maintenance of individual batch histories, using information from other activity levels.

Some data logging and reporting tasks are common to batch and continuous processes. They include periodic/on-demand reports of current process data and logs of alarms. There are other requirements that are specific to batch processes, such as batch reports and batch/material tracing.

Batch reports provide a record of what occurred during the course of the processing of an individual batch in a concise format. They enable production and engineering personnel to identify the reasons for a bad batch, recurring problems or the potential for future difficulties.

In the pharmaceutical industry, regulatory bodies like the DoH and FDA require evidence of proper production procedures for individual batches. This requirement is becoming increasingly common in other sectors in order to meet standards for quality assurance (e.g. BS 5750 and ISO 9000), safety and the environment. The batch report must therefore contain the complete process history of the batch. The task of keeping track of materials and batches in a single product/stream plant is not difficult. In multi-product/stream and multipurpose plants the task is more complex; several batches exist at any one time and a batch may be switched from one train in a process cell to another during processing. In such cases, separate log files must be maintained with each batch. As the batch moves through the process, it may come under a different local controller, so the log file must move with the batch. The problem is simplified if only one global database is associated with the batch control system.

Process management. The functions performed at this level include:

- Selecting control recipes and transforming them into working recipes.
- Managing resources necessary for batch execution, including allocating batches to process units and resolving conflicting demands on shared resources.
- Initiating and supervising the execution of batches.

In allocating batches to process units, account must be taken of constraints. These may range from equipment availability to potential cross-batch contamination and the need to avoid changeover operations such as cleanouts.

Conflicting demands on shared resources may be solved on a simple 'first-come, first-served' basis. In other cases, different priorities may be recognized, e.g. one batch of product may be more valuable than another batch of product; the more valuable product may secure the use of a common resource even though the other batch is ready to use the resource first.

In the high-level activities outlined above the focus is on co-ordination control. In the remaining levels the emphasis is on procedural and basic control.

Unit supervision. When the time comes to cook a meal, the recipe itself does not contain information to how to operate the cooker. It is up to the cook to interpret the instruction 'bring to the boil' in terms of the appropriate switch settings, etc. The same is true when a production schedule calls for the manufacture of a single batch by the execution of a control recipe. The procedural elements, e.g. operations, phases, etc., defined in the procedure section must first be linked to the equivalent equipment control functions that actually cause the equipment assigned to the batch to operate. This linking is a major activity at the unit supervision level. Other tasks include inter-unit co-ordination (e.g. of material transfers between units) and the collection of production information relevant to the batch.

Process control. This level involves both procedural control and basic control. Recipe phases have already been translated into the equivalent equipment phases by unit supervision. Process control involves executing these phases in sequence by sending commands to the interlock, discrete and regulatory functions that make up basic control.

Interlock functions typically specify that an output to a process element or device is not possible unless the inputs from other elements or devices indicate that a desired status exists.

Safety interlocks prevent abnormal process actions that would endanger personnel, harm the environment or damage major equipment, e.g. a reactor may be shut down if its temperature exceeds a high limit because this could cause a runaway or a decomposition. For maximum safety, this level should be physically separate from all other levels. Indeed, it is common for such functions to be implemented as 'hard-wired' systems separate from any computer control.

Process interlocks protect product quality or prevent minor equipment damage, e.g. the discharge from a reactor may be stopped if the outlet valves are not set to a correct destination. This could cause the product to be ruined if the transfer is made, but would not necessarily pose a safety hazard.

If provision is made for purely manual operation, this should be above the safety interlock level.

Discrete control involves maintaining the outputs of a process to a target value chosen from a set of known stable states. In practice, the choice usually lies between two such states: 'OPEN' or 'CLOSED', 'ON' or 'OFF' etc. One distinctive feature is that the 'normal' status of devices such as on/off valves or motors may change within the processing of a batch. During a charging operation, the normal status of a reactor agitator may be 'STOPPED', whilst during the reaction operation itself the 'normal' status will be 'RUNNING'.

Regulatory control involves maintaining the outputs of a process as close as possible to their respective set-point values despite the influences of set point changes and disturbances. Two features distinguish regulatory control as applied in batch processing from its use in controlling continuous processes. The first is the need for ease of reconfiguration as discussed earlier in this chapter. The second is that setpoints are often deliberately changed or 'profiled' during a batch in order to obtain a desired product or product property.

Two levels in a batch control system – the process itself and process I/O – do not appear in the SP88 model. Whilst they do not involve any control functions as such, they have an important role to play.

Emphasis should be placed at the design stage on making the process itself as inherently safe as possible, e.g. by using a less hazardous material or reducing the inventory of hazardous material to a minimum. Because there are always possibilities of control failure at any higher level, process equipment must be designed with appropriate safety margins to minimize the consequences of such a failure. Provision may need to be made for 'last-ditch' protection systems such as bursting disks. Key aspects of process safety are discussed in Chapters 1 and 9.

Process I/O begins with sensors that generate analogue and digital inputs representing the current status of the process. A typical analogue input might be from a thermocouple measuring temperature. Digital inputs from, say, limit switches might indicate whether a valve is open or closed, a motor is on or off, or a tank is empty or full. Pulse trains from turbine meters or revolution counters and serial data transmitted from 'intelligent' instruments are other forms of digital input. Output signals implementing control actions demanded by the control system are generated at this level. Typically, analogue outputs adjust flows by opening or closing valves, whilst digital outputs energize solenoids, motors, pumps, etc. Clearly, the proper functioning of this level is essential to the execution of the control functions at other levels.

The aim of the ISA SP88 standard is to provide a set of models and terminology that can be applied regardless of the degree of automation. In practice, it is likely that batch control will be implemented by a computer-based system.

8.5 Computer control

8.5.1 Introduction

Computers have been used to control both continuous and batch processes for over 25 years. The advantages and disadvantages of computer control are therefore fairly well understood. In the case of batch processes, the major benefits include [2, 3]:

- Increased production arising from reduced batch cycle times, better equipment utilization and higher yields.
- Improved consistency in product quality, which reduces the need for blending.

- Cost reductions resulting from better use of materials, utilities and equipment.
- More flexibility in changing operations and products whilst maintaining repeatability.
- Improved batch reporting and traceability.
- Greater safety for plant, personnel and consumers of the end product.
- A reduced impact on the environment through tighter control and improved operating procedures.

These and other advantages of computer control must be set against the disadvantages. The disadvantages include:

- A substantial investment in application software development and management.
- Higher purchase and installation costs for field instrumentation.
- Greater levels of technical and engineering effort that offset reductions in process manning levels.
- Duplication and redundancy of systems in order to guarantee reliability and safety.

The interactions between these and other advantages and disadvantages are summarized in Figure 8.6 [4].

In general, the technology now exists to bring any batch process under computer control. Justifying this level of control involves evaluating the various benefits in terms of economic gains that can be set against the costs involved. Tangible benefits such as increased production, improved quality and better use of resources can be directly evaluated. Those relating, say, to safety and the environment are less tangible and more difficult to take into account. The methods available for justifying computer control for new or existing processes range from qualitative risk versus pay-off assessments to quantitative costs versus benefits studies.

8.5.2 Systems architecture – hardware and software

The SP88 batch control models described in this chapter define a number of general functional requirements for batch control. Individual applications will differ in their demands for these functions. Low level functions – I/O, interlocking, discrete/regulatory control and sequencing – are likely to be common requirements. Here, the differences are largely ones of scale (e.g. I/O count and type) or of technical detail (e.g. provision for intrinsic safety). The demand for higher level functions such as recipe management and scheduling will depend on the complexity of the processes involved. A computer-based control system must have an architecture – a mix of hardware and software – that allows these demands to be met.

8.5.3 Choosing an architecture

The choice of an appropriate architecture for computer-based control of batch processes is not simply a matter of deciding between alternative computer systems as such. The experience of end-users, contractors and vendors is that the choice is project dependent; in other words 'the architecture of the control system needs to reflect that of the plant' [5].

The relationship between plant and control system architecture is discussed in general terms by Carlo-Stella [6]. Rosenof and Ghosh [3] describe an analysis based on a mapping of the process units and the need for intercommunication between their control systems against the various functional levels of a batch control model. Figure 8.7(b) illustrates this procedure for the simple batch plant of Figure 8.7(a) comprising two reactors sharing common charge, preparation and finishing units. The communication paths between process units required at each level are indicated by arrowed lines. For clarity, the destinations of some of the paths at level 5 are indicated by the appropriate symbol for the process unit concerned. Alternative architectures can be represented by 'envelopes' that enclose units and levels and cut some of the communication paths between them.

The analysis starts by recognizing that architectures whose envelopes cut fewer paths are more 'natural'; they match more closely the organization of the process and its controls. On this basis, the best architecture would be made up of the largest possible subsystems. This 'centralized' approach also has certain 'natural' advantages in other areas such as data management and the synchronization of parallel operations [7]. Other factors, however, favour a 'distributed' approach, e.g. the need for independent safety functions to minimize common failures means that the controls at level 2 should be separated. Similarly, at levels 3 and 4 a 'distributed' approach reduces the potential impact of computer system failures on process operations. Flexibility and ease of expansion are other advantages of distributed systems.

In practice, an evaluation of alternatives involves establishing relationships between architectures and their associated penalties or benefits. The procedures used to measure the degree of centralization and to determine the penalty or benefit values will inevitably be somewhat subjective; they are also likely to vary from project to project. The end result, however, will often resemble that shown in Figure 8.8 for the simple batch plant of Figure 8.7(a). The best architecture corresponds to a distributed approach to implementation at levels 2 to 4, coupled with a centralized system from level 5 upwards. An alternative that incurs a modest additional penalty extends the distributed approach to level 5.

So far, the discussion has focused on the relationship between the distribution of control functions and the structure of the process. As in the case of a building, there is a 'third dimension' in the architecture of a control

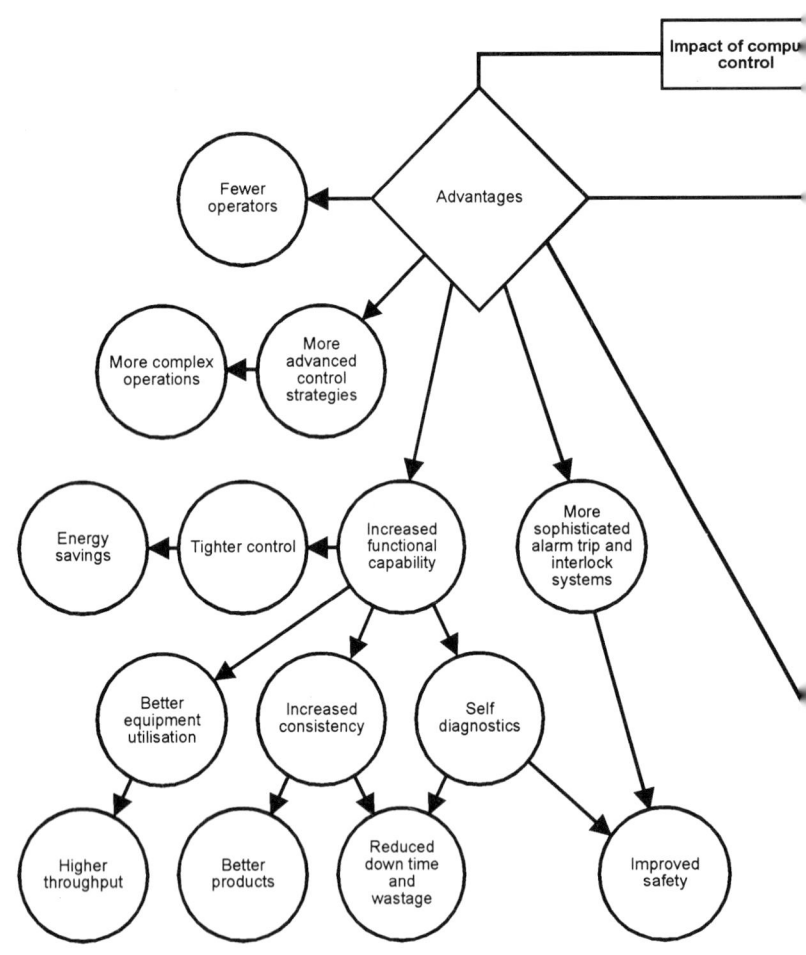

Figure 8.6 Batch computer control advantages and disadvantages. (Redrawn with permission from Sawyer [8], © 1993 IChemE.)

system, that of the geographical distribution of the hardware, e.g. the hardware of a control system that is functionally distributed may need to be sited centrally in a safe environment remote from process hazards.

8.5.4 Hardware

From a functional point of view, either a centralized or a distributed architecture can be implemented using four main hardware components:

- I/O subsystems.
- Processors.

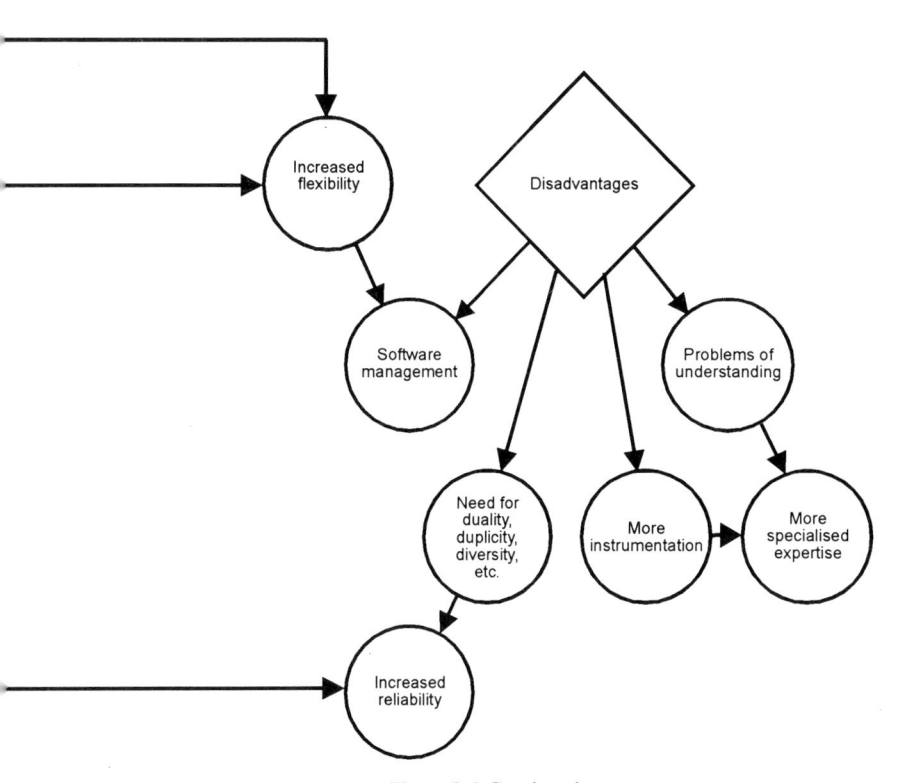

Figure 8.6 Continued.

- Consoles.
- Data highways.

I/O subsystems can be located remotely from the processors they service. Process inputs and outputs connected to the I/O subsystem are handled by modules that offer a wide range of I/O conversions. High- and low-level analogue signals, discrete (digital) inputs, pulse trains and other inputs can be read, validated and converted to engineering units where appropriate. Analogue and time-proportional signals can be output to regulate process variables and discrete outputs to open/close valves or start/stop equipment. Most I/O modules also provide for serial digital communication to and from the growing number of 'intelligent' transmitters and actuators now to be found mounted out in the field.

Signal conversion is only one of the tasks of the I/O modules. They must also offer:

- Isolation, noise rejection, transient suppression and over-voltage protection.
- Supply of power to field-mounted devices.

240   HANDBOOK OF BATCH PROCESS DESIGN

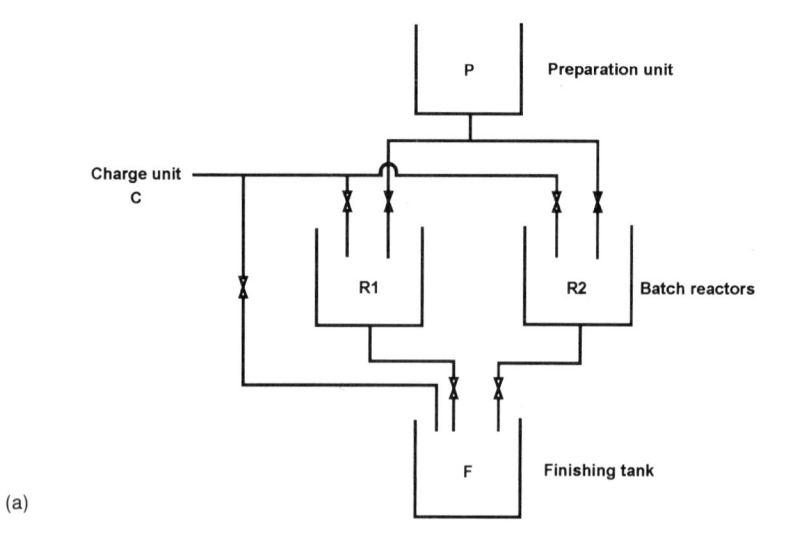

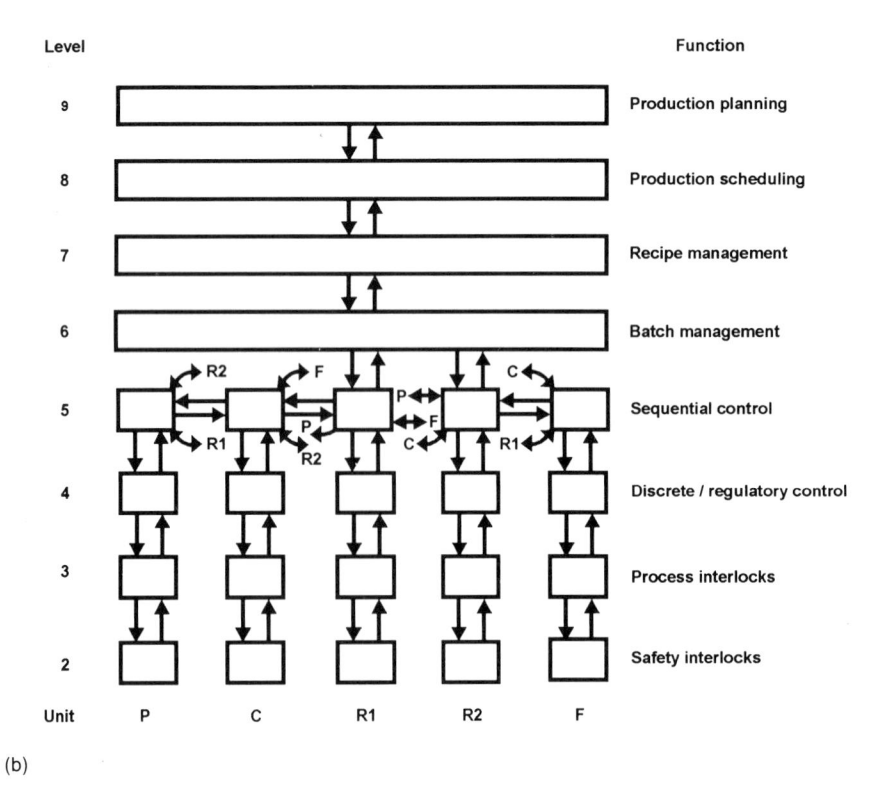

Figure 8.7 (a) A simple batch plant. (b). Mapping of process structure to batch control functions. (Redrawn with permission from Rosenof and Ghosh [3], © 1987 Van Nostrand Reinhold.)

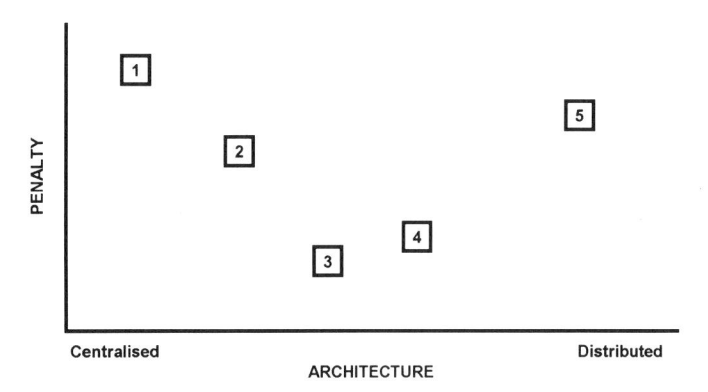

Figure 8.8 Penalties for various architectures. (Redrawn with permission from Sawyer [8], © 1993 IChemE.)

- Environmental ruggedness.
- Energy limitation to or isolation from field devices in hazardous areas; as an option this intrinsic safety (IS) isolation can be achieved by the use of separate safety barriers.
- Security of operation via self-diagnostics and redundancy.

Processors provide the functions required at various levels of control activity. Within process control, for example, they implement basic regulatory control by accessing process data provided by I/O subsystems, executing control algorithms and transmitting the required commands back for output as control signals. Procedural control is implemented by processors that execute the equipment phase logic derived from recipe phases. At a higher level, they may implement statistical process control (SPC) as part of the management of production information.

Consoles provide operators and engineers with access to the system. Plant personnel can monitor and control the process by interacting with graphical displays. Typical activities undertaken in this way include acknowledging alarms, initiating control actions and changing controller parameters. As well as providing for operator interaction, operator consoles also support the facilities needed to provide instantaneous, historical, alarm, batch, and other displays and logs. Similar consoles allow process and control engineers to configure and modify control systems and displays, construct master recipes, and perform other system maintenance and development functions.

Data highways are used to transmit information between the various functional modules and to integrate them into an overall control system. At first sight, they appear similar to the local area networks (LANs) widely used in commercial MIS applications. There are, however, some significant differences. The need to achieve secure, reliable and accurate high-speed data transmission and to guarantee access to these data in real-time has led to the use of a variety of protocols and access control techniques involving proprietary hardware and software. A common feature, however, is the provision of 'gateways' to other communication systems ranging from simple asynchronous serial interfaces to industry-standard networks such as Ethernet, Token Ring and MAP.

A variety of hardware platforms are available which can be used as a basis for implementing the various levels of a batch control strategy. They include:

- Digital controllers.
- Personal computers/workstations (PCs).
- Programmable logic controllers (PLCs).
- Distributed control systems (DCSs).

Digital controllers include the 'single loop' controllers traditionally used in continuous applications. These can be used for blending applications in batch processing as well as for control of pressure, level, flow and temperature. Other examples are controllers dedicated to basic batch operations such as weighing and mixing.

PCs/workstations – industrialized versions of the familiar desktop models – are now readily available. They provide a powerful and cost-effective platform for supervisory control and data acquisition (SCADA) systems. Such PC-based systems appear very attractive for other real-time applications such as batch control. Apart from the familiarity and low cost of the hardware and systems software, applications are now available that make developing a control system (almost!) as easy as using a spreadsheet. However, a number of issues need to be taken into account, including:

- The capacity for expansion and reliability of hardware.
- The robustness of systems software.
- The integrity of the applications software.
- Long-term support for hardware and software.

A PLC is a solid-state control system designed and packaged for high reliability in an industrial environment. PLCs were originally designed as stand-alone replacements for electromechanical devices – relays, cam and drum sequencers – used for sequencing operations in manufacturing industries. The ability to handle large numbers of discrete I/O signals is one advantage that has been claimed in the past for PLCs. Another has been their capacity for implementing interlocking and sequencing. This led to the view that their primary role in batch process control was the control of individual items of mechanical plant equipment such as filters and driers. Limited capacity for regulatory or supervisory control and lack of suitable software and programming methods were amongst the disadvantages quoted as precluding their use for overall batch process control. These 'rules of thumb' no longer hold good. A number of vendors now offer PLCs as powerful as the processors in a distributed control system and capable of being networked in a similar way.

PLCs are also playing a growing role in another important area of batch process control – that of safety. They are being used in place of 'hardwired' relays to implement independent protection systems.

DCSs were first introduced in the mid 1970s for controlling continuous processes. Since then, they have developed significantly, reflecting the changes in the underlying technology – microprocessors, memory and storage, displays and communications – that have occurred.

The DCS concept aims to exploit not just this underlying technology but also to secure the advantages gained by distributing the functions of a control system throughout a process facility rather than concentrating them at a single central location. This flexibility makes them attractive as platforms for batch control where the structure of the system needs to mirror that of the process. Also in their favour are reductions in entry-level system size and cost coupled with increased functionality.

Ongoing developments make it more and more difficult to choose between the PC, PLC and DCS for implementing batch control on the basis of cost and functionality. Systems based on either platform are becoming more powerful, cheaper (to buy, run or maintain), more open and wider in scope. In the future, issues such as the supplier's industrial experience, application know-how and ability to offer life-cycle support are likely to be the deciding factors.

8.5.5 Software

The software required to implement computer-based batch control schemes can be divided into three categories, described below.

- System software – standard software provided by the system vendor. This includes regulatory and batch control packages as well as utilities such as editors, compilers, database generators, etc.
- Application software – generated by the end-user, the vendor or a third party to define the specific requirements for a particular batch process. Sequence logic, databases, displays, reports come in this category.
- Operating systems – again, supplied by the vendor. The operating system supervises and controls the execution of a combination of system and application software in real-time so as to control the process.

8.6 Procedural control

8.6.1 Introduction

A number of specifications will be generated during the course of a project aimed at implementing a computer-based batch control system. They will originate both from the purchaser and the supplier of the equipment. These specifications, their content and role in the 'life-cycle' of the project are discussed. The procedural control requirements of the system will form a major part of these specifications. Case studies [8] show that the specification of these requirements and their subsequent coding or configuration in a batch control language account for the majority of the resources used in developing applications software.

8.6.2 Identifying procedures

The first step is to identify the processing actions involved in the manufacture of each product. Information in the form of:

- The flowsheets and P&I diagrams for the process.
- The actual or proposed operating instructions.
- The knowledge and experience of personnel from production, engineering, quality control, maintenance and other relevant departments.
- Advice and assistance from system vendors.

can be collated and structured into the operations, phases and steps that make up the procedure for each product.

As an example, consider the simple batch plant shown in Figure 8.1. A simplified set of process instructions for the manufacture of a batch of product in this plant is set out in Table 8.1. A possible breakdown of the corresponding unit procedure into operations, phases and control steps is presented in Table 8.3. The control steps are not, however set out in detail (e.g. with tag numbers for instruments).

In a batch control system that is in line with the SP88 standard, a high-level description in terms of operations and phases is all that is required for recipe configuration by production personnel. During the production of an individual batch, the unit supervision activity maps these generic operations and phases into the corresponding operations and phases required for the control of the equipment actually assigned to the batch. These will be coded in a batch control language by applications specialists working from detailed low-level specifications.

8.6.3 Specifying procedures

Procedural control in batch processing is a specific example of sequencing. This involves a number of stages or 'states'. Progress from stage to stage – 'transition' – depends on events or actions which show that one stage is finished and the next can begin. In addition to sequences made up of stages in series, subsequences can occur in parallel; the simultaneous charging of two components called for in Table 8.1 is an example. This implies the need to co-ordinate the completion of subsequences in some way before proceeding with the rest of the sequence.

A number of tabular and graphical methods are available for representing sequences in these general terms. They include:

- State charts and decision tables.
- State transition diagrams.
- Petri nets.
- Sequential function charts.
- Flowcharts.

Another general method involves the use of 'structured English' or 'pseudo-code' to produce a written rather than a graphical specification.

These methods of specification are reviewed by Fisher [9], Rosenof and Ghosh [3] and Mallaband [10]. Three methods emerge as most suitable for use in specifying sequences in batch processing:

- Flowcharts.
- Sequential function charts.
- Structured English.

Flowcharts are widely used to document sequences. However, they have a number of drawbacks. Amongst those quoted [3, 9, 10] are:

(i) It is difficult to ensure a consistent style and level of detail.
(ii) They are difficult to read.
(iii) They are inefficient for use as checkout tools.
(iv) Considerable effort is required to draw them up and to maintain them.
(v) Parallel, dependent sequences are difficult to document.

It is possible to overcome these difficulties and to formalize, standardize and structure flowcharts. The measures suggested [3, 9, 10] include the following:

(i) A standard set of symbols is used, such as ISO Standard 5087.
(ii) Flowcharts are supplemented by recipes, plant unit descriptions, etc., to form a complete batch/sequence specification.
(iii) Flowcharts are drawn up at several levels, corresponding to the operation/phase/control step hierarchy of the procedural control model.
(iv) Parallelism is allowed for where necessary and shown on flowcharts by the use of appropriate symbols (the ISO standard provides for such a symbol – a double horizontal line).
(v) Alarms and failure actions are flowcharted separately from normal operations.
(vi) Flowcharts are supplemented by structured English, logic diagrams, etc., when it is more appropriate to specify detailed requirements by another method.
(vii) Use is made of the software packages now readily available for developing flowcharts on PCs and workstations.

The examples of flowcharts shown in Figure 8.9(a–c) correspond to the sequences of operations, phases and control steps set out in Table 8.3.

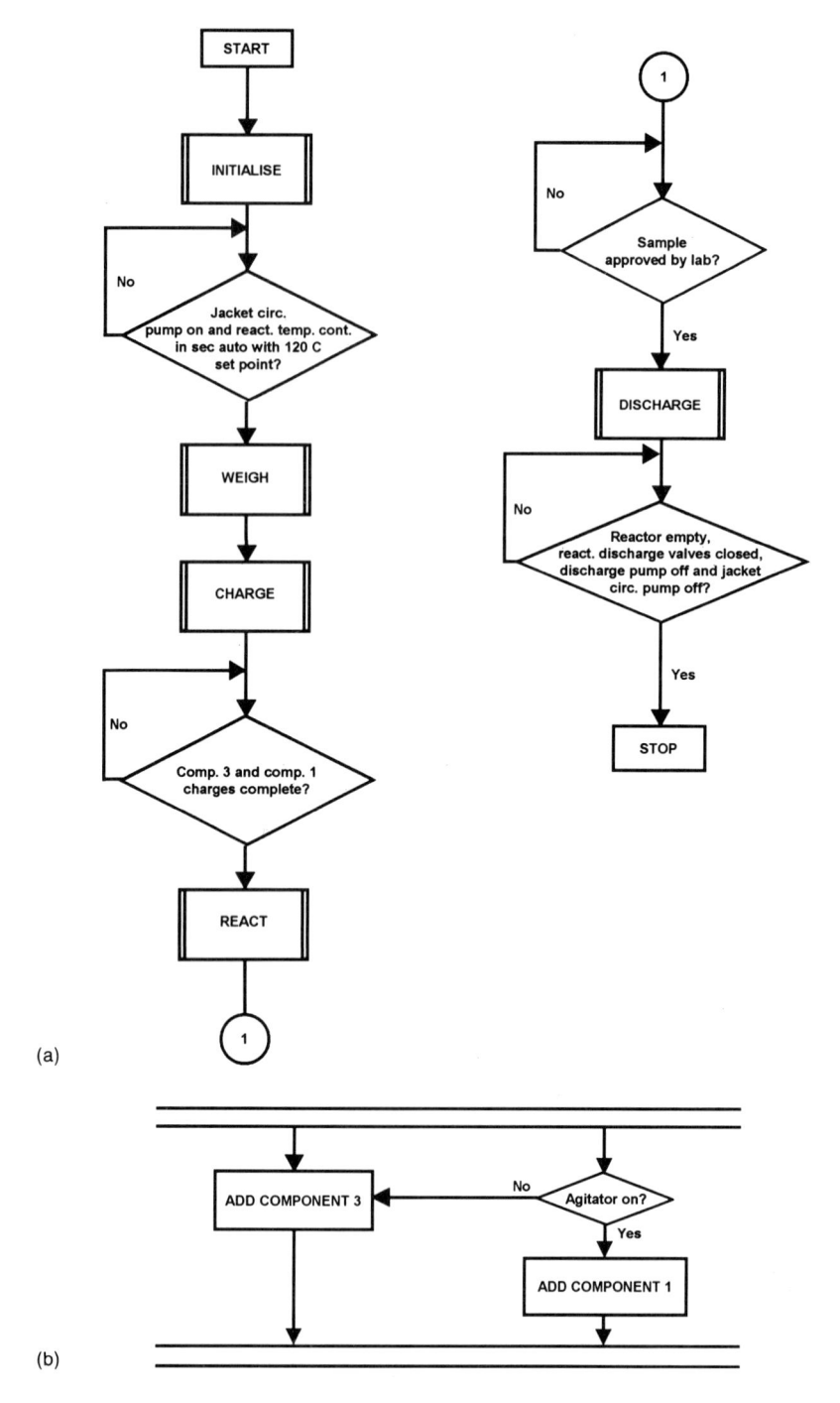

Figure 8.9 (a) Flowchart for operations. (b) Phase flowchart for CHARGE operation ('double line' symbol represents synchronization of parallel operations). (c) Control step flowchart for 'add component 3' phase. (Redrawn with permission from Sawyer [8], © 1993 IChemE.)

CONTROL

[Flowchart: Initialise → All interlocks satisfied? (No → Print message 'interlocks not satisfied') → Open outlet valve from weigh tank → Enough of Comp. 3 charged to cover agitator? (No loops back) → Yes → Start agitator → Amount of comp. 3 charged = preset? (No loops back) → Yes → Close outlet valve from weigh tank]

(c)

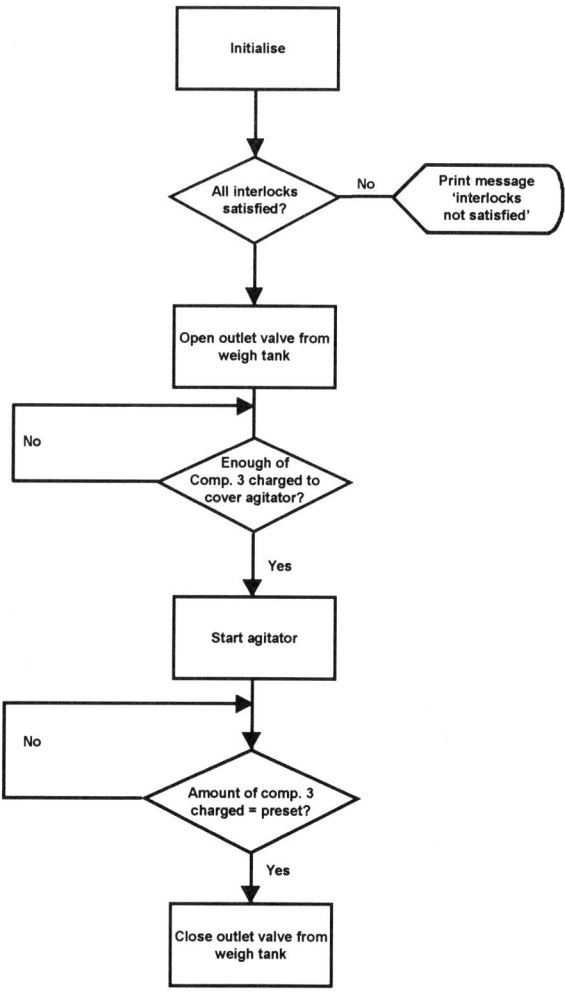

Figure 8.9 Continued.

Sequential function charts have been developed specifically for describing sequential control systems. They do so in terms of three basic elements: steps, transitions and directed links. Steps represent commands or actions that are either active or inactive; each step is followed by a conditional transition. If the condition is satisfied control passes to the next step which becomes active; the previous step then becomes inactive. Directed links tie steps and transitions together to form sequences.

The standard symbols and conventions used to construct sequential function charts are shown in Figure 8.10(a). They can be used to define simple serial sequences and also to allow for alternative or parallel paths and loops as shown in Figure 8.10(b).

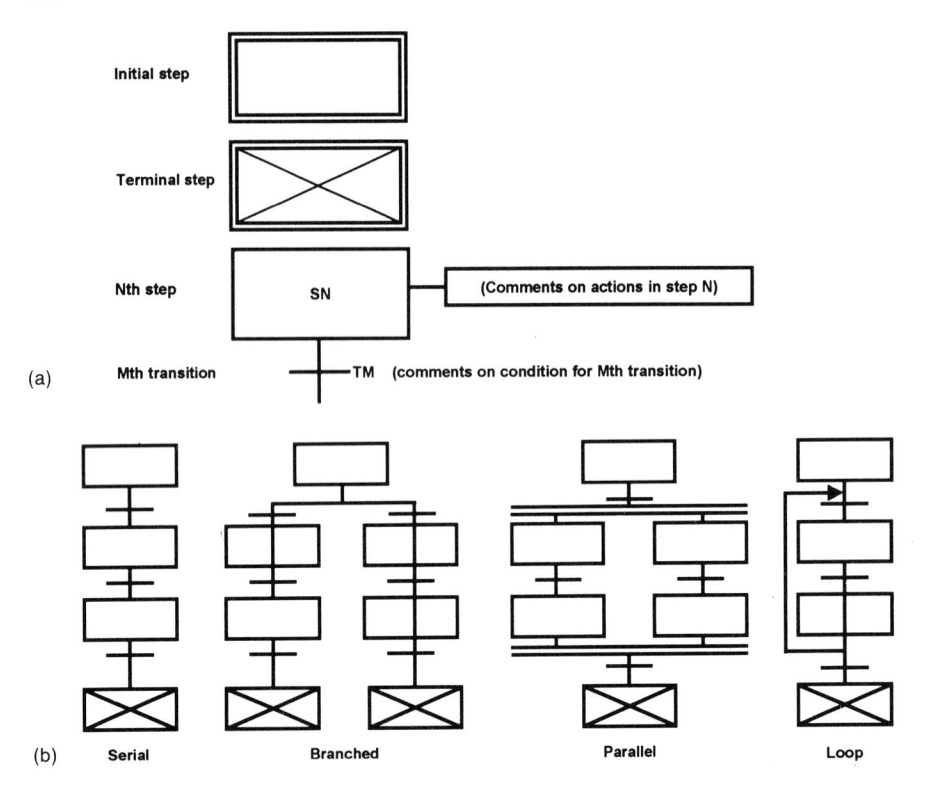

Figure 8.10 (a) Basic symbols for sequential function charts. (b) Sequential function chart structures. (Redrawn with permission from Sawyer [8], © 1993 IChemE.)

Sequential function charts have considerable potential for use in specifying sequences in batch processing [9]:

- They represent sequences graphically.
- They have an in-built structure which caters for hierarchy and parallelism.
- They are flexible enough to handle procedure changes in recipes.
- They are defined by an international standard (IEC 848).
- Like flowcharts, they can be generated by software on PCs and workstations.
- Vendors of distributed/PLC control systems offer facilities for the direct generation of sequence control software configured using such charts.

The examples of sequential function charts in Figures 8.11(a–c) correspond to the operations, phases and steps of Table 8.3.

Figure 8.11 (a) Sequential function chart for operations. (b) Sequential function chart for CHARGE operation. (c) Sequential function chart control steps in 'add component 3' phase. (Redrawn with permission from Sawyer [8], © 1993 IChemE.)

CONTROL

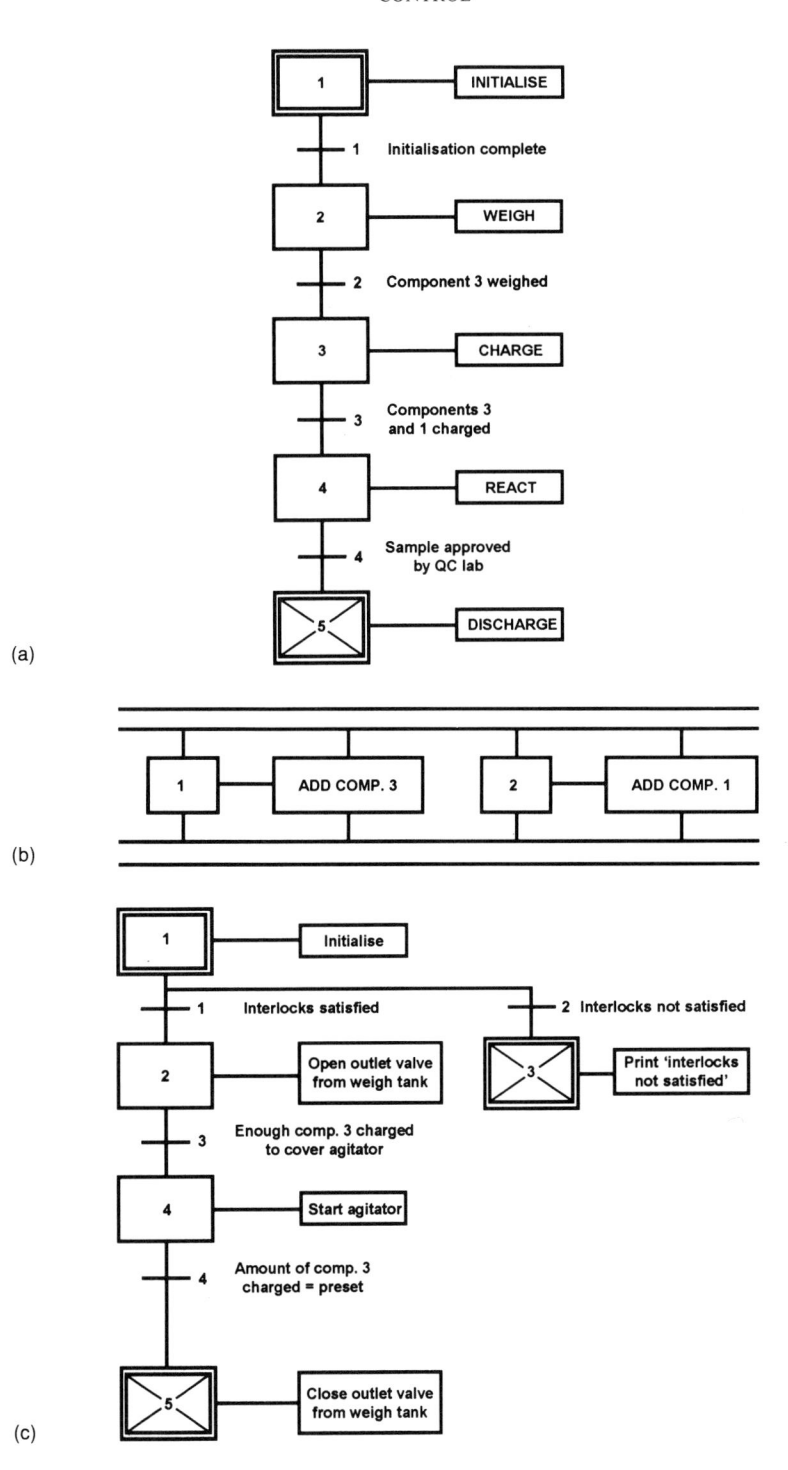

(a)

```
[1] ── INITIALISE
 │
─┼─ 1  Initialisation complete
 │
[2] ── WEIGH
 │
─┼─ 2  Component 3 weighed
 │
[3] ── CHARGE
 │
─┼─ 3  Components 3 and 1 charged
 │
[4] ── REACT
 │
─┼─ 4  Sample approved by QC lab
 │
[5] ── DISCHARGE
```

(b)

```
[1] ── ADD COMP. 3        [2] ── ADD COMP. 1
```

(c)

```
[1] ── Initialise
 │
─┼─ 1  Interlocks satisfied ────────── 2  Interlocks not satisfied
 │                                      │
[2] ── Open outlet valve              [3] ── Print 'interlocks
         from weigh tank                        not satisfied'
 │
─┼─ 3  Enough comp. 3 charged
 │     to cover agitator
 │
[4] ── Start agitator
 │
─┼─ 4  Amount of comp. 3
 │     charged = preset
 │
[5] ── Close outlet valve
         from weigh tank
```

Structured English describes logic in plain language but with certain rules and conventions. These have yet to be standardized – one disadvantage of the method – and those stated below are based on the recommendations of Fisher [9] and Rosenof and Ghosh [3].

Statements preceded by five hyphens are comment statements. Statements preceded by a single hyphen are a continuation of the line above. This may be a continuation of either a comment (or other non-executable statement) or an executable logic function. Subroutines are identified by labels with a leading double asterisk. The label marks the entry point to the subroutine. Conditional branching in the logic uses the IF, THEN, ELSE format. Multiple levels of branching are allowed and indentation is used to indicate this. Operator messages are shown in quotation marks. AND and OR functions can be used in the logic; NAND and NOR functions should be avoided as they are not as easily understood.

Rosenof and Ghosh [3] give some general guidelines for writing in structured English:

- Use simple statements.
- Clearly define the required function in a statement.
- Identify the plant hardware address wherever possible.
- Indent where necessary.
- Avoid negative logic.
- Avoid excessive nested logic.

A structured English description of the CHARGE operation in Table 8.3. is set out in Table 8.4.

The choice of a method for specifying sequences needs to be made according to its suitability for the primary purpose – that of forming a basis for the development of control software. However, two subsidiary but important purposes also need to be taken into account. The specification should be readily usable both as a process description at all stages of a project and as an operating guide once a process is in operation. This means that it must be in a form that is understood by people who are not computer programmers or who have no control systems background. The method should also [9]:

(i) Clearly show the overall sequence while maintaining the operation/phase/control step hierarchy.
(ii) Represent both combinatorial and sequential logic.
(iii) Be able to represent discrete variables, describe alarm status, compare analogue variables and switch control instrumentation modes.
(iv) Specify sequences briefly, intuitively and accurately.
(v) Produce a description that can be used 'as is' to code application software in a batch sequencing language.
(vi) Be self-documenting and easy to maintain.
(vii) Adaptable to generation on a engineering workstation or PC.

Table 8.4 Structured English description of the CHARGE operation of the model process

```
** CHARGE OPERATION
-----CHARGE OPERATION MAIN LOGIC
 -THIS OPERATION CHARGES COMPONENTS 3 AND 1 TO THE
 -REACTOR
 -THE AMOUNT OF EACH COMPONENT IS A RECIPE VARIABLE
** ADD COMPONENT 3 PHASE
 OPEN OUTLET VALVE FROM WEIGH TANK
 IF REACTOR LEVEL SWITCH IS ACTUATED
   THEN START THE AGITATOR
 IF AMOUNT OF COMPONENT 3 CHARGED EQUALS PRESET
   THEN CLOSE OUTLET VALVE FROM WEIGH TANK
** ADD COMPONENT 1 PHASE
 RESET COMPONENT 1 FLOW TOTALISER TO ZERO
 IF REACTOR AGITATOR IS ON
   THEN OPEN OUTLET VALVES FROM HEAD TANK TO FLOWMETER
   -AND FROM FLOWMETER TO REACTOR
 IF COMPONENT 1 FLOW TOTAL EQUALS PRESET
   THEN CLOSE OUTLET VALVES FROM HEAD TANK TO
   -FLOWMETER AND FROM FLOWMETER TO REACTOR
 IF AMOUNT OF COMPONENT 3 CHARGED EQUALS PRESET
   -AND COMPONENT 1 FLOW TOTAL EQUALS PRESET
     THEN GOTO **REACT OPERATION
```

8.7 Acknowledgements

This chapter is based on material that first appeared in *A Guide To Computer-Controlled Batch Processing* (Sawyer, 1993). I am grateful to the Institution of Chemical Engineers for allowing me to quote extensively from Chapters 1, 2 and 3 of this publication. I am also indebted to the International Society for Measurement & Control (ISA) for their permission to quote from the ISA SP88 Standard (ISA, 1995). Other sources are acknowledged in the text or cited as references. Finally, thanks are due to Mr Chris Hart for his skill and patience in preparing the illustrations.

References

1. ISA (1995) *ISA-S88.01-1995 Batch Control Part 1: Models and Terminology.* ISA, Raleigh, NC.
2. Mehta, A. (1983) *Chem. Engng Progr.*, October, 47–52
3. Rosenof, H.P. and Ghosh, A. (1987) *Batch Process Automation – Theory and Practice.* Van Nostrand Reinhold, New York.
4. Love, J. (1987) *Chem. Eng.*, June, 34–35.
5. Love, J. (1988) *Chem. Eng.*, April, 24–26.

6. Carlo-Stella, G. (1982) *Intech*, March, 35–39.
7. Simmons, A. (1989) *Process Engng,* May, 47–51
8. Sawyer, P.E. (1993) *Computer-controlled Batch Processing.* IChemE, Rugby.
9. Fisher, T.G. (1990) *Batch Control Systems – Design, Application and Implementation.* ISA, Raleigh, NC.
10. Mallaband, S. (1988) In *Advances in Process Control II.* IChemE, Rugby.

9 Hazards from chemical reactions and flammable materials in batch reactor operations
R. ROGERS

9.1 Introduction

Many of the chemicals used in industry can present safety, health and environmental problems, particularly if either their inherent hazards or those arising from specific operations have not been identified, evaluated and a basis for safe operation of the process developed and implemented. This chapter is primarily concerned with chemical reaction and fire and explosion hazards that may arise during batch or semi-batch processing. The underlying cause of such hazards is associated with energy release and is thus intrinsically linked to the balance between the rate of heat generation, be it from a runaway reaction or an incipient flame kernel in a flammable mixture, and the rate of heat loss.

Safety in chemical production requires not only a knowledge and understanding of the hazardous properties of the materials being handled but also an appreciation of how these are affected by the equipment or process in which they are being used. In contrast to physical chemical properties, hazard data are often not inherent characteristics of a substance, but rather depends on and are changed by the interaction of the substance with the plant situation. This influences not only the experimental methods used to measure these properties but also the interpretation of the data obtained and its relationship to plant-scale operations.

The hazards arising during the operation of reactors can arise from two main areas:

- Uncontrolled chemical reactions, i.e. chemical reaction hazards.
- Ignition of flammable atmospheres, i.e. operational hazards.

It is important that any possible operational or chemical reaction hazard that could arise is considered at an early stage of reactor design. This is not only to ensure provision of an effective basis of safety, but also in order to design a reactor system such that the hazard is prevented from occurring. This minimizes other expensive additional systems needed to protect against the consequence of the hazard.

The hazards from batch reactor operation can arise both from the hazardous properties of the materials being handled, e.g. flammability, explosibility and also from the reactions themselves. A systematic procedure is therefore required to identify and evaluate any potential hazards and also to develop a safe method of operation or 'basis of safety'.

9.2 Hazard identification

Batch chemical manufacture usually involves the production of relatively small quantities of (kilograms to hundreds of tonnes per year) of chemicals that are designed and produced for their effect, e.g. pharmaceuticals, agrochemicals, dyestuffs, etc. The process chemistry is often complex, sometimes incompletely understood and requires multistage synthesis. This invariably leads to a high frequency of change as processes are improved and developed and, in order to be economic, the plants are necessarily reusable or multiproduct. Naturally, to meet marketing requirements a rapid response to change is required. A systematic procedure to ensure safe manufacturing in a manner that is compatible with production, engineering, financial and marketing requirements needs to be used. The systematic procedure should assess the hazards that may arise from both chemical reactions and process operations. The essential stages of a hazard assessment include:

- Characterization of materials.
- Identification of sources of hazard.
- Assessment of the risk in the specific manufacturing process.
- Definition/design of safety measures most appropriate to the process.

In response to this problem ICI Fine Chemicals Manufacturing Organization developed such a procedure over a period of 10–15 years [1–3]. This procedure, which is carried out in addition to and interrelates with other hazard studies, e.g. hazard and operability studies [4], provides a technical evaluation and understanding of the hazards of a process and ensures that safety measures are precisely but simply specified and implemented.

The systematic assessment procedure relies on three key elements: a defined procedure, selective technical examination and specification of a 'basis of safety':

- *A defined procedure.* This specifies the what, when and how of the assessment and details the responsibilities for the various stages.
- *Selective examination.* This involves the technical assessment of the hazards of the process leading to a specified safe envelope or boundary for its operation.
- *Specification of a basis of safety.* The basis is specific to the proposed plant and operation conditions.

This assessment procedure is applicable to the evaluation of both chemical reaction and fire and explosion hazards of batch reactor operation.

9.2.1 The defined procedure

The essential stages in the procedure are show in Figure 9.1. It relies on the clear definition of responsibilities for initiating the assessment, evaluating the hazards, implementing the recommendations, and monitoring and maintaining the specified safety measures.

Naturally, the different stages in the procedure cannot be carried out in isolation as each stage must take account of the other considerations and interact with them. In addition, a framework has to be developed to ensure that the procedure is carried out. Potential hazards must be identified and appropriate actions are taken to ensure safety at all stages in the life of a process, i.e. from initial development work through to pilot-plant and full-scale production, including when any modifications are made to the plant/process.

Examination of incidents that have occurred during chemical manufacturing operations shows several common factors. These have been taken into account in developing the procedure. The key points which emerge are the following.

- Every process at every scale needs to be examined. Incidents are not the prerogative of large-scale plant manufacture but often also occur in laboratories and semi-technical/pilot-plant units.
- All process changes and modifications should be assessed.

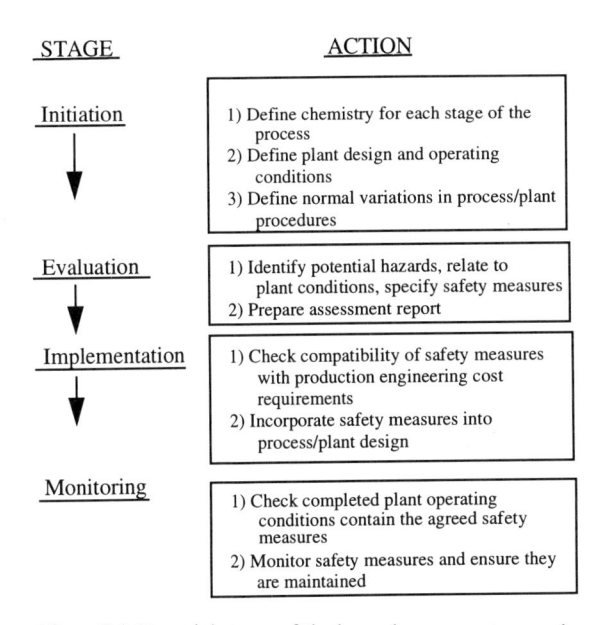

STAGE	ACTION
Initiation ↓	1) Define chemistry for each stage of the process 2) Define plant design and operating conditions 3) Define normal variations in process/plant procedures
Evaluation ↓	1) Identify potential hazards, relate to plant conditions, specify safety measures 2) Prepare assessment report
Implementation ↓	1) Check compatibility of safety measures with production engineering cost requirements 2) Incorporate safety measures into process/plant design
Monitoring	1) Check completed plant operating conditions contain the agreed safety measures 2) Monitor safety measures and ensure they are maintained

Figure 9.1 Essential stages of the hazard assessment procedure.

- The whole process needs to be examined including raw materials, the effect of materials of construction and the means of effluent disposal. The methods proposed for dealing with unsuccessful batches and clean out after campaigns must not be forgotten.
- There is an intrinsic link between the process specification and hazard evaluation. It has to be recognized that it is not feasible to assess every conceivable abnormality, maloperation or change which could occur in a process. The hazard evaluation must, however, take into account credible maloperations. More importantly this boundary or envelope of credible maloperations must be recognized. Similarly, it is not sufficient to know that the process operates safely as specified if a minor deviation, which could lead to a runaway situation or the ignition of a flammable material, has not been identified.
- The safety measures must be precisely but simply specified and recorded and they must be implemented and maintained.

Hazards can occur at any stage of a process life from laboratory work through to full-scale manufacture. The defined procedure must take this into account and must specify responsibilities to ensure (and promote an awareness) of safety at all stages.

Naturally the amount of experimental testing and the depth of hazard assessment or investigation into possible hazards depends on the scale of the process. There is little justification for accurately measuring the rate of temperature rise of a runaway reaction if the process is only being operated at the gram scale in the laboratory. Hence, a differing degree of hazard assessment is required at the different stages in process development.

Laboratory scale. To ensure safety at the laboratory scale it is usually sufficient for the chemist responsible to ensure that the materials being handled will not detonate or deflagrate and that the process does not involve a violently exothermic reaction. However, a more detailed evaluation of the potential chemical reaction hazards is often valuable at this stage to aid in route selection and process development.

Pilot plant. Once the process leaves the laboratory and is carried out at the pilot-plant scale it is essential that a formal hazard assessment is carried out. The depth of investigation required will depend on a variety of factors including scale, familiarity with the type of process and the level of competent supervision. As a minimum the assessment at this stage should demonstrate that the process can be operated safely both as defined and with the normal variations which could occur in practice, e.g. variations in temperature.

Normal manufacture. The assessment of the chemical reaction hazards associated with full-scale production should start at an early stage of the project. This early identification of potential hazards provides an opportunity for the process and plant design to be changed in order to design out the hazard where practicable, and achieve an inherently safe process. Minor changes are often all that are needed if the hazard is identified at this stage. In contrast, if the assessment is not carried out until the design is fixed or construction has started, expensive add-on protective systems are often the only way to achieve safe operation. Where a formal assessment procedure has been adopted, information will already be available from the laboratory and pilot-plant stages. This will need to be re-evaluated and further testing may well be required to relate the data to the specific plant and operating conditions that occur at this scale of manufacture.

In addition, the effect of any maloperations or incidents (such as agitation failure, rapid charging, loss of cooling, extended reaction times, etc.) that have not previously been considered should be evaluated. The integration of the process with other manufactures may also need to be considered.

Modifications. Many incidents that occur in the chemical industry follow modifications that have been poorly or incompletely thought through. It is essential, therefore, that any assessment procedure includes a provision to ensure that all modifications are assessed. If a hazard assessment of the plant and process is available and the modification does not invalidate the basis for safe operation, then the formal recording of this decision and the reasons is all that is required. Alternatively, a detailed evaluation may be necessary as well as alternative or additional recommendations to ensure safe operation.

A key aspect of the procedure is the formal recording of the results of the investigations and the recommendations/conclusions that have been made to ensure process safety. The report should be circulated to all relevant parties (e.g. production and safety departments) and a copy kept with the process file. The contents of the hazard assessment report will naturally vary with the type of assessment, e.g. with laboratory-scale work a simple statement may be all that is required. However, in order to ensure that all relevant factors are considered a standard format is recommended.

9.3 Chemical reaction hazards

The evaluation of chemical reaction hazards requires the identification of potential thermal instability in the reactants, reaction masses and products; the measurement of the heat of reaction; and the detection of any gas evolution. The key aspect in any hazard evaluation is the relationship between the experimentally determined data and the conditions that actually occur on a plant scale [3, 5]. Unfortunately, no single experimental technique is able to produce all the data required to specify safe operating conditions. In most investigations some form of small-scale thermo-analytical method is generally used for the initial detection of exothermicity and/or gas generation. However, the use of such data is limited by the sensitivity of the techniques which range from 0.5 to 5 W kg^{-1}. In comparison, a typical value at which heat generation can be observed in a stirred 500 gallon reactor with a filled jacket is from 0.04 to 0.08 W kg^{-1}. It can be seen, therefore, that small-scale tests can only be used to give an indication of the possibility of exothermic activity and an approximate value of the minimum onset temperature. They are useful for quickly screening large numbers of reactions, but in order to specify safe operating conditions either the results from such small-scale tests have to be scaled or extrapolated, or more sensitive tests that more accurately simulate the plant-scale heat loss conditions have to be used.

The consequences of violent exothermic runaway reaction can be as severe as those from the ignition and explosion of a fuel–air mixture. Chemical hazards principally arise from:

- Thermal stability of reactant mixtures and products.
- Rapid exothermic reactions that can raise the temperature to the decomposition temperature or cause violent boiling of the reactants.
- Rapid gas evolution that can pressurize and possibly rupture the plant.

These chemical hazards must be considered in assessing whether a particular chemical process can be operated safely on the manufacturing scale. A knowledge of the heat change associated with the desired reaction, ΔH_r, and information about thermal stability are essential to evaluate the hazards. The thermal stability may be characterized by the temperature at which any decomposition reaction may occur on the plant scale and the magnitude of that decomposition. In addition to the above, the consequences of possible process maloperation must be considered, e.g. overcharging or omitting one of the reactants, agitation failure or poor temperature control.

A number of inter-related parameters govern the hazards associated with a process. These include:

- Chemical constitution.
- Process temperature.
- Process pressure.
- Thermochemical characteristics of the reaction.
- Reaction rate.
- Reactant ratios.
- Solvent effects.
- Batch size.
- Operational procedures.

The assessment of the hazards of a particular process requires a technical examination of the effects of these parameters by suitable experimental work, the interpretation of the results in relation to the plant situation and the definition of a suitable basis for safe operation.

9.3.1 Thermal explosions

The occurrence of a runaway reaction or thermal explosion depends not only on the rate of heat generation from a chemical reaction but also on the rate of heat loss from the system. As mentioned above, it is therefore not possible to determine and assign a stability temperature to a substance as one can with melting points or flash points. A material which is stable at some temperature in one situation may runaway from the same temperature if the system, in particular the rate of heat loss, changes.

There are two extreme cases which can be considered in describing heat loss from a system (Figure 9.2). In the first, originally discussed by Semenov [7], the temperature is assumed to be uniform throughout the reactant mass. This situation occurs in gaseous and well stirred liquid systems where the rate of heat loss is governed by heat transfer at the boundary.

In the second case, considered by Frank-Kamenetskii [8] the temperature distribution is non-uniform and heat loss is controlled by heat transfer through the bulk. This occurs in large, unstirred liquid masses, powders and solids.

Chemical reactions are often carried out in the liquid phase and the Semenov heat transfer conditions can be applied as follows. Semenov assumed a pseudo-zero-order exothermic reaction following an Arrhenius type rate law, that is the rate of reaction (and therefore the rate of heat production) increases exponentially with temperature. Thus, for an irreversible reaction $A \rightarrow R$ at constant volume, V, the rate of heat production, Q_r, is given by

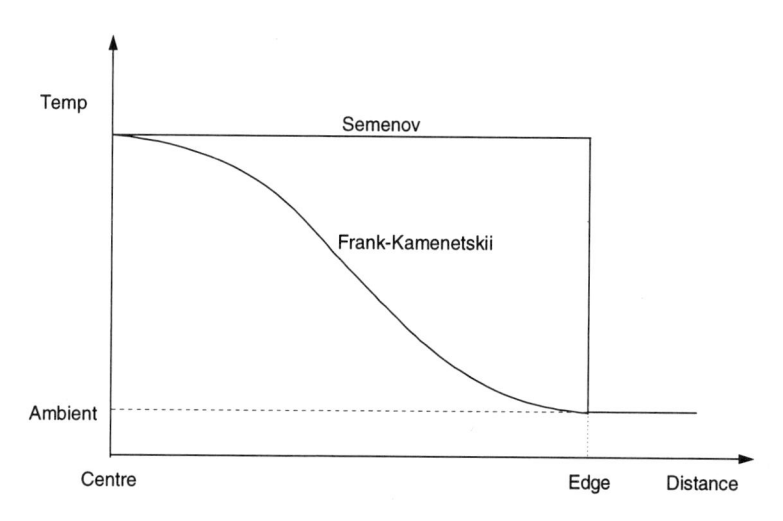

$$Q_r = V(-\Delta H_r)k_0 C_A \exp(-E_a/RT) \tag{9.1}$$

Figure 9.2 Semenov and Frank-Kamenetskii temperature profiles.

where ΔH_r is the heat of reaction, C_A is the initial concentration which is assumed to remain constant for a limited time and k_0 is the initial rate constant for the reaction with activation energy E_a.

The rate of heat loss, Q_c, is assumed to be governed by Newtonian cooling, i.e. it is linearly dependent on the temperature difference, the heat transfer coefficient, U, and area, A:

$$Q_c = UA(T - Ta) \qquad (9.2)$$

Three cases for difference ambient coolant temperatures can be identified (Figure 9.3).

In the first case the rate of heat loss (line 1) intersects the exponential heat production curve at two points A and B where the chemical heat production rate is balanced by the heat removal capacity. The low temperature point A represents a stable situation which can be illustrated by considering an increase in temperature to point C. At this temperature the rate of heat loss is greater than the rate of heat production and the temperature will return to point A. In contrast, point B is unstable as any slight increase in temperature will cause an increase in the rate of heat production not matched by the rate of heat loss and an accelerating runaway will occur. Line 3 represents the situation where the rate of heat loss from the system is always less than the rate of heat production and a runaway reaction will always occur. Line 2 describes the critical situation where the heat production is just equal to the heat removal.

Since the rate of heat loss is dependent on the heat transfer coefficient and area, a decrease in either of these will lead to a decrease in the slope of the line and a reduction in the rate of heat loss from the system.

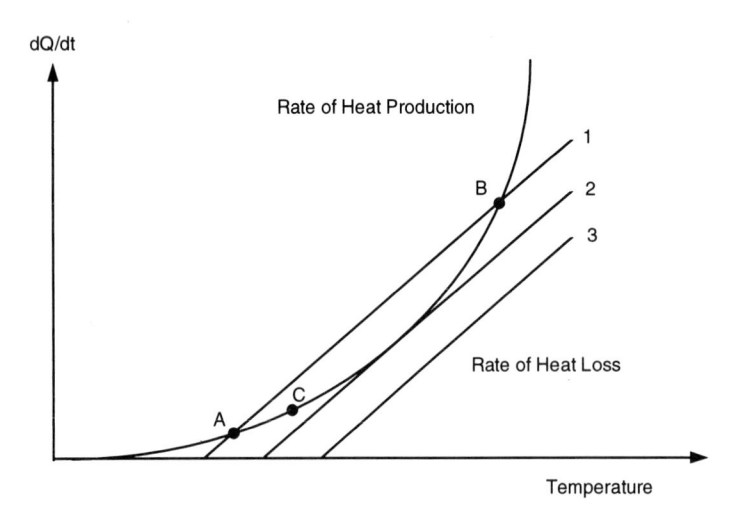

Figure 9.3 Heat balance for Semenov-type systems.

HAZARDS IN BATCH REACTOR OPERATIONS

Table 9.1 Typical rates of heat loss for reactors and laboratory equipment

		Heat loss (W kg^{-1} h^{-1})	$t_{1/2}$ Std. (h)	Time for 1°C fall in temperature with $\Delta T = 60°C$
Reactor	(2.5 m^3)	0.054	14.7	21 min
Reactor	(5 m^3)	0.027	30.1	43 min
Reactor	(12.7 m^3)	0.020	40.8	59 min
Reactor	(25 m^3)	0.005	161.2	233 min
Glass tube	(10 ml)	5.91	0.12	11 s
Glass flask	(100 ml)	3.68	0.19	17 s
Glass Dewar	(1 l)	0.018	43.3	62 min

Typical natural rates of heat loss from different sized reactors (i.e. without jacket cooling) are given in Table 9.1.

Table 9.1 also shows the large difference between rates of heat loss from industrial-scale reactors and laboratory equipment. This causes many problems during process scale-up. Reactions carried out in the laboratory are often believed to be non-exothermic and therefore to pose no hazard when operated on the plant.

As an example, consider an exothermic reaction with a heat of reaction ΔH_r of -34 kJ mol^{-1} which is to be operated as a semi-batch process, i.e. the second reactant is added over a period of 1 h. Assuming a relative molecular weight of 200, the heat of reaction can be expressed as 170 kJ kg^{-1}. This will result in an adiabatic temperature rise of 100 K assuming a specific heat C_p of the reaction mixture of 1.7 kJ kg^{-1} K. This temperature increase will occur over the 1 h addition period, thus the rate of temperature rise will be 100 K h^{-1}.

Table 9.2 shows the difference in behaviour between the laboratory and what would happen when the reaction is carried out in a 2.5 m^3 reactor on the plant.

The rate of heat production per unit reaction mass (which is scale independent) is the same in both cases. It can be seen that the rate of heat loss when the reaction is performed in the laboratory is greater than the rate of heat production and therefore the reaction will need to be heated to maintain

Table 9.2 Heat losses from an exothermic reaction at small and large scale

	Laboratory (100 ml)	Plant (2.5 m^3)
Heat loss (W kg^{-1} h^{-1})	3.68	0.054
Reaction temperature 80°C		
Reaction time 1 h		
Rate of heat loss (°C h^{-1})	210	2.8
Rate of heat production (°C h^{-1})	100	100
	Needs HEATING	Needs COOLING

the desired temperature of 80°C. However, on the plant scale the rate of heat loss is much smaller and is far less than the rate of heat production of the reaction. Thus when operated on the plant, a runaway reaction will occur unless the reactor is cooled.

9.3.2 Characterization of the desired reaction

As has been described above, the key parameter in evaluating the hazards of the primary or desired chemical reaction is its heat of reaction, ΔH_r. An initial estimate of ΔH_r can be obtained from the literature or calculated from the heats of formation (ΔH_f) of the reactants and products according to Hess's law:

$$\Delta H_r = \sum \Delta H_{f_{products}} - \sum \Delta H_{f_{reactants}} \qquad (9.3)$$

The exothermic nature of many industrial reactions is illustrated by typical values for their heats of reaction given in Table 9.3.

The heat release of a specific reaction will often have to be measured and the following calorimetric methods may be used to determine the heat of reaction:

- Differential thermal analysis DTA/DSC.
- Dewar calorimetry.
- Isothermal reaction calorimetry, i.e. heat flow or heat balance calorimetry.

Differential thermal analysis should only be used when the reaction components form a homogeneous mixture at room temperature or lower and when the reaction begins at a higher temperature.

Dewar calorimetry is one of the most sensitive and absolute calorimetric methods and is particularly suitable for batch processes, i.e. where all the reactant components are charged at the beginning of the reaction [9].

Table 9.3 Typical values of the heat of reaction for some common industrial reactions [6]

Reaction	Typical ΔH_r (kJ mol^{-1})
Diazotization	−65
Sulphonation (SO$_3$)	−105
Nitration	−130
Epoxidation	−96
Hydrogenation (nitroaromatic)	−560
Amination	−120
Neutralization (HCl)	−55
Neutralization (H$_2$SO$_4$)	−105
Heat of combustion (hydrocarbons)	−900
Diazo decomposition	−140
Nitro decomposition	−400

As no heat is lost from the Dewar when it is operated in the adiabatic mode, the heat of reaction is directly proportional to the temperature rise measured.

Isothermal reaction calorimetry has the advantage that it uses a normal laboratory reactor and is particularly suitable for use with semi-batch processes. Calibrations must be carried out to measure either the heat flow between the reactor contents and the cooling/heating medium or the heat change occurring in the cooling/heating medium in heat balance calorimetry. The heat of reaction is then determined by integrating the heat release curve obtained.

(a) Reaction kinetics

It is not necessary to have a complete description of the formal kinetics of a chemical reaction in order to evaluate its potential reaction hazards. However, it is important to determine whether, particularly for a semi-batch process, the reaction proceeds rapidly or whether reactant is accumulated during the dosing period. Similarly sufficient data must be obtained to assess the effect of possible maloperations, e.g.

- Loss of agitation.
- Temperature/pressure deviations.
- Reactant charging errors – omissions/over charging/wrong order.
- Extended reaction times.

Such data can be obtained from reaction calorimetry experiments and need to be interpreted and related to the particular plant situation.

(b) Measurement of gas evolution

The rate of gas evolution during the normal process and under any envisaged maloperations is required to ensure adequate vent and/or scrubber sizing. The rate of gas evolution is not dependent on scale. Data obtained from small-scale experiments can therefore be directly related to the plant scale.

9.3.3 Characterization of exothermic decomposition reactions

(a) Chemical composition

Certain chemical groups are known to reduce the stability and possibly confer explosive properties on a compound [10]. These include the groups listed in Table 9.4.

However, not all organic compounds containing, for instance, nitro groups and nitrate esters possess explosive properties. The possession of such properties is dependent on the oxygen balance. This is a measure of a compound's inherent 'self-oxidation' ability and can be calculated, ignoring

Table 9.4 Chemical groups associated with chemical instability and/or possible explosive behaviour

Aromatic Nitro	Azo	Hypochlorite
Aliphatic Nitro	Azide	Chlorate
Nitrate ester	Peroxide	Perchlorate
Nitramine	Ozonide	Acetylenic

any atoms other than C, H and O, from the substance's empirical formula as follows.

$$C_aH_bO_c + (a + b/4 - c/2)O_2 \Rightarrow aCO_2 + b/2H_2O \qquad (9.4)$$

The oxygen balance is then given by:

$$\text{oxygen balance} = -[1600(2a + b/2 - c)/\text{mol. wt}] \qquad (9.5)$$

Compounds that have oxygen balances greater than -100 are likely to be detonating explosives and those with balances between -100 and -150 may show detonation properties under severe confinement. Compounds with oxygen balances less than -200 are not likely to possess explosive properties though they may still be thermally unstable.

If the presence of one of the groups listed above or an oxygen balance of more positive than -200 suggests that a reactant or the reaction mixture may possess explosive properties then in addition to the evaluation of its thermal stability, it should be tested for explosive properties [11].

(b) Thermal stability
The thermal stability of a reactant or reaction mixture gives a measure of the maximum temperature at which a process can be operated. It can also be used to determine the effects of adding or omitting a solvent, varying the reactant ratios and consequences of possible process maloperations such as overcharging or omitting one of the reactants.

Some form of small-scale scanning calorimetry is generally used for the initial detection of any decomposition exotherm and gas generation. DSC or DTA using pressure resistant sealed sample cells can be used but these techniques may be limited by their small sample size (i.e. milligrams) and the difficulty of obtaining a representative sample.

Experimental determinations of thermal stability can be made by heating about 10 g of the mixture under test in a glass tube sealed with a pressure transducer and fitted with a re-entrant thermocouple pocket, so that the temperature at the centre of the sample can be recorded continuously. Materials of plant construction should be added to the sample which is then typically heated at $2°C \, \text{min}^{-1}$ in an electric furnace.

The exotherm onset temperature is dependent on the sensitivity of the equipment but on a 10–20 g scale exothermicity can generally be detected at a self heating rate of 2–10°C h^{-1} or approximately 3–10 W l^{-1}. This means that self heating in a 5 m^3 vessel will occur at a temperature approximately 60–100°C lower than that observed in the small-scale test provided that there is no induction period for the decomposition.

Depending on the results from the screening tests (i.e. exotherm size, proximity of decomposition onset temperature to process temperature) secondary testing may be required to determine more accurately:

- The minimum temperature above which the reactor will be unstable on the scale used and the time available to instigate safety measures.
- The consequences of the exotherm – heat of reaction, adiabatic temperature rise/pressure developed/venting requirements.

Such secondary testing usually involves some form of adiabatic calorimetry in order to minimize the heat loss from the sample during the test and therefore to detect the low rates of generation which may occur on the plant scale. These can be less than 1 W l^{-1} or 1–2°C h^{-1} for a large-scale reactor.

In small-scale testing the sample container often represents a substantial proportion of the system heat capacity and this will abate both the temperature rise and the total exotherm rise. This is indicated by the 'phi' factor of the system [12] which is given by:

$$\phi = \text{heat capacity of sample and container/heat capacity of sample} \quad (9.6)$$

A sample of about 10–20 g in a normal container will have a phi factor of about 1.5 compared to a phi factor of about 1 on the plant scale. Thus, the magnitude of any exotherm seen in the laboratory test will be only two-thirds that which will actually occur during a runaway on the plant.

(c) Characterization of the runaway reaction

Characterization of the runaway reaction or decomposition is required where the safety of a process is to be based on coping with the consequences of the runaway rather than preventing it occurring. It is self evident, therefore, that this evaluation need not be carried out for all cases, e.g. if previous testing has shown that the minimum exotherm temperature on the plant scale cannot be reached by any means then it is not necessary to determine the runaway reaction parameters.

The data required depend on the chosen basis of safety but usually involves the measurement of the rate and magnitude of changes in temperature, pressure and gas evolution that could occur in the plant situation. The critical information required to characterize a runaway reaction is shown in Figure 9.4 which shows the thermal history of a runaway scenario.

The occurrence of an upset condition during the operation of an exothermic reaction will result in the temperature rising from the process temperature to the maximum temperature of the synthetic reaction

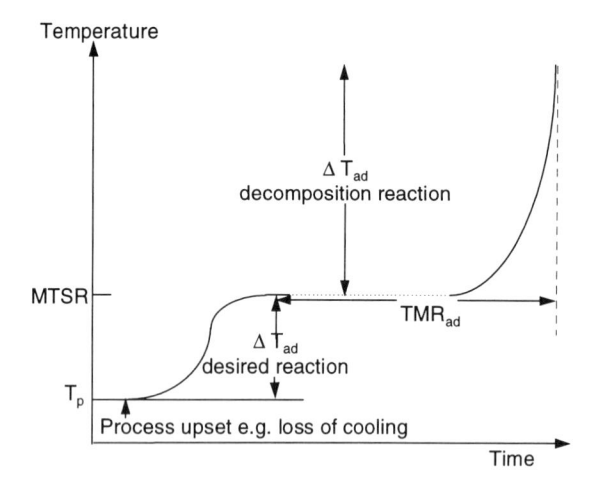

Figure 9.4 Schematic of the temperature profile of a runaway reaction.

(MTSR). This temperature equals the process temperature T_p plus the adiabatic temperature rise ΔT_{ad} resulting from the continuation of the desired reaction. The maximum adiabatic temperature rise which could occur can be calculated from the heat of the desired reaction ΔH_r and the specific heat, C_p, of the reaction mixture:

$$\Delta T_{ad} = \Delta H_r / C_p \qquad (9.7)$$

The resulting temperature rise from this runaway of the desired reaction may be sufficient in itself to cause an incident, particularly where the desired reaction has a high heat of reaction as is the case for polymerization reactions. In addition, the temperature reached, i.e. the MTSR, may be sufficient to initiate a secondary decomposition reaction. Such decomposition reactions are often highly energetic and an indication of their severity can again be obtained by calculating the resulting adiabatic temperature rise using, in this case the heat of decomposition, ΔH_d.

In order to evaluate the potential hazard of a decomposition reaction it is also necessary to know whether it will be initiated at a particular temperature. Unfortunately, it is not possible to quote a specific temperature at which a particular reaction will runaway since the majority of chemical reactions follow an Arrhenius rate law. Thus, their rate is exponentially dependent on the temperature and the temperature at which a runaway will occur is therefore critically dependent on the environment, i.e. size and rate of heat loss. However, an indication of the probability of such a decomposition reaction being initiated can be obtained from the adiabatic time to maximum rate or TMR_{ad}. This can be estimated [12] from:

$$\text{TMR}_{ad} = C_p R T_0^2 / q_0 E_a \quad [s] \qquad (9.8)$$

where C_p (J kg^{-1} K^{-1}) is the specific heat capacity, R is the gas constant (J mol^{-1} K^{-1}), T_0 is the initial temperature (K), q_0 is the heat output at T_0 (W kg^{-1}) and E_a (J mol^{-1}) is the activation energy. The time to maximum rate gives an indication of how much time is available to introduce emergency measures once a process disturbance has occurred.

From an experimental point of view any technique that is used must be able to simulate not only the normal process and plant operations but also the effect of any maloperations, e.g. loss of agitation, loss of cooling, rapid addition, etc. Various adiabatic calorimetry techniques have been used. Rogers and Wright [13] have described an adiabatic pressure Dewar system which allows the direct measurement of pressure, temperature and gas evolution changes under the low heat loss conditions that occur on the plant scale. There are many papers describing the application of data from the ARC adiabatic calorimeter in thermal hazard evaluation [12, 14]. In addition the VSP apparatus which uses a pressure equalization system to allow the use of a weak test cell with low thermal mass has been recently developed to produce the data required to design reaction relief systems [15].

It should be recognized that numerical simulation methods which use thermodynamic and kinetic data for the desired reaction that have been obtained under isothermal conditions or at lower temperatures may not accurately reflect the behaviour of the runaway reaction on the plant scale.

9.3.4 *Selection of safety measures*

The safe operation of a chemical process can be achieved by either preventative or protective measures. The option chosen will depend on the process detail but both require a thorough understanding of any potential hazard. It is a common fallacy that plants fitted with protective safety devices, e.g. trip systems, vents, etc., will deal effectively with any possible process deviation and therefore do not need a vigorous hazard assessment. In reality a protective system is only satisfactory for the particular situation for which it has been designed.

(a) *Preventive measures*

The use of preventive measures to achieve safe operation requires the early identification of process hazards in order that both the plant design and processing conditions can if necessary be altered.

Inherent safety. The ideal is an inherently safe process, i.e. one in which no disturbance, whatsoever, can cause an incident. An inherently safe process is one in which the potential hazards have been identified and the process and plant designed such that they are eliminated. This can be achieved by for example changing reactants, reaction conditions, solvent, catalysts, etc. [16]. Although it is seldom possible to completely eliminate all hazards, the major hazards can often be eliminated or minimized by process design provided they have been identified at an early stage.

Process control. In practice such 'absolute' inherent safety is rarely attainable. However, if the hazards have been identified and the safety measures are incorporated at the design stage, it is often possible to approach this ideal. Examples are the selection of heating medium to ensure that the temperature cannot reach the plant minimum decomposition temperature; specification of charge vessel sizes to limit the quantity that can be added at once, therefore limiting the temperature rise due to exothermic reaction, etc.

The most common critical parameters requiring control are the following:

- *Temperature* – definition of minimum temperature at which uncontrolled exotherms will start under plant conditions and the safety margin between operating temperature and exotherm temperature; identification of minimum or maximum temperatures for desired reaction to prevent accumulation.
- *Additions* – specification of maximum rate (e.g. limit by orifice plate) commensurate with the capacity of the cooling system.
- *Agitation* – identification of possible two phase system and reactant accumulation on failure.
- *Scrubbing system/vents* – identification and quantification of rate and magnitude of gas evolved or the need for a permanently open vessel.
- *Key reagents* – e.g. catalyst, phase transfer agent, ensuring addition at appropriate time and prevention at other times.

Safe manufacturing by prevention involves, therefore, the specification of boundary conditions or an envelope within which the process can be operated safely, and the provision of measures to ensure that the process remains within this envelope. This should be achieved where possible by the design of the process and plant. In addition, organizational procedures such as rigorous instruction, strict enforcement of appropriate operating conditions and provision of adequate maintenance are required. In some cases suitable trips and alarms may also be needed.

(b) Protective measures

Protective measures to ensure safe operation have to either deal with or mitigate the consequences of the runaway reaction. The options include containment, reactor venting, crash cooling, drown out and reaction inhibition.

Since the initial conditions markedly affect the course of a runaway reaction, the critical aspect for all protective systems is the interrelationship between their design and the definition of the worse case, i.e. the conditions leading to the runaway. The features of some protective measures are as follows.

Crash cooling. In this system, forced cooling is applied when the reaction deviates from set limits. It is important that the magnitude and rate of any hazard is quantified in order that reaction control can be maintained with the applied cooling.

Dumping/dilution/inhibition. In this system the contents of a reactor are rapidly emptied into a dump tank when the reaction deviates from set limits. The dump tank usually contains a liquid (often water) which acts as a heat sink and effectively stops the reaction. The liquid used in the dump tank should not react with the reactor contents. Dilution and inhibition involves the rapid addition of a diluent or inhibitor to the reactor contents to stop the reaction. The choice of diluent or inhibitor and the efficiency of mixing require careful investigation.

Protective systems such as crash cooling, drown out and reaction inhibition involve the detection of the onset of the runaway reaction and subsequent corrective action to prevent it occurring. In addition to suitable detection methods and the availability of process compatible systems, these techniques need time to act and are of limited use when the runaway is caused by a sudden event.

- *Containment.* In its broadest sense, this is a good principle for the design and operation of chemical reactors. Its use is becoming more common as environmental pressures on the release of chemicals increase. Both the reactor and all ancillary equipment have to be designed for the peak pressures/temperatures which may be reached. The main limitations, as well as cost, are problems associated with design and operation of such equipment.
- *Emergency reactor venting.* This is a widely used effective safety measure particularly in polymerization reactions. Although it is used in fine chemicals manufacture, it suffers from two major problems; containment of the ejected material and its limited effectiveness for low vapour pressure systems [17]. Venting normally involves the provision of either a relief valve or bursting disc which opens at a set pressure. The venting of material from the reactor by either single- or two-phase flow tempers the runaway reaction by removing heat and therefore maintaining the temperature and also the pressure in the case of a vapour-phase system. For purely gas systems the vent is sized to cope the peak gas generation rate and, therefore, prevents pressure increase.

Historically, the sizing of vents for emergency relief of chemical reactors has been carried out using either in house techniques or empirical standards such as the Factory Insurance Association methods. These have now been superseded by the DIERS methodology (the Design Institute for Emergency Relief Systems) which requires thermal data, flow regime and viscosity characterization to be measured under runaway conditions [15].

In addition to calculating a suitable vent size, the use of venting as a basis of safety requires the provision of a safe discharge area. This will often involve a dump or containment tank [18], particularly where the reactants or products of decomposition are flammable or toxic.

Protective systems are rarely used on their own and some preventive measures are usually included to reduce the demand on the protective system. It should be recognized that it is not always possible to design a protective system to cope with the consequences of a runaway reaction. In such cases, if preventive measures are also unsuitable to ensure safe operation then the process may need to be radically redesigned or even abandoned.

9.4 Fire and explosion hazards

Fire and explosion hazards during batch operations can arise from the formation and subsequent ignition of flammable gas, vapour or dust atmospheres. The systematic procedure outlined above can be used to identify the hazard, assess the risk and define an appropriate basis of safety to ensure the process is free from operational hazards.

Safe operation of processes and plants handling flammable materials requires that due consideration is given to the potential fire and explosion hazards that can arise and the procedures that can be implemented to avoid them. The hazards are determined by the flammability characteristics of the materials involved, the process conditions, the design of the plant and the operating procedures. From a consideration of all these aspects a system of safety can be devised.

Most organic materials are combustible under certain conditions. Combustion is a chemical process; an oxidation reaction which is accompanied by the evolution of heat and light. For a combustion reaction to take place, the presence of both fuel and oxidant are required, together with an ignition source of sufficient energy to initiate the reaction. The 'Fire Triangle' provides a useful means of illustrating this (Figure 9.5). Remove any one component to break the triangle and the combustion reaction will not take place.

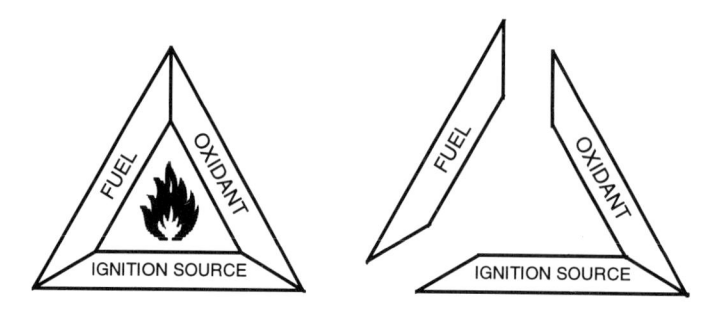

Figure 9.5 The fire triangle.

It is the mixture of fuel and oxidant which together constitute the flammable atmosphere. A flammable mixture is defined as one through which flame will propagate, away from the influence of the ignition source. Sources of ignition can be categorized as:

- Those external to the specific manufacturing process (e.g. naked flames, welding, electrical equipment).
- Those associated with the plant operations that form part of the manufacturing process.

External ignition sources should be controlled/eliminated as a consequence of good plant house keeping and associated in-house regulations. They are not considered in detail in this chapter.

Ignition sources that may arise from chemical plant operations need to be evaluated. These include for example the generation of static electricity during reactor charging and materials handling, friction during the grinding of powders and the development of exothermic decomposition during drying operations.

In unconfined situations, ignition of the flammable mixture generally results in a fire. Under confined conditions significant increases in pressure can be generated in a matter of milliseconds leading ultimately to an explosion. The severity of the explosion is influenced by a number of factors including the initial pressure, the mixture composition and the degree of confinement.

Whenever possible, it is generally preferable to operate processes outside the range of flammability. This can be achieved either by inert gas blanketing or by use of appropriate process control to prevent formation of the flammable atmosphere by, for example, restricting the temperature or introducing well defined ventilation. It is also possible to base safety on the avoidance of ignition sources if all potential sources can be identified and eliminated. When neither of these options can be reliably implemented, suitable protective measures, e.g. explosion containment, venting, suppression, etc., have to be installed to deal with the consequences of any explosion. To implement such precautions satisfactorily, the flammability characteristics of the materials must be well defined and clearly understood.

9.4.1 Flammability characteristics of materials

In order to assess the operational hazards which may arise in a plant it is necessary to have information on the flammability characteristics of any gases, liquids and powders used. This includes:

- Flammability – does the material support combustion and under what conditions?
- Ignition sensitivity – temperature/energy required for ignition.

- Ignition consequences – rate and type of combustion/flame spread, pressure development during an explosion.

Data on the flammability characteristics of common liquids and gases are readily available in the literature. Information for powders is, however, rarer. This is because the physical form, i.e. particle size, moisture content of solids markedly affect their flammability characteristics. It is often necessary, therefore, to measure these parameters using a sample of the powder being used [19].

Parameters that provide flammability characteristics are described below. It should be recognized that in specialist industries (e.g. the explosives industry, peroxide manufacture) these parameters will need to be supplemented by data obtained using test methods specific to these industries.

(a) Limits of flammability

Mixtures of combustible materials in air can only be flammable if the concentration of fuel falls within a particular range. This range is bounded by the lower and upper flammability or explosive limits (LEL and UEL). These are the minimum and maximum concentrations of gas or vapour in air respectively that will permit flame propagation. Below the LEL, the mixture is too 'weak' in fuel to support combustion, above the UEL, the mixture is too 'rich' for the flame to propagate (Figure 9.6).

For many gases and vapours the flammable concentration range is quite small (1–15% v/v), there are however notable exceptions such as hydrogen and methanol. The LEL for dust–air mixtures is 10–$60 \, \text{mg} \, \text{l}^{-1}$. Upper limits are less well defined but are of little importance because of the difficulty in maintaining high concentration dust dispersions.

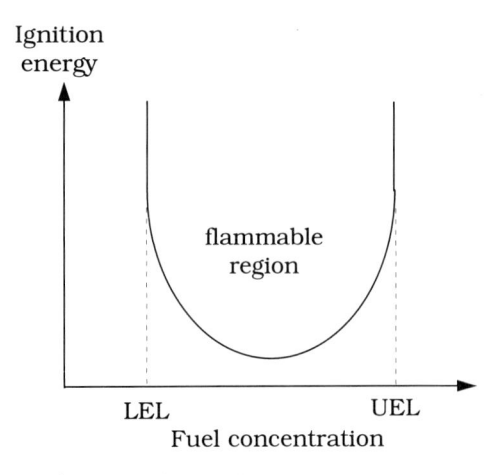

Figure 9.6 Diagram showing flammability limits.

Limits of flammability are influenced by a number of factors which need to be taken into account. Limits are widened by increases in temperature and pressure; the effects of temperature increases being greatest. The changes in limits are most noticeable in the UEL.

A great deal of data on flammability limits is available in the literature. These values refer to normal flames initiated by high temperature sources with little or no delay between the application of the source and the appearance of flame. Under some conditions, what are called 'cool' flames can occur which increase the upper limit to well within the region which was non-flammable to normal flames.

Cool flames are initiated by lower temperature sources applied for some time, are only slightly exothermic, and result from the early stages of the combustion process. Cool flames can transform into normal flames to produce explosions, but this only occurs within the normal flammable range. Although in themselves cool flames do not produce an explosion problem, their possible occurrence and transition into normal flames has to be examined.

(b) Flash point of liquids

The flash point is the lowest temperature to which a combustible liquid must be raised before it will form a flammable mixture of vapour in air. It provides an indication of the conditions under which flammable atmospheres will occur and is equivalent to the lower temperature limit of flammability. Flash point is influenced by a range of factors such as ambient pressure, equipment design, size of sample, operator bias, etc., but standard procedures have been developed that enable the variation due to most of these to be eliminated, although pressure corrections are still necessary.

The standards allow flash point to be determined in what are termed 'closed-cup' or 'open-cup' apparatus. For 'closed-cup' measurements the sample is heated in a closed vessel and an ignition flame applied at intervals until the lowest temperature for ignition is found. 'Open-cup' measurements use an open vessel. Because the sample is open to atmosphere, the vapour concentration above the liquid will be less than in the 'closed-cup' apparatus at a given temperature. Consequently, flash points measured in 'closed-cup' apparatus are normally several degrees lower than those measured in 'open-cup' equipment.

Where a flammable liquid is handled above its flash point an ignitable concentration of vapour in air will be formed. It is, of course, possible to specify an upper temperature limit of flammability but this is used only rarely.

Provided the process operation does not produce a mist of fine droplets the atmosphere will be non-flammable if the operating temperature is at least 5°C below the flash point.

(c) Fire point/combustibility

The fire point of a liquid is the temperature to which a liquid must be raised in order that the liquid continues to burn when an external flame is applied to the surface. It is usually a few degrees above the flash point.

There is no comparable standard test method for powders. However, information is required on the possibility of local ignition propagating through bulk material or along powder layers. The Train Firing Test for determining the flammability of powders, used for the Notification of New Substances Regulations and for Transport Regulations, provides an indication of whether this will occur. The test essentially consists of a train of powder (typically 25 cm long by 2 cm wide by 2 cm deep) which is ignited at one end and the rate of burning away from the ignition source evaluated [20].

(d) Flammability of a dust cloud

The flammability of a dust cloud is determined by dispersing it around hot coil, spark and flame ignition sources [21, 22]. If the cloud ignites and the flame propagates away from the ignition source then the material is classified as flammable.

In the UK such dusts are termed Group (a) [23]. A dust cloud that does not ignite is considered to be non-flammable and is classified as Group (b). These tests are carried out at ambient temperatures and Group (b) dusts may become flammable at temperatures greater than 110°C [24]. Additional testing may be required in this situation.

An alternative system of classifying the flammability of dust clouds uses the 20 l sphere test which is also used to measure the explosion violence of a dust cloud explosion (Section 9.4.1.i). In this test, which results in a St 0, St 1, St 2 or St 3 classification, a 10 kJ chemical ignitor is used as the ignition source as it is believed that this is much larger than would ever be encountered on a chemical plant [25]. Some dusts which propagate an explosion in this test may be non-explosible when tested with a more representative ignition source. Table 9.5 presents the criteria which dusts are allocated to the St categories.

(e) Minimum oxygen concentration (MOC)

For combustion to occur the supporting atmosphere must contain a certain amount of oxygen. This is termed the minimum oxygen for combustion (MOC). The MOC depends on the individual material, and the conditions of temperature and pressure. Under normal atmospheric conditions few materials have MOC values below 5% v/v (e.g. hydrogen). Certain exceptional materials (e.g. ethylene oxide) do not require oxygen to 'burn' or decompose.

The addition of inert gases to the supporting atmosphere modifies the limits of flammability by reducing the available oxygen, and can suppress

Table 9.5 St classification of dust explosions (based on strong ignition source of 10 kJ and 1 m³ test apparatus)

Dust explosion class	KSt (bar m s^{-1})	Characteristics
St 0	0	no explosion
St 1	0 < KSt < 200	weak/moderate explosion
St 2	200 < KSt < 300	strong explosion
St 3	>300	very strong explosion

flame propagation. The effect of various inert gases on the limits of flammability depends on the gas being used. The main changes occur in the upper limit, and the effectiveness of the inert gases is in order of their molar heat capacities, i.e. $Ar < He < N_2 < H_2O < CO_2$. Most fuels have MOCs in the range 6–14% v/v, but these values decrease with increasing temperature or pressure.

Table 9.6 gives typical values for the flammability characteristics of some common materials.

(f) Minimum spark ignition energy

The minimum spark ignition energy of fuel–air mixture depends markedly on the electrical characteristics of the test circuit (i.e. the inductive, resistive and capacitative elements). The spark ignition energy is used to assess the risk from two types of sparks – those from electrical equipment and those from static electricity. Sparks from electrical equipment are produced by relatively low voltage, high inductance sources and data on the ignition energy of gas/vapour air mixtures are available in the relevant standards on specifying electrical equipment. The major hazard from static electricity is a spark from an insulated conductor (e.g. metal plant). This is essentially a high voltage pure capacitative spark and the minimum ignition energy is determined by passing such sparks of known energy through the flammable atmosphere.

Gases and vapours have with few exceptions minimum spark ignition energies in the range 0.2–1 mJ. Dust/cloud spark ignition energies have values varying from about 1 to 5000 mJ. This wide range should be taken

Table 9.6 Flammability characteristics of some commonly used materials

Material	Flash point (°C)	Flammability limits [% (v/v)] Lower	Flammability limits [% (v/v)] Upper	MOC with N_2 (inert)	Auto-ignition temperature (°C)
Hydrogen	–	4.0	75	5.0	400
Methanol	+11	6.7	36	10.3	385
Acetone	−18	2.6	13	13.8	465
Cyclohexane	−17	1.3	7.8	11.2	245
Benzene	−11	1.4	7.1	11.2	560

276   HANDBOOK OF BATCH PROCESS DESIGN

into account in the assessment of risk and definition of appropriate safety measures.

(g) Auto-ignition temperatures
This is the temperature at which a material will ignite in a heated environment in the absence of other ignition sources. Test methods for liquids and dust clouds are well established. It is important not to confuse the auto-ignition temperature of a dust cloud with the temperature required for ignition of a powder in bulk or layer form which is sometimes also referred to as the auto- or self-ignition temperature.

Auto-ignition temperatures for liquids and vapours cover a very wide range, but few materials ignite below 250°C and most have ignition temperatures in excess of 350°C. They are significantly reduced by increases in temperature and pressure, and can be complicated by the materials in contact with the gas or vapour and the presence of cool flames. It becomes necessary when working in regions where cool flames can occur to base determination of auto-ignition temperature on the appearance of the flame.

(h) Ignition temperature of bulk powders and layers of powders
The temperature at which powder in bulk or layer form will ignite is dependent on the quantity and physical shape of the pile of powder. These systems cannot be described by a single temperature, rather the temperature is a point function depending on the three spatial co-ordinates. The problem of analysing heat transfer and hence of deriving the critical conditions for thermal explosions therefore becomes extremely complex.

In essence, the onset of runaway reactions in such systems depends not only on their size but also on their shape or symmetry. An increase in sample size reduces the heat loss from the reaction zone at the centre of the sample by acting as additional insulation. Since heat loss from the system occurs at the surface this change is proportional to the surface area whilst the heat generation is proportional to the mass or volume of the substance. Thus the ignition temperature decreases as the size increases.

Most workers base their analysis on the treatment by Frank-Kamenetskii with the basic equation:

$$\ln(\delta_{cr} T_i^2 / r^2) = M - N/T_i \qquad (9.9)$$

where δ_{cr} is the dimensionless critical parameter that depends on the size and shape and boundary conditions of the material, T_i is the minimum ignition temperature for the particular system, r is the radius, and M and N are constants that characterize the material such that

$$M = \ln[p(-\Delta H_r)EZ/LR] \qquad (9.10)$$

$$N = E/R \qquad (9.11)$$

where Z is the Arrhenius pre-exponential factor, p is the density and L the thermal conductivity.

From these equations it is evident that a plot of $\ln(\delta_{cr} T_i^2/r^2)$ against $1/T_i$ for a given substance in a series of containers of similar shape but different sizes, should be a straight line with slope of $-E/R$. Further details can be found in the reviews by Gray and Lee [26] and Bowes [27].

A simpler treatment is given by Leuschke [28] which results in a relationship between the ratio of the volume, V, to the surface area, SA, and the ignition temperature such that:

$$\ln(V/SA) \propto 1/T_i \quad (9.12)$$

This allows a quick and approximate correlation to be made for samples of different shapes and sizes. Figure 9.7 illustrates this relationship for the ignition of activated carbon. It can be seen that moderate increases in scale markedly reduce the ignition temperature of the bulk material.

In contrast to liquid reaction masses which contain all the components involved in their decomposition, air is often an essential ingredient in the exothermic activity of powders. Factors such as packing density and the distribution of interparticle voids that are not a function of the chemical constitution of the powder often control the initiation and rate of decomposition.

The majority of substances handled in the chemical industry are complex molecules or mixtures of substances and decomposition can involve consecutive and/or parallel reactions. The rate of heat production depends on their chemical constitution and reactivity. Small changes in formulation can often markedly affect the decomposition process. Autocatalysis in

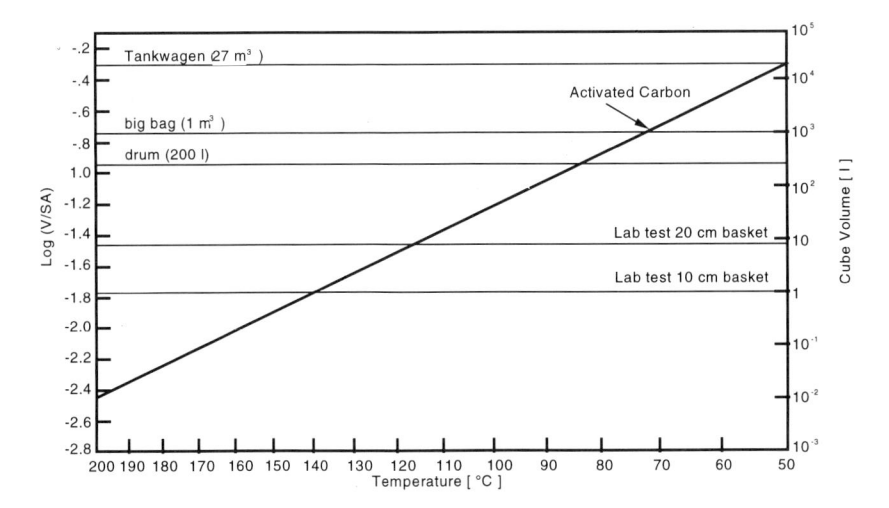

Figure 9.7 Relationship between ignition temperature and size.

which the first stage of a reaction produces a catalytic component in sufficient concentration to activate the second may also occur.

In addition, contaminants can have a marked effect on thermal behaviour. For example, an explosion occurred during spray drying of a material for which the lowest temperature at which ignition of 10 mm layers of material (the thickest that would occur during operation) was measured as 350°C, well above the maximum temperature in the dryer of 220°C. Detailed examination of the material in the dryer showed contamination by the lubricating oil from the inlet atomizer. This contamination reduced the ignition temperature to 215°C and initiated the explosion.

It can be seen that many factors affect the stability of substances. The material tested must therefore be representative of the substance handled in full-scale manufacture and the tests used should simulate the practical conditions including the temperatures and time cycles that will occur [29]. Ignition temperature values are therefore quoted with the conditions used in the measurement (e.g. 5 mm layer for specification of electrical equipment [30]). The data appropriate to a specific situation need to be obtained by simulating the process conditions with suitable sample form (i.e. bulk, layer or fluidized) and heat exposure combination.

(i) Explosion violence
The potential violence of a vapour or dust cloud explosion is determined by measuring the rate of pressure rise and maximum pressure developed when the flammable atmosphere is ignited in a cloud vessel. Maximum pressures are of the order of 900 kN m^{-2} (120 p.s.i.) when the mixture is initially ignited at atmospheric pressure. The data are required if safety is to be based on containment or venting. In the case of dust clouds it is important that the parameters, especially rates of pressure rise, are measured using conditions relevant to the plant and appropriate to the method used for vent sizing.

The 20 l sphere test procedure has now become widely accepted as a means of measuring the explosion violence of a dust cloud for the design of dust explosion protection systems and also as a method for dust explosion classification. VDI 3673 [31] for the design of dust explosion vents based on 1 m³ tests can be implemented using 20 l sphere results.

The 10 kJ chemical ignitor was chosen to ensure that a larger ignition source than is likely to be encountered in chemical plants is used in the standard test. For the majority of dusts the results obtained are independent of the size of the ignition source. However, with some weakly explosible dusts, e.g. PVC, this high ignition energy initiates a different and more vigorous decomposition/combustion mechanism. In such cases, provided that it can be guaranteed that only smaller ignition sources will be present in practice, results from tests using more representative smaller ignition sources may be used [32, 33].

The results obtained in the 20 l sphere are used to estimate the explosion violence to be expected in larger vessels by using the 'cube root' law:

$$(dP/dt)_{max} V^{1/3} = KSt \qquad (9.13)$$

where dP/dt is the maximum rate of pressure rise determined in a vessel of volume V and KSt is a constant characteristic of a particular dust with units $bar\,m\,s^{-1}$. The classification of dusts into the different explosion classes is shown in Table 9.5.

(j) Flammability in atmospheres other than air or oxygen
The flammability characteristics described above have been considered when air or oxygen is the oxidizing component. It should be recognized, however, that any oxidizer mixed with a fuel may produce a flammable mixture. Table 9.7 lists typical values obtained when the oxidant is chlorine.

The data illustrate that flammability characteristics vary considerably depending on the oxidant present in the mixture. Flammability data need to be determined for the atmosphere actually present.

(k) Flammability of mists
Fine droplet mists (or aerosols) of flammable liquids are ignitable and will be so at temperatures well below the flash point of the parent liquid. The temperature at which a mist becomes ignitable is determined by the droplet size and the size of the source of ignition. As the droplet size reduces the flammability characteristics of a mist approach those of a vapour and at sub-micron sizes are indistinguishable from a vapour.

(l) Hybrid mixtures
Atmospheres containing both flammable gas or vapour and dust are termed 'hybrid mixtures'. Their flammability parameters are difficult to define; in most cases falling between the dust at one end, and the gas or vapour at the other. However, it is possible that a synergy could occur between the materials that would produce entirely different characteristics. A very important feature of hybrid mixtures is that they can be flammable when

Table 9.7 Flammability characteristics in chlorine

Fuel	Auto-ignition temperature (°C)			Flammability limits (mol %)		
	Air	Oxygen	Chlorine	Air	Oxygen	Chlorine
H_2	400	–	207	4.0–75.6	3.9–95.8	3.5–89
CH_4	537	–	<300	4.0–16.0	5.0–61.8	5.5–63
CH_3Cl	618	–	215	7.0–17.4	8.0–66.0	10.2–63
C_3H_6	455	423	150	3.0–15.4	3.0–67.0	4.9–59

both components are below their respective LELs. In cases where the gas or vapour content is significant (more than say 25% of its LEL) the mixture should be assumed to have the flammability characteristics of the gas or vapour.

In addition to flammability characteristics other properties of materials may need to be known or measured in order to identify potential hazards and specify a safe operating procedure, e.g. a measure of the resistivity of any liquids being used may be needed to prevent hazards arising from static electricity.

9.4.2 Sources of ignition

The presence of a flammable atmosphere does not on its own constitute a hazard. An ignition source is also required that is capable of igniting the atmosphere. The main ignition sources (excluding electrical equipment, for which detailed standards are available) associated with chemical plant operation are summarized below.

(a) Flames and hot surfaces

Heat is the most common form of energy input and sources of heat can directly or indirectly lead to ignition. Direct ignition of a flammable atmosphere can occur when its temperature exceeds the auto-ignition temperature. The temperature of the atmosphere can be measured and this risk readily specified.

In powder manufacturing, ignition of a flammable atmosphere is likely to be more indirect and be caused by exothermic decomposition of bulk or layer material developing to red heat. The temperature of bulk material and surfaces where layers can form needs to be established and related to the thermal decomposition temperatures of the material (see Section 9.4.1.h).

(b) Friction

Friction can be present in many operations ranging from high speed sustained contact to single impacts between materials. Mechanical friction can arise from malfunction in bearings, binding of moving parts, tramp metal, etc., all of which are capable of igniting flammable atmospheres.

A potent source of ignition is the thermite reaction that can result from impacts involving aluminium, titanium or magnesium and rust.

(c) Static electricity

Static electricity can be generated in virtually all industrial operations. Electrostatic spark discharges can occur from conductors, personnel, insulating materials (i.e. plastics), liquid and powder surfaces and fine droplet mists. Seven main types of electrostatic discharge need to be considered:

Corona	discharge from or to a pointed conductor.
Brush	discharge between a charged insulator and a conductor.
Propagating brush	discharge of a charge bound across a dielectric.
Cone	discharge from bulked powder.
Human	discharge from human body.
Spark	discharge from isolated conductor.
Lightning	lightning discharge.

Corona discharges. Although definitive results on the incendivity of corona discharges are not available, it is generally assumed that they are only capable of igniting particularly sensitive gas–air mixtures, e.g. hydrogen, acetylene, carbon disulphide with a MIE of less than 0.025 mJ. Thus they are non-incendive for common gas and solvent vapour–air mixtures and dust air mixtures.

Brush discharges. Brush discharges have been shown to ignite flammable gas and solvent vapour–air mixtures with energies of 1–4 mJ. The maximum energy available in a brush discharge has been estimated as 5–15 mJ from measurement of the field strengths before and after the discharge. Brush discharges are therefore capable of igniting almost all gas and vapour–air mixtures.

Despite extensive laboratory testing by many workers, it has so far not been possible to ignite pure dust–air mixtures by brush discharges. This may be partly due to the fact that the nature of brush discharges are changed by the dust cloud, but mainly due to the dust being attracted to the charged surface leading to much weaker brush or corona discharges. Because of the lack of definitive evidence that brush discharges could not ignite flammable dust clouds, and since they have sufficient potential energy to ignite sensitive dust–air mixtures, it has previously been recommended that they should be considered capable of igniting dust–air mixtures with a MIE of less than 25 mJ. This value is frequently being revised and it is likely that the European Standard on electrostatic hazards will indicate that brush discharges should be considered capable of igniting dust–air mixtures with an ignition energy of 3 mJ or less. However, it is recommended that, if a dust with a MIE of less than 10 mJ is being handled, the system is critically examined to be certain that only brush discharges are likely to be present.

Propagating brush discharges. These discharges occur as a result of the build up of polarized layers of charge occurring on an insulator. Typical situations in which this can occur are where an insulator that is exposed to a charging mechanism is backed by an earthed metal substrate, or highly charged powder is held in an insulating container and the outer surface becomes charged by corona. The short circuit of these charges by the approach of an earthed object, or breakdown of the insulator, will initiate the discharge. Such discharges are highly incendive and are capable of igniting all gas and solvent vapour–air mixtures and dust–air mixtures with a MIE of up to 1 J if not higher.

Cone discharges. The incendivity of such discharges is uncertain at present. An equivalent energy of about 100 mJ based on theoretical considerations has been proposed. Recent work is indicating that the igniting power of cone discharges depends on the amount of material present and that discharges of several hundred millijoules may be possible in large containers.

Human discharges. The human body has a capacitance of about 80–300 pF depending on its insulation from earth. Sparks from charged humans have been shown to ignite both flammable gas and solvent vapour–air mixtures and sensitive dust–air mixtures. Not all the energy stored on the body is released in the spark discharge, mainly because some is absorbed by the body resistance. Approximately 2–3 times the MIE of the flammable atmosphere is required to be stored on the human body to cause ignition. Accordingly it has been recommended that the maximum equivalent energy of sparks from the body should be considered as 30 mJ, i.e. such sparks are capable of igniting all flammable gas and solvent vapour–air mixtures and dust–air mixtures with a capacitative ignition energy of up 30 mJ.

Spark discharges. The energy of spark discharges depends on the capacitance of the conductor and the voltage to which it is charged. The stored energy is therefore limited only by the magnitude of these values and if sufficiently large such sparks are capable of igniting all flammable atmospheres.

Lightning discharges. These are capable of igniting all flammable mixtures.

(d) Material properties

Certain materials are pyrophoric (e.g. Raney Nickel catalysts) or have a tendency to self heat or undergo rapid decomposition. Under certain conditions they may ignite themselves or act as an ignition source for any other flammable atmosphere that may be present.

(e) Other sources of ignition

In addition to the above, individual processes may contain sources of ignition specific to the process. An essential part of process definition is the assessment of this possibility, e.g. in processes involving solid amines, it is important that they do not come into contact with nitrous fumes as exothermic reaction and ignition can occur.

9.4.3 Assessment of hazards and definition of appropriate safety measures

Every process, reactor or operation requires a defined basis of safety appropriate to its hazard. The objective is to provide an acceptable level of safety consistent with the manufacturing, engineering and economic requirements of the process as well as satisfying all applicable regulations.

Safe operation in chemical plants using flammable materials can be achieved by:

- Avoidance of ignition sources
- Elimination of flammable atmospheres
- Provision against consequences of ignition.

(a) Safety based on avoidance of ignition sources

In addition to the control of external sources of ignition (e.g. electrical equipment, welding, use of flame) and chemical reactivity, this basis for safe operation requires the identification and elimination of *all* ignition sources associated with plant operations under normal and maloperation conditions. Detailed precautions to achieve this objective will depend on process and plant operating conditions but the common essentials are as follows.

Control of heat sources. Heat applied to the process (e.g. hot air for drying, steam in reactor heating coils) and that generated during manufacture (e.g. energy input from grinding) must not lead to temperatures capable of initiating auto-ignition or uncontrolled exothermic decomposition that could develop into a fire or explosion. Material parameters that provide guidance on this are given in Table 9.8.

A good link exists between the measured auto-ignition temperature and the plant temperature that would initiate auto-ignition and an explosion.

It must be emphasized that ignition temperature, particularly of powders, depends on the physical form of the material and the chemical constitution. A change in formulation may markedly reduce the decomposition temperature. This aspect must be fully evaluated before the avoidance of ignition sources can be used as the basis for safety.

Control of friction. Friction can lead to both the generation of heat and sparks. To address heat from mechanical friction, a programme of regular and careful maintenance should be carried out paying close attention to bearings and moving parts where binding friction could occur. Precautions

Table 9.8 Relevant flammability parameters for control of heat sources

Liquids	Powder/dust clouds
Flash point	Flammability
Combustibility	Auto-ignition temperature of dust cloud
Auto-ignition temperature	Ignition temperature of powder in bulk and layer form

should also be taken to avoid the ingress of tramp metal when it can become caught up in moving parts. (e.g. by the fitting of suitable screens to scroll feeds, agitators, mills, etc.)

To minimize the risk from friction sparks, equipment made from aluminium, magnesium, titanium or light alloys containing these metals should not be used if flammable atmospheres can be present – unless it can be ensured that rusty ferrous metal is excluded or the equipment is located where the impact risk is low.

Static electricity. Precautions for the avoidance of electrostatic hazards are given in relevant standards [33]. For liquid systems where a flammable vapour or gas atmosphere can be present the complete elimination of ignition sources from static electricity requires that:

- All conducting items are earthed – this includes drums, scoops, wire, reinforcement in hoses, metal clamps on QVF glass lines, etc.
- All personnel are earthed – use of conductive or antistatic footwear and conductive floors as well as suitable gloves.
- Electrostatic charge generation is minimized – by restricting flow velocities to less than 5 m s^{-1} for single phase liquids and less than 1 m s^{-1} for liquids containing immiscible phases (liquids or solids).
- Charge accumulation on bulk liquids and mists is prevented – by earthing the contents of the vessels and minimizing the production of mists by suitable vessel filling using dip legs.
- Use of plastics is controlled – insulating plastics should not be used.

For processes only involving flammable dust clouds, the precautions required to eliminate electrostatic ignition sources are dependent on the ignition sensitivity of the dust cloud as measured by its minimum spark ignition energy. Typically with about 55% of powders the only precaution necessary is the earthing of conducting items, 30% require that personnel should also be earthed and 10% require a restriction on the use of non-conductors or plastics. Some dusts, perhaps 5% of the total, are very sensitive and the elimination of ignition surfaces will not provide an adequate degree of safety and alternative methods need to be used.

Other sources of ignition. A process may contain specific sources of ignition not considered above (e.g. pyrophoric catalyst). It is important that the depth of the hazard assessment is sufficient to identify these and eliminate them.

The ignition of flammable mixtures vented from reactors should be avoided by venting vapours above roof level, away from sources of ignition and protecting vent lines against flame propagation should ignition occur (e.g. by lightning) by using appropriate flame traps or inert gas purging.

(b) Safety based on elimination of flammable atmosphere

The elimination of flammable atmospheres can be achieved in two ways – operating outside the flammable limits or operating below the critical oxygen level.

Operating outside the flammable limits. This requires that the fuel component of a mixture is maintained below the LEL or above the UEL. Provided account is taken of the effect of pressure and temperature on the limits then the published data on gases and vapours give the information required for defining appropriate gas concentration control. For liquids, the flash point prescribes the temperature for the lower limit.

In powder operations the inhomogeneity of dust clouds usually precludes this method as a basis of safety.

Operating below critical oxygen level/inert gas blanketing. Control of oxygen below the level required to support combustion is achieved by the use of inert gas. MOC data provide the necessary design information, provided they were obtained with the inert gas to be used and account is taken of the effect of pressure and temperature. Processes where highly flammable hydrogen–air atmospheres may be present are usually carried under a carefully controlled inert gas blanket.

In inert gas blanketing systems the following recommendations should generally be applied.

- Blankets generally should reduce oxygen contents to less than 5% v/v for hydrocarbon vapours and dusts, and to less than 2% for hydrogen.
- The inert gas supply should be reliable and the gas itself should not contain more than 1% v/v of oxygen (0.5% v/v where hydrogen may be present).
- Some processes (e.g. where a pyrophoric catalyst is used) may need high reliability blanketing. This can require the use of monitoring systems and procedures to cover failures of the inert gas supply or the control equipment.
- Inert gas blankets should be set up before flammable materials are introduced, and should be maintained throughout the process whenever flammable atmospheres can occur.
- Inerting by pressurizing and then relieving a vessel is preferable to the use of flow purging as it is more efficient, and generally uses less inert gas. The oxygen content in a vessel blanketed by pressurization and relief can be calculated from $C_p = C_0(P_1/P_2)^n$, where C_p is the oxygen content after purging (%), C_0 is the oxygen content before purging (%), P_1 is the ambient (usually atmospheric) pressure (absolute), P_2 is the purging pressure (absolute) and n is the number of purges.
- At least two purges should be used in blanketing vessels. Because of the need to mix the gases properly satisfactory blanketing is unlikely to be obtained from a single purge.
- Purging by flow requires a volume of gas equal to about five times the volume of the space to be blanketed. Inlet and outlet lines used for flow purging must be kept as far apart on the vessel as possible to aid gas–air mixing.
- Satisfactory blankets cannot be maintained in open vessels. Consequently, re-purging becomes necessary after any operation, such as sampling, is

carried out via a chargehole into a blanketed vessel and also after liquid loading by vacuum or blowing, unless this is effected in a way that avoids the ingress of air (e.g. blowing by inert gas pressure). This also means that it is impracticable to maintain an efficient blanket on a vessel with an open chargehole.
- Because of the extreme ignition sensitivity of hydrogen, vessels in which it is present should be purged with inert gas to reduce the concentration to less than 1% v/v before air is admitted.
- Flame arresters are of limited used with hydrogen so that vent lines that carry flammable concentrations to atmosphere should be purged with inert gas.
- It should be appreciated that the inert gases are simple asphyxiants and operators working with blanketed vessels must be suitably protected.
- In general terms 50% of the bulk of any powder is air, and this must be taken into account in calculating purge rates during powder loading to vessels. It should be recognized that rotary valves and similar equipment can have a pumping action by taking air into a vessel and removing gas from it.

(c) Provision against consequences of an ignition

Vapour and dust cloud explosions develop pressure and the objective of protection techniques is to prevent damage arising from this. The techniques that can be used are containment of the explosion, explosion vents and explosion suppression. The salient features of each technique are as follows.

Containment of an explosion. This technique requires that the equipment withstands the maximum pressure developed in the explosion. In a simple enclosure this is unlikely to exceed $800 \, \text{kN} \, \text{m}^{-1}$ but in long ducts and interconnected vessels the phenomena of pressure-piling and detonation can induce pressures greatly in excess of this value. This technique is best used for the protection of relatively small single volume units that can be mechanically isolated from other plant units.

Explosion venting. A common form of protection against explosion in which a vent is opened by the pressure rise in the early stages of an explosion. This releases the combustion products at such a rate that the pressure developed remains below the strength of the plant and can be released safely to atmosphere.

Explosion suppression. This method of protection consists essentially of detecting an explosion in its early stages with a pressure monitor that activates a system capable of injecting a suppressant throughout the volume of the protected unit within a few milliseconds. This stops combustion by one or both of the following mechanisms: either diluting the flammable mixture so that it is outside the flammable range and providing a heat sink, or removing the radicals necessary for flame propagation.

9.4 Conclusions

A careful evaluation of the hazards that may occur is essential to safety in chemical reactor operations. This requires a systematic assessment of the complete process and involves the measurement and collation of relevant safety data on the materials used, its interpretation with respect to the particular operation, in particular to the scale, and the specification of a suitable basis of safety.

References

1. Rogers, R.L. (1989) The systematic assessment of chemical reaction hazards. In *Proc. Int. Symp. on Runaway Reactions.* CCPS/AIChemE, New York, p. 578.
2. Gibson, N., Rogers, R.L. and Wright, T.K. (1987) *IChemE Symp. Ser.*, **102**, 61.
3. Barton, J.A. and Rogers, R.L. (eds) (1993) *Chemical Reaction Hazards – A Guide.* IChemE, Rugby.
4. *Guidelines for Hazard Evaluation Procedures.* CCPS, New York.
5. Regenass, W. (1984) *IChemE Symp. Ser.*, **85**, 1.
6. Rogers, R.L. (1990) Fact-finding and basic data – Part 1: hazardous properties of substances. In *IUPAC Workshop on Safety in Chemical Production.* Blackwell, London.
7. Semenov, N.N. (1928) *Z. Phys. Chem.*, **48**, 571.
8. Frank-Kamenetskii, D.A. (1969) *Diffusion and Heat Transfer in Chemical Kinetics*, 2nd edn. Plenum Press, New York.
9. Rogers, R.L. (1989) *IChemE Symp. Ser.*, **114**, 47.
10. Bretherick, L. (1990) *Handbook of Reactive Chemical Hazards*, 4th edn. Butterworths, London.
11. Cutler, D.P. (1986). *IChemE Symp Ser.*, **97**, 133.
12. Townsend, D.I. and Tou, J.C. (1980). *Thermochim. Acta*, **37**, 1.
13. Rogers, R.L. and Wright, T.K. (1986) *IChemE Symp Ser.*, **97**, 121.
14. Wilberforce, J.K. (1984). *IChemE Symp Ser.*, **85**, 329.
15. Fisher, H.G. et al. (1992) *Emergency Relief System Design using DIERS Technology.* AIChemE, New York.
16. Rogers, R.L. and Hallam, S. (1991) *Trans. IChemE*, **69B**, 149.
17. Regenass, W. (1984). *IChemE Symp Ser.*, **85**, 1.
18. Hermann, K. and Rogers, R.L (1995) Design of quench tanks using jet condensers. In *Proc. Int. Symp. on Runaway Reactions and Pressure Relief Design.* AIChE, New York.
19. IChemE (1990) *User Guide to the Prevention of Fires and Explosions in Dryers*, 2nd edn. IChemE, Rugby.
20. UN (1989) *Recommendations on the transport of dangerous goods.* United Nations, New York.
21. Field, P. (1983) *Explosibility Assessment of Industrial Powders and Dusts.* HMSO, London.
22. Field, P. (1982) *Dust Explosions Handbook of Powder Technology*, Vol. 4. Elsevier, Amsterdam.
23. Raftery, M.M. (1974) *Explosibility Tests for Industrial Dusts (Fire Research Technical Paper 21).* HMSO, London.
24. Gibson, N. and Rogers, R.L. (1980) *IChemE Symp Ser.*, **58**.
25. Bartknecht, W. (1981) *Explosions*, 2nd edn. Springer-Verlag, Berlin.
26. Gray, P. and Lee, P.R. (1967) *Oxidat. Combust. Rev.*, **2**, 1.
27. Bowes, P.C. (1984) *Self Heating: Evaluating and Controlling the Hazards.* HMSO, London.
28. Leuschke, G. (1976) Self ignition of powdered materials. In *Proc. VFIB 5th Int. Fire Protection Seminar*, p. 1.
29. Gibson, N., Harper, D.J. and Rogers, R.L. (1985) *Plant/Operations Progr.*, **4**, 181.

30. BS 6467. *Electrical Apparatus with Protection by Enclosure for use in the Presence of Combustible Dusts.* Part 1, 1985. Part 2, 1988. British Standards Institution.
31. VDI 3673 (1979). *Guidelines on Venting of Dust Explosions.* Verein Deutsche Ingenieure, Dusseldorf.
32. Gibson, N., Maddison, N., Rounsley, J.S. and Stokes, P.S.N. (1986) *IChemE Symp Ser.*, **97**, Rugby.
33. BS 5958. *Code of Practice for the Control of Undesirable Static Electricity.* Part 1: General Considerations, 1980. Part 2: Recommendations for Particular Industrial Situations, 1983. British Standards Institution.

10 Environmental protection and waste minimization
C. JONES

10.1 Introduction

Wherever a process is operated it will be subject to controls on its environmental performance. These usually regulate emissions to the environment, and increasingly force operators to reduce waste generation by process and equipment modification. While the legal framework differs from country to country, the impacts on process design and operation are similar. In this chapter, batch plant environmental performance will be considered in the context of UK regulations, although the approach presented applies equally well under other regulatory frameworks.

In the UK, the Environmental Protection Act 1990 (EPA 90) [1] provides a mechanism and a legal basis to examine the impact that a process has on the environment as a whole. This holistic approach ensures that substances that are unavoidably released to the environment are released to the medium to which they will cause least damage. Every prescribed process must meet a number of conditions to be authorized to operate.

- The process must apply the best available techniques not entailing excessive cost (BATNEEC) to prevent or, if that is not practicable, to minimize the release of prescribed substances into the medium for which they are prescribed, and to render harmless both any prescribed substances which are released and any other substances which might cause harm if released into any environmental medium.
- The releases from the process must not breach any agreed environmental quality standard or similar control.
- Where the process is likely to involve releases into more than one medium, the best practicable environmental option (BPEO) must be achieved, i.e. the releases from the process are controlled through the use of BATNEEC so as to have the least effect on the environment as a whole.
- Waste minimization is a key element in achieving good environmental performance. While abatement technologies (which will be dealt with briefly later in the chapter) can reduce emissions, the most significant improvements are often found by design and operation of processes so as to avoid the generation of wastes in the first place. The primary focus of this chapter is thus the application of waste minimization to batch processes.

The definition of waste given under Section 75 of Part II of EPA 90 [1] is:

(i) Any substance which constitutes scrap material or an effluent or other unwanted surplus substance arising from the application of any process.
(ii) Any substance or article which requires to be disposed of as being broken, worn out, contaminated or otherwise spoiled.

The definition of waste minimization in the IChemE's *Waste Minimisation Guide* [2] is:

> Waste minimisation involves any technique, process or activity which either avoids, eliminates or reduces a waste at its source, usually within the confines of the production unit, or allows reuse or recycling of the waste for benign purposes.

Once waste has been created we cannot destroy it. All we can do is concentrate it, dilute it or change its physical or chemical form. Wastes fall into two main types, process and utility. The process waste generated is dependent on synthesis route, reactor designs, separation processes, recycle systems, control systems and operating procedures. The generation of utility waste is related to heating and cooling loads, the heat transfer efficiencies and the methods used to generate the heating and cooling streams.

A battery of waste minimization tools has been assembled from existing chemical engineering methods, e.g. by Smith and Petela [3–7]. The application of many of these methods is dependent on a good understanding of the reaction chemistry and thermodynamics. However, the industrial sectors which commonly use batch processes must often respond rapidly to changing customer needs and market opportunities. The tendency is therefore to be product rather than process driven. The result is a frequent lack of accurate physical and chemical property information, making the use of many techniques based on simulation and modelling difficult or impossible.

Batch reactions are predominantly carried out in the liquid phase, usually in a solvent. The use of batch processing and solvents often results in the generation of large quantities of liquid wastes during the product separation. Typically the quantity of liquid waste generated would be in the range 1–50 times the quantity of product produced in each reaction stage [8]. Filtration is one of the operations prominent in the generation of liquid waste in batch processes. Other major sources include batchwise solvent extraction operations and the cleaning of plant between batches or campaigns [8–10].

Waste can be minimized by: developing and using new processes/equipment that produce less waste; by changing operating conditions; by improving the process control; or by altering the mode of operation of the system [4]. Increasing the percentage conversion/recovery at any stage in the process will mean less raw materials are needed and potentially a smaller number of batches being required, again reducing waste.

Methods of undertaking waste minimization programmes can be found in a number of texts including the IChemE's *Waste Minimisation Guide* [2].

Having identified target areas two useful approaches to waste minimization in batch processes are the use of checklists [9, 11] and the use of methods which rely on restricted measured data [12]. The key factors affecting the generation of waste in different unit operation and a number of checklists are presented here. These are designed to highlight the areas of concern, possible causes and possible solutions. They provide a starting point for a technologist embarking on a waste minimization programme or the optimization of a unit's performance.

10.2 Batch reactor waste minimization

Probably the most important parts of a batch plant for waste minimization are the reaction stages. They are at the heart of the process and their demands and performances will affect the waste produced by the whole system. There are five root causes of process waste generation in batch reactors [13]:

- Low conversion (if recycling is difficult).
- Waste by-products from primary reaction.
- Waste by-products from side reactions.
- Feed impurities undergoing reaction and/or requiring separation from products.
- Catalyst waste

10.2.1 Process chemistry

The generation of reactor waste is dependent primarily on the fundamental chemistry of the process, i.e. the chosen synthesis route. Greater understanding of the chemical equilibria and kinetics of a reaction creates greater scope for reactor performance optimization and therefore waste minimization.

10.2.2 Heat effects

The mixing and heat transfer characteristics of the reactor systems can also affect the overall efficiency of the system due to the formation of dead spaces and hot spots. For example, consider an exothermic reaction being carried out batchwise in a stirred tank reactor. There are two reactants A and B. A is charged to the reactor and B is then dosed in slowly over a period of several hours to ensure the reaction temperature is not exceeded. The primary reaction is

$$A + B \rightarrow \text{product} \tag{10.1}$$

However, the main source of waste has been identified as a secondary reaction

$$B + \text{product} \rightarrow \text{waste by-product} \tag{10.2}$$

Laboratory tests indicate that by-product formation increases with temperature. Using this information a number of options are available to reduce waste generation:

- Improving heat transfer, e.g. using a colder cooling agent, therefore minimizing the temperature rise caused by the exothermic reaction.
- Improving the agitation, thereby reducing the creation of local hot spots in which by-product formation will be increased.
- Reducing the rate of addition of B, or by adding it in dilute solution, to minimize the excess available in solution to form the by-product.

10.2.3 Mixing and contacting pattern

The choice of reactor type will determine the fluid dynamics of the system and influence waste formation. For example, consider a pair of parallel reactions:

Feed 1 + Feed 2 → product

$$\text{rate}_1 = k_1 (C_{\text{Feed 1}})^{a_1} (C_{\text{Feed 2}})^{b_1} \quad (10.3)$$

Feed 1 + Feed 2 → by-product

$$\text{rate}_2 = k_2 (C_{\text{Feed 1}})^{a_2} (C_{\text{Feed 2}})^{b_2} \quad (10.4)$$

To minimize the formation of the waste by-product the following ratio must be minimized [4].

$$\frac{\text{rate}_2}{\text{rate}_1} = \frac{k_2}{k_1} (C_{\text{Feed 1}})^{a_2 - a_1} (C_{\text{Feed 2}})^{b_2 - b_1} \quad (10.5)$$

In a batch plant a number of strategies could be used to minimize the formation of unwanted by-product, the selection depending on the relationship between a_1 and a_2 as well as b_1 and b_2 Some of these are listed in Table 10.1. Note that detailed knowledge of the kinetics is not required to assess strategy – simple factorial experiments could provide sufficient information to judge which situation prevails.

Table 10.2 shows a summary of some reactor intensification options aimed at matching reaction kinetics and reactant mixing.

10.3 Equipment for the production of solid products

The final product or intermediates in batch processes are often solids. When considering opportunities for waste minimization in both the design and operation of batch solid production processes it is important to consider the solid formation system as a whole. Consideration needs to be given not

Table 10.1 Reactor operation strategies to minimize by-product formation in competing reactions

Situation	Strategy
$a_2 > a_1, b_2 > b_1$	Minimize the concentration of both feeds. Use continuous addition of both into batch. Consider pre-dilution of feeds.
$a_1 > a_2, b_2 > b_1$	Minimize the concentration of Feed 1. Dose Feed 1 into batch of Feed 2. Consider pre-dilution of Feed 1.
$a_1 > a_2, b_2 > b_1$	Minimize the concentration of Feed 2. Dose Feed 2 into batch of Feed 1. Consider pre-dilution of Feed 2.
$a_1 > a_2, b_1 > b_2$	Maximize the concentration of Feed 1 and Feed 2. Rapid addition and mixing desirable. Consider use of 'T' pipe or static mixer reactor if reactions are fast.

only to the number and type of waste streams that will be produced but also how the design and operation at one stage will affect the generation of wastes in others.

The decisions made at one stage will have a knock-on effect on the subsequent ones. For example, in reducing a crystallizer cycle time to remove a production bottleneck, the product crystal size may be reduced. This may increase the waste generated in a downstream filtration stage by changing the product's filtration properties. In order to minimize drying heat requirements, a low initial moisture content is desirable. This, however, will increase the separation problem in a filter or centrifuge and may increase overall waste generation. Such operations should be considered together, and not in isolation, as only in this way can the most economic and efficient process to be selected.

The following sections cover the three main steps in the production of solid products:

- Solid formation, crystallization/precipitation.
- Solid–liquid separation, filtration/sedimentation.
- Drying.

Table 10.2 Some reactor intensification options

Type of reaction	Mixing required for maximum productivity and yield	Possible intensified reactor system
Fast, liquid phase reaction	Rapid mixing to bring reagents together	Motionless mixer, ejector or rotor-stator mixers.
Liquid reaction with solids present	Sufficient mixing to ensure adequate mass transfer	Stirred tank reactor with the reagents fed in near the impeller.
Precipitation	Rapid, low shear mixing	Pipeline 'T' mixers.

The approach taken here is to list possible waste minimization options in checklists. These are likely to be the starting point for modelling or experimental investigations.

10.3.1 Crystallization processes

The particle size distribution achieved during a crystallization process will have an effect on the efficiency of the down stream processes and hence the associated waste generation. During batch crystallization, the purity, shape, size distribution, degree of agglomeration, structure, voidage and slurry density of the crystalline material are among the properties that need to be considered. Uniform and narrow crystal size distribution (CSD) and a large mean crystal size are often desirable attributes [14, 15]. This makes solid–liquid separation easier and reduces solid waste (fines loss) and liquid waste (wash liquor requirements).

There are two main areas of crystallizer operation open to waste minimization that affect the unit itself. These are reduction of encrustation on the heat transfer surfaces and the reduction of the vapour losses from evaporative systems. Table 10.3 is a checklist for the minimization of waste in crystallization processes.

Any crystallization process will inevitably be a balance of the maximum possible yield of the largest possible crystals, against an acceptable batch time and overall process productivity.

10.3.2 Precipitation processes

Solids are also formed by precipitation where two liquids are brought together and solid particles are formed rapidly either by a chemical reaction or by displacement of a dissolved substance from the solution. The important parameters which can be varied in a stirred tank precipitator are concentration, pH, temperature, flow rate and agitator speed.

It is important for the formation of large particles that the amount of solids to be formed is small compared to the amount of solids already present, so that the newly formed solids can agglomerate on existing particles. Precipitation ideally requires rapid, low shear mixing, this is best achieved in pipeline 'T' mixers (see Figure 10.1). The crystals produced by precipitation are usually smaller and more difficult to filter than those generated by crystallization.

10.3.3 Solid–liquid separation

Solid–liquid separation may be achieved by filtration, use of hydrocyclones or sedimentation. The efficiency of these operations not only determines the product recovery efficiency but also the energy that will be required to

Table 10.3 Checklist for the crystallizer optimization and waste minimization

Problem	Possible solutions	Effects
Crystal size too small (decreasing downstream efficiencies)	The degree of secondary nucleation depends on the power dissipated per unit mass of suspension, so slow down the mixer to reduce number of nuclei.	Decrease in waste generation and cycle time in subsequent stages. May increase crystallizer cycle time.
	Slow down the rate of cooling to decrease spontaneous nucleation.	As above.
	Temporarily increase the temperature at the end of crystallization [15] to remove small crystals.	As above. Utility waste will increase from crystallizer.
	Reuse part of the suspension left in the crystallizer from the preceding batch to seed the next batch.	Decrease in waste generation and cycle time in downstream stages. Crystallizer cycle time may improve.
	Increasing the size of the seed nuclei.	As above.
	Change the concentration or the temperature profile of the solution in the crystallizer.	Change in waste generated and cycle time in subsequent stages.
	May be due to the impurities, consider the purification of feed.	As above, but may increase upstream waste.
Crystal growth too slow (bottleneck)	Impurities inhibiting crystallization: consider the purification of the feed prior to crystallization.	Decrease in waste and time in subsequent stage, but may increase upstream waste. May decrease overall batch time and utility wastes.
	Increase the degree of agitation.	As above, but will increase the electrical power required.
Crystal size distribution too wide	Consider changing the power dissipated per unit mass of suspension.	Less waste in subsequent stages.
	Consider changing the concentration or the temperature profile of the solution in the crystallizer.	As above.
Poor heat transfer/ encrustation	Large initial temperature differences on cooling can lead to heavy encrustation on the cooling surfaces and therefore decrease the heat transfer. Try different cooling profiles. Consider temporary heating of cooling surfaces to remove encrustation during batch.	Reduction in cleaning wastes. May change overall batch time and amount of utility waste.

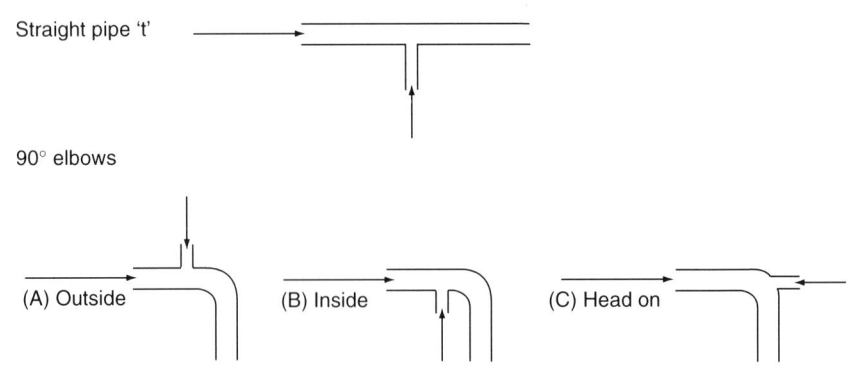

Figure 10.1 Pipeline and 'T' mixers.

produce the final dried product and the quantity of washing liquor required to ensure product quality.

(a) Filtration
Much work has been carried out on filtration as a unit operation [16–19]. Cake dewatering and washing have also been the subject of research [17, 18]. In batch processes the recovery of a valuable solid from a suspension with the liquid being discarded or recycled is common. The opportunities for waste minimization in this type of filtration are often consistent with optimal operation of the filtration; methods of improving the filtration may also lead to the generation of less waste. Equally, some process improvements, such as the use of filter aids, may ultimately add to waste production.

The liquid effluents arising directly from batch filtration are the residual liquid (mother liquor) removed from the solid during the initial filtration and any wash liquor used to remove impurities and mother liquor from the filter cake. A large number of factors can influence the quality and quantity of these wastes.

- The particle size, particle size distribution, particle shape and the degree of agglomeration of the solid will all influence the filtration properties of the cake, e.g. voidage, resistance, compressibility and tendency to crack.
- The solid will also influence the choice of wash solvents and restrict the type of wash techniques which can be used.
- The liquid phase will have an impact on both the filterability and the washing efficiency of the cake. The composition of the filtrate will affect its recovery potential.
- The washed cake must meet the product specifications for sale or further processing; this will be a function of the amount and composition of the liquid remaining in the cake after the last wash.

The features of batch filtration have already been described in Chapter 6. Typical batch filters are the simple Nutsche (vacuum and pressure), automated Nutsche, leaf filters and tubular filters. Table 10.4 is a checklist for waste minimization and process optimization opportunities that may arise in the operation of a filtration system. The key targets for waste minimization in filtration processes are: to minimize the amount of wash liquor required and to maximize the recycle or recovery potential of the wash liquors.

Two main washing techniques are commonly used – slurry and displacement. In slurry washing, the wash liquor is added to the filter and the material in the filter is agitated to mix thoroughly the liquid and suspend the solid. The slurry is then filtered in a similar way to the initial filtration. The key areas that affect the volume and concentration of waste from slurry washing are:

- The amount of liquid required to suspend the solid.
- The adsorption of solute on to the surface of the solid.
- The proportion of material not suspended.

Table 10.4 Opportunities for waste minimization and process optimization for filtration

Area of concern	Possible solutions	Effect on waste
Initial fines pass through the filter medium	Recirculation of initial filtrate.	Less product lost. Increased pumping costs.
Slow filtration due to small particle size	Production of larger crystals at the crystallization stage.	Depends on the crystallization, but usually a slower process, therefore longer batch time.
	Use of flocculation/coagulation	The flocculant or coagulant appears in waste.
Slow filtration due to small medium pore size	Use a large mesh size and increased recirculation of the initial filtrate.	Some fine product may enter waste streams.
Slow filtration due to filtrate properties	Dilution of the feed to reduce the viscosity.	Increase amount of filtrate waste and decrease in its recycle potential.
	Heating the feed to reduce the viscosity.	No increase in the volume of filtrate. Increased heating losses. Increased loss of product.
Slow filtration due to particle form	Change the crystallization process.	Depends on the crystallization process.
High resistance filter cake	Change solids concentration prior to filtration, may give a lower resistance cake.	Will change amount and recycle potential of filtrate waste.
Slow filtration due to pore plugging	Decrease the initial feed rate.	

In displacement washing, the liquid is passed through the cake without agitation. The wash liquor displaces the residual liquid. Typically, a displacement wash will remove 90% of the liquid held up in the cake [16]. The remaining liquid is retained in isolated pockets in the cake, and cannot be displaced. Instead, it is removed by diffusion washing as the residual liquid diffuses into the wash liquor, as it passes through the cake. Diffusion washing is usually inefficient, producing large quantities of low strength waste. The key factors in the generation of waste in displacement washing are:

- Solute being retained inside particles or adsorbed on to the solid surface.
- Uneven distribution of wash liquor at the surface.
- Axial dispersion as the wash passes through the bed.
- Cake cracking (which allows bypassing of wash liquor).

The filter design will determine the maximum allowable wash volume, the type of washing that can be undertaken, the efficiency of wash liquor distribution on the cake in displacement washing, and the efficiency of the mixing of wash liquor and cake in slurry washing.

The number and type of washes used will be constrained by equipment and by the cycle time allowed for the whole washing cycle. The recycling of washes, e.g. by retaining the final wash for use as the first wash for the next batch, can significantly reduce waste. This will be achieved not just by the reduction of the quantity of wash liquor required but also by the filling in of any small cracks with fines carried over in the final wash. Table 10.5 presents a qualitative checklist for waste minimization and process optimization opportunities during cake washing.

The use of different filters and different filter wash strategies can create great variations in the quantity and quality of waste arising from batch filtration. There is a trade off between the capital costs of a filtration (e.g. additional receivers, filter facilities) and the waste generation. Some general principles are:

- Blowing the cake dry before washing has substantial benefits, both in the quantity of high strength waste recovered and in reducing the overall volume of waste.
- The use of cake smoothing is effective in reducing bypassing where cake cracking occurs.
- Recycling of later washes for use as initial washes on the next batch reduces waste substantially (but does require additional process vessels).
- Slurry washing, while inefficient at high concentrations, has benefits over displacement as a final wash.
- Filters which allow flexibility in operation are likely to give the greatest opportunities for waste minimization. Features which contribute to this are the ability of perform both slurry and displacement washes, the ability to smooth the cake and the ability to blow through the cake to dry it.

Table 10.5 Opportunities for waste minimization during batch filter cake washing

Area of concern	Possible solutions	Effects
Increased washing required due to high initial moisture content	Increasing the final filter pressure (this could lead to cake compression problems).	
	Cold blow-through prior to the start of washing to remove liquid (this could lead to channelling problems).	Less waste produced and it will be less contaminated and therefore have a higher recycle potential.
High loss of fines in wash liquor	Recycle initial wash liquor.	Decrease in the fines lost. Increased pumping requirements and cycle time.
Slow wash rate due to cake porosity or compression	Lower final filter pressure.	Leads to increased initial moisture content, therefore increased contaminant level in wash.
	Change the form of the cake particles and the overall structure.	
Poor wash efficiency due to channelling	Lower the pressure drop.	Leads to increased initial moisture content, therefore increase contaminant level in wash.
	Recycling of fines to plug the holes.	Less wash liquor required. Increased pumping cost.
	Cake smoothing.	Increased batch cycle time. Less, more concentrated waste.
High wash liquor requirements or too many washes required to fit the cycle time	Changing the wash regime.	Less wash liquor.
	Improving the distribution of the wash liquor on to the cake (displacement wash).	As above.
	Improving the mixing of the wash liquor and the cake liquid hold-up (reslurry wash).	As above.
	Using an immiscible wash liquor (displacement washing).	Will depend on the liquor and its recyclability.
Poor recycling potential of the used wash liquor	Changing the wash regime.	Less wash liquor.

(b) Centrifugation

Centrifuges can be classified by their separation method as either sedimentation or filtration devices. Centrifugal filters are used to collect the solids from a solid–liquid slurry. The feed slurry concentration is very important since the mechanism of separation is by liquid flow through a bed of solids and this liquid flow is the rate controlling step. A suspension to be separated by centrifugal filtration must be free of extremely small particles, to prevent the top layer of the sediment being impermeable. A pre-treatment using a hydroclassifier to remove very small particles may be required (desliming).

Variables in batch centrifugation performance are rotation speed, liquid viscosity and density, particle diameter and density, bowl radius, and volume of suspension in the batch. There are a number of areas of centrifuge operation which present opportunities for waste minimization and process optimization. They are summarized in Table 10.6.

10.3.4 Batch drying

There are two types of moisture to be removed in the drying process – bound moisture and free moisture. Bound moisture differs in that its equilibrium vapour pressure is less than that of the pure liquid. There are also two stages in the drying process: the constant rate period (where the rate of liquid removal is constant) and the falling rate period (where the drying rate decreases with time at constant temperature). Under conditions where diffusion is not important, the constant rate period corresponds to the removal of free moisture. The primary environmental concerns associated with dryers are contaminated gas streams and energy usage. Often, these

Table 10.6 Checklist of opportunities for waste minimization in batch centrifugation

Area of concern	Possible solutions	Effects
High cake compaction	Lowering the centrifugal acceleration.	
High cake resistance	Lowering the speed of rotation.	
	Reducing the number of peeling operations before cleaning.	More washing waste produced but shorter cycle time.
	Trying to reduce crystal breakage during discharge.	May improve product quality.
	Desliming to remove the fines before the centrifuge.	The fines can be recycled.
Slow cake drainage	Lowering the viscosity of the liquid.	
	Using a surfactant to decrease the surface tension of the liquid (particularly with water).	Surfactant will end up as waste.
	Changing the porosity, permeability and height of the cake.	

are linked – dryers requiring high gas flows will have greater heat requirements for the gas and will produce greater volumes of contaminated gas.

The thermal efficiency of batch drying increases with the load and decreases with drying time, and the normal efficiency range is 50–60% [20]. The most efficient way of drying a batch, in terms of heat utilization, involves using different drying techniques at different stages of the process. The first stage of the drying process should be a cold blow through to decrease the initial moisture content by liquid entrainment. During the second stage (constant rate drying), the gas mass velocity, and thermal and physical states of the product are important. The most efficient way to supply heat during this stage is hot gas and agitation, since the drying takes place at the saturated surface of the material and pick up depends on temperature and humidity. Hot gas recirculation during this period is optimal at between 80 and 90% [21]. During the third stage, the falling rate period, the cake properties and the heat input are the controlling factors. The use of heated surfaces combined with vacuum drying will produce a steep falling rate curve [22].

Improving the thermal efficiency of a drying system offers a good opportunity for waste minimization. Table 10.7 summarizes the different methods available for increasing the thermal efficiency of the various types of batch dryers and provides some suggestions for efficient operation.

Selecting the appropriate dryer for a particular process can be difficult, since classification of the product according to its physical state does not provide a simple way to identify the optimum type of dryer. Usually, drying trials will be necessary to select the most effective type of dryer. It is also important to note that appropriate control and instrumentation of dryers can bring substantial savings in cycle time, energy usage and waste generation.

10.4 Fugitive and other minor emissions

Fugitive emissions are minor leaks and emissions from process equipment. They are not emergency or abnormal events and equally are not usually thought of as effluent streams leaving a process. Indeed, they are seldom considered in any detail by the process designer. Such emissions are from multiple sources, primarily flanges, valve seals, pump seals and vents. Other typical sources of minor emissions (sometimes classified as fugitives) are storage tanks and vessels, drum transfers, spillages and process drains [24]. Typical ways of reducing emissions arising from outbreathing of vessels during transfers include, vapour collection and treatment, use of floating roofs, membranes, and back venting (piping the tank vent back to the feed tank).

There are a number of factors that will affect the fugitive emissions from valves, pumps and flanges. These include the type and size of the process,

Table 10.7 Opportunities for increasing the thermal efficiency and waste gas generation of batch dryers

Type of dryer	Possible method	Comments
Tray (atmospheric)	Recycle the hot gas.	Closed loop substantially reduces emissions (only viable if the vapour can be recovered). If vapour recovery is possible at a temperature greater than ambient, heat load will also reduce.
	Change drying medium from hot gas or gas to superheated steam 'airless drying' [23].	Improves the recovery of sensible heat from the gas outlet stream.
	Pre-form the feed to increase the effective surface area.	Depends on the feed. Can reduce heat requirement and make emissions lower in volume but more concentrated.
	Use rotating shelving.	Involves more complex equipment and depends on the feed properties. Can reduce heat requirement and make emissions lower in volume but more concentrated.
	Improve gas distribution, by using vane bends, flow smoothing grids or swirl reduction techniques.	Involves more complex equipment. Can reduce heat requirement and make emissions lower in volume but more concentrated.
	Improve gas distribution by using back to front circulation instead of side to side.	May involve more complex equipment. Can reduce heat requirement and make emissions lower in volume but more concentrated.
Vacuum	Minimize the leaks into the system.	Reduces both heat requirement and contaminated off-gases. There is a balance between cost and benefit.
	Improve the lagging on the external hot surfaces.	Cost optimization should be considered.
General	Use an initial cold blow-through to minimize the initial moisture content.	Suitability depends on the dryer system. May increase thermal efficiency at expense of greater amount of dilute gaseous effluent.
	During the constant rate drying period supply the heat as hot gas.	During this phase the gas velocity, thermodynamic and physical state of the product are important.
	During the falling rate period, supply the heat using heated surfaces and remove the evaporated liquid by vacuum.	During this phase the cake properties and the heat input are the controlling factors.

the operating conditions and the type/size of the valves, pumps or flanges. Pumps and compressors tend to leak through their seals [24]. Retrofitting pumps with double or tandem mechanical seals will reduce emissions, for problem services, canned-motor (seal-less) and magnetic drive pumps may be applicable. The releases from pumps and valves can account for over 50% of the fugitive emissions from a plant.

Fugitive emission rates are considerably lower from valves than those from pump seals. However, valves are the main contributor to fugitive emissions from a plant due to their large number. The main route for a potential leak from a valve is along the valve stem. The emission rate from a valve increases as the size increases. New developments in technology have led to valve packings which incorporate an anti-extrusion ring and live loading. This can result in low leakage levels and long seal lives.

Flanges tend to leak through their gaskets [24], the larger the flange, the higher the release rate. Other factors which may affect the emission rate are, the gasket, flange design, bolting procedures, flange face surface condition, both specifications, both tightening procedures and the stresses in each bolt.

10.5 Cleaning wastes

The minimization of cleaning waste can be broken down into three main areas:

- Improved methods of washing.
- Reuse of solvent washes.
- Improved scheduling.

Two methods that can improve washing are jet and spray washing. Jet washing is quicker than other cleaning methods and so saves on plant downtime. It produces less wastes than solvent cleaning. Spray washing uses a hollow sphere drilled with a large quantity of holes, enabling the cleaning fluid to be distributed over the tank or vessel's internal surface. They operate at low liquid velocities and because they require larger flow rates than the jet washing systems are less efficient.

Solvent washes are often used for the cleaning of vessels after a batch. The size, number and quality of these washes can be modified to minimize the quantity and maximize the quality of them. It is possible to modify

- The number of individual washes.
- The effluent strengths.
- The choice of extractant.

A single-stage batch wash process is very inefficient; if time allows it is more efficient to use two or more smaller washes. Significant savings can also be made by recycling less contaminated material from the second wash for use in the first wash after the next batch.

10.6 Waste treatment and solvent management

The treatment of waste streams once they have been produced is also important in a waste minimization programme. Process waste streams are often mixed in an attempt to render pollutants as dilute as possible ready for biological treatment. However, the mixing of streams prematurely can often inhibit recovery of useful materials. The collection and segregation of waste can help in recovery and recycling.

During design there are four main ways of maximizing the reuse and recycling of materials (particularly solvents) [25]: segregation, substitution, downgrading and recycling. Waste stream recycling (direct reuse) is by far the simplest way of reducing waste, if it is possible within the constraints of the process and the relevant regulations. A typical example here would be recycling of solvent evaporated from one batch as a raw material in the next [5].

Downgrading the use of a solvent may involve the use of pure solvent in a reaction step. After the solvent has been boiled off and condensed it may be directly recycled to an extraction step, where traces of another solvent will not impact on the extraction.

The methods of recycling or reclaiming materials will be determined by: considering the chemical compositions, identifying the data requirements for the treatment, generating a list of potential technologies and identifying which technologies can be used. It is always important to consider whether trace quantities of highly toxic materials are more significant than larger quantities of relatively inert materials.

When batch distillation is used to recover spent solvent materials, it is usually only economic if the mixture is greater than about 50% solvent by weight. The separation of batches of solvent using heat can be divided into simple and fractional distillation. Simple distillation involves no reflux and is good for quick, crude separation of solvents. The degree of separation is not high but the energy requirements are low.

There are two main modes of operation in batch fractional distillation; fixed reflux and variable reflux. Operating at a fixed reflux ratio will lead to the purity of the top product steadily decreasing over the duration of the batch. This is easy to achieve and gives much better separation of solvents than simple distillation. The boil-up requirements for low degrees of separation are higher than for simple distillation. Operating with an increasing reflux ratio will help maintain the top product composition. This is more difficult to achieve, although modern control technology makes it possible. The ability to specify the degree of purity of the overhead product can be a great advantage. Nevertheless it can be more difficult to achieve low levels of the volatile component in the still. This strategy will use the least energy to produce a high level of overhead product purity. The recovery of solvents by the use of batch distillation and other techniques has been thoroughly described by Smallwood [27].

10.7 Environmental protection

The ever increasing environmental pressures on industry means waste minimization and minimizing the impact of any remaining emissions are integral in the design and operation of manufacturing processes. In the UK this is embodied in the establishment of which processes or abatement systems are considered to be BATNEEC and BPEO.

A waste minimization programme will identify and quantify the sources of waste associated with a process. Once these discharges have been minimized steps needs to be taken to render them harmless by the application of BATNEEC. To evaluate what constitutes BATNEEC for the abatement of an emission, its frequency, type and concentration need to be established. Data relating to the allowable discharge limits, the toxicity and odour detection thresholds of the various components of the stream are also required. Having identified the allowable discharge rate, abatement systems capable of meeting these limits have to be identified.

There are an increasing number of technologies available to treat emissions, e.g. there are now at least 12 different technologies available for the treatment of gaseous VOC emissions. Seven are standard technologies: thermal oxidation, catalytic oxidation, condensation, adsorption, absorption, flaring and boilers/process heaters. Three others have recently become commercially available: membrane separation, UV oxidation and biofiltration, while corona discharge destruction and plasma technology are moving towards commercialization.

Abatement systems will require raw materials and produce secondary emissions either directly (by conversion of the feed contaminants into other substances) or indirectly (due to the use of heating, cooling and power). Therefore consideration has to been given not only to the abatement equipment that can be used to reduce an emission to one medium the most effectively, but also to the impact on the other media, e.g. a thermal oxidation unit will reduce the concentration of VOCs in a gaseous discharge but there will be an associated release of combustion products, including NO_x and carbon dioxide, that also have impacts on the environment. Thermal oxidation units also often require support fuel, which will also generate combustion products.

There will inevitably be some iteration during the evaluation of abatement equipment, since the design and operation of gaseous abatement systems will impact of the wastewater generated on the site and the design and operation of the wastewater treatment system will impact on the gaseous emissions generated by the site. The transfer of pollutants from one medium to another needs to be carefully assessed to ensure that the BPEO is achieved.

10.8 Conclusions

Undertaking a waste minimization programme during the design or operation of a plant should focus attention on reducing the impact of the process and also increasing productivity, due to increased efficiency, representing a positive effect on the industry. The guidance and checklists in this section should provide a basis from which to achieve this.

References

1. UK Government (1990) The Environmental Protection Act.
2. Crittenden, B.D. and Kolaczkowski, S.T. (1994) *Waste Minimisation Guide.* IChemE, Rugby.
3. Smith, R. and Petela, E. (1991) Waste minimization in the process industries, Part 1: the problem. *Chem. Eng.*, **506**, 24.
4. Smith, R. and Petela, E. (1991) Waste minimization in the process industries, Part 2: reactors. *Chem. Eng.*, **509/510**, 17.
5. Smith, R. and Petela, E. (1991) Waste minimization in the process industries, Part 3: separation and recycle systems. *Chem. Eng.*, **513**, 25.
6. Smith, R. and Petela, E. (1991) Waste minimization in the process industries, Part 4: process operations. *Chem. Eng.*, **517**, 21.
7. Smith, R. and Petela, E. (1991) Waste minimization in the process industries, Part 5: utility waste. *Chem. Eng.*, **523**, 32.
8. Caughlin, T.R. (1993) A study of waste and pollution management in the fine chemicals manufacturing sector. MSc Dissertation, University of Manchester Institute of Science and Technology.
9. Jones, C.E. (1993) Opportunities for waste minimization in batch processes. MSc Dissertation, University of Manchester Institute of Science and Technology.
10. Krichphiphat, A. (1992) Waste minimization in speciality chemicals production plant. MSc Dissertation, University of Manchester Institute of Sceience and Technology.
11. USEPA (1991) *Guide to Environmental Assessment Compliance.* USEPA, Washington, DC.
12. Sharratt, P.N. (1993) Waste minimization for batchwise extraction. *IChemE Symp. Ser.*, **132**.
13. Smith, R. and Petela, E. (1992) Waste minimization in the process industries, In *Proc. IChemE NW Branch Symp.*, 20–21 May. IChemE, Rugby.
14. Rohani, S. and Bourne, J.R. (1990) Self-tuning control of crystal size distribution in a cooling batch crystalliser. *Chem. Engng Sci.*, **47**, 3457.
15. Garside J., Davey R.J. and Jones A.G. (1991) *Advances in Industrial Crystallization.* Butterworth–Heinemann, Oxford.
16. Coulson, J.M. and Richardson, J.F. (1987) *Chemical Engineering*, 3rd edn, Vol. 2. Pergamon, Oxford.
17. Wakeman, R.J. (1985) In *Mathematical Models and Design Methods in Solid–Liquid Separation. NATO ASE Series E, No. 88* (Rushton, A., ed.). Martinus Nijhoff, Dordrecht.
18. Hermia, J. (1985) In *Mathematical Models and Design Methods in Solid–Liquid Separation. NATO ASE Series E, No. 88* (Rushton, A., ed.). Martinus Nijhoff, Dordrecht.

19. Svarovsky, L. (1990) *Solid–Liquid Separation*, 3rd edn. Butterworth, London.
20. Backhurst, J.R. and Harker, J.H., (1972) *Process Plant Design*. Heinemann, London.
21. Nonhebel, G. and Moss, A.A.H. (1971) *Drying of Solids in the Chemical Industry*. Butterworth, London.
22. Perlmutter, B.A. (1991) Combined filtration and drying. *Chem. Engng Progr.*, **87**, 29.
23. Butcher, C. (1993) All steamed up. *Chem. Eng.*, **539**, 16.
24. Shiel, H.H., Siegell, J.H. and Jones, A.L. (1992) Modelling plant hydrocarbon emissions. *Chem. Eng.*, **502**, 21.
25. ERL (1991) *Pollution Control for Chemical Recovery Processes*. DoE report No. DoE/HMIP/RR/92/029.
26. Mukhopadhyay, N. and Moretti, E.C. (1992) *Current and Potential Future Industrial Practices for Reducing and Controlling Volatile Organic Compounds*. Centre for Waste Reduction Technologies, AIChemE, New York.
27. Smallwood, I.M. (1993) *Solvent Recovery Handbook*. Edward Arnold, London.

11 Future developments in batch process design and technology

P.N. SHARRATT

This chapter does not set out to present a definitive (and thus almost certainly wrong) prediction of the future. Nor does it set out to list and explain the ever-growing range of technologies being sold to the batch sector. Instead, it discusses the challenges that must be faced and identifies a few of the more important new approaches to problems.

Many of the preceding chapters have already identified equipment or techniques that are novel or emerging in the batch process industries – static and 'T' mixers, combined reactor filters, merging control standards, use of supercritical fluids, etc. These chapters perhaps provide the best predictions of the immediate future in each of the fields covered.

11.1 Influences for and against change

In many ways the batch process industries are well adapted to change. They specialize in bringing novel compounds to market quickly and possibly using unfamiliar chemistry. However, the core technology, the stirred tank reactor, has changed little in the past 50 years. The failure of various new techniques to bring significant change to batch processing results partly from their inappropriateness, but also from the barriers that must be overcome. The barriers to change include the following.

- There is a large investment in existing technology. Because equipment is frequently reused, new technologies are competing against equipment that has already been paid for.
- The implicit requirement to reuse equipment means that technologies that are not versatile are less acceptable.
- A new technology may require the entire process to be redesigned in an innovative way. For example the use of supercritical solvent for a process would also make substantial changes to mechanical design, control, sampling, materials introduction and removal, etc. The cumulative risk of multiple pieces of innovative technology together may be unacceptable.
- New technologies may require extended development periods that are not acceptable in the normal rapid framework of process development and whose costs cannot be borne by any individual product.

- There are traditional boundaries between academic disciplines, when the skills required for process development lie between the areas of chemistry and chemical engineering.

These factors lead to technical conservatism and risk-adversity. However, these barriers are as much perceived as real, and a growing body of pressures on the industry sectors involved is forcing re-evaluation of the underlying assumptions. Pressures suggesting that change is required include the following.

- A growing environmental awareness among many groups in society. The large quantities of 'hazardous' waste generated by the batch organic chemicals sector has become a growing burden. Disposal routes such as landfill and incineration are becoming unavailable or increasingly costly as a result of tightening regulations and pressure from various interested parties. Some intrinsically dirty processes are being replaced with cleaner syntheses. Doing this presents new technological challenges.
- The competitive commercial situation that places ever growing emphasis on speed to market with new products.
- Safety concerns that could be alleviated by reduced inventories of material in process.
- The need for better levels of containment for ever more active products.
- A desire to reduce the working capital tied up in large materials inventories.
- Concerns about the efficiency and effectiveness of production and supply change planning and management.
- The production of bulk active materials and generic pharmaceuticals puts pressure on margins and process efficiency.

It seems likely that the forces for change will have a substantial impact in the medium term (10–20 years), bringing new chemistry, new technologies and new design methods.

11.2 New technologies

New technologies of interest for batch processing include new or enhanced unit operations, new plant layout philosophies, intensified operations and new process chemistry.

11.2.1 *New or enhanced unit operations*

Developments in unit operations take many forms. Some are simply the enhancement of well-known operations to improve their performance in existing duties. Pressure enhanced centrifugation extends the range of solid–liquid separation by reducing the cycle time for what by other means would be prohibitively long operations. Other developments are new ways of carrying out operations. Microwave drying replaces hot gas with microwaves that are used to introduce the energy required for drying a damp solid. This has the potential to reduce substantially the generation of gaseous effluent.

Technology developments may open up new process possibilities. Membranes have been available for many years, but recent developments to increase their stability to extremes of pH and to solvents [1] could allow substantial process improvements, e.g. they could allow the recovery of caustic from wash liquors, an operation that would seldom be considered, reducing effluents and saving raw material costs.

The use of 'power fluidics' as a means to carry out high turbulence intensity operations can extend the range of mixing intensity that can be applied. In mixing-sensitive processes, the turbulence intensity can be key to obtaining good performance (Chapter 5). Bowe has discussed the use of such systems for crystallization [2].

Process-scale chromatography is becoming a widespread operation, especially in pharmaceuticals manufacture. The use of chiral packing opens up the possibility of separating optically active molecules. Reactions involving racemic mixtures, where only one of the product isomers is desired, is wasteful of reactants and produces substantial quantities of waste. Separation of racemic mixtures before reactions to avoid these problems is a good example of clean technology [3].

11.2.2 New plant designs

As well as new technologies, there may be significant advantage to be gained from the use of innovative plant layouts. The concept of modular plant has already been mentioned in Chapter 7. Cellular plant design has been considered by several organizations. Broadly, this involves the design of 'generic' plant in the form of 'cells' that provide a group of process functions. Most often this will be in the form of a stirred tank reactor with the associated peripherals – heat/cool loop, overhead condenser, charging facilities. The cell has pre-designed piping layouts that link to the edge of the cell at standard connection points. When a new plant is to be designed for a given process, the process is fitted to the pre-designed cell, deleting any items or lines that are superfluous for that particular process. Once this is done, mechanical design is essentially instantaneous as the layout, steelwork, piping isometrics, etc., are already designed and need merely be printed. The potential savings in design time are substantial, but unless a lot of new plant is being constructed this approach may not be beneficial.

Pipeless plant is another concept that lends itself to the versatility required by many batch process industries. Process plant is made mobile – vessels on wheels – so that instead of having to erect pipework to change plant configuration, the vessel is moved [4].

11.2.3 Process intensification

Process intensification involves operations which give the same productivity using smaller plant with lower hold-up of materials. It can reduce working capital requirements, improve safety and may even improve yields. Consideration of a typical batch process reveals that much of the highly expensive capital equipment spends most of the time idle, or carrying out trivial operations such as charging, discharging, heating and cooling. The time spent in carrying out the key operations – reaction and separations – may only be a small proportion of the cycle. Furthermore, for intrinsically fast reactions, the reaction time itself may not be set by kinetics, but by the time taken to charge reactants. Thus, the theoretical time and volume required to carry out the processing steps may be only a tiny proportion of the time used to make the batch.

Perhaps the best example of process intensification would be the replacement of a large stirred tank reactor with an in-line or static mixer using continuous feed of reactants. On the surface it may seem that if such a possibility exists it should always be used. However, the saving may actually be rather small. If the operations upstream and downstream are carried out batchwise, then the upstream unit will have to act as a feed vessel and the downstream vessel as a receiver. Thus, the continuous reactor ties up two other units that may be greater in volume than the original reactor would have been.

In order to obtain the greatest benefit from a small, intensified, continuous operation it is necessary to link together two or more such operations in series, e.g. a continuous crystallizer could feed directly to a (continuous) rotary vacuum filter and a continuous drier. Of course, the design of continuous plant usually requires more accurate experimental data and more careful design – remembering that the capacity of all continuous equipment in series is limited by the slowest operation in the chain, so all operations need to be designed robustly.

Another problem generated by intensification is the need for more accurate and particularly more rapid control. Clearly, it is not possible to place the same reliance on sampling and analysis for process control. This raises the need for on-line or in-line chemical composition measurement, pH measurement, particle size measurement, etc. Of course, new measurement techniques are being developed all the time. For example, 'artificial nose' technologies bring the prospect of on-line measurements that can directly be linked to subtle changes in the chemical composition of solutions by 'sniffing' the vapour above the liquid [5].

The control of processes would, using intensified continuous processing, become more similar to the control problem of continuous plant. However, the need for versatility, plant reuse and rapid set-up in multipurpose plant would provide a significant challenge.

11.3 New processes

New approaches to industrial synthetic chemistry are emerging, particularly in response to environmental pressures. Broad themes include:

- Replacement of homogeneous catalysts with heterogeneous ones to facilitate catalyst recovery and reuse.
- Identification of alternatives to the commonly used solvents.
- More elegant and efficient chemistry.

All of these areas have been widely considered, e.g. Clark [6] has reviewed in depth the issues above, providing numerous examples. Some of the new chemistry can be used in existing technologies, while some would require the development of new technologies to deal with new problems in fluids handling, separations, containment, instrumentation or control.

11.4 New design methods

The first problem to address in the area of batch design methods is to understand what elements belong in a batch process design method. More is needed than sophisticated scheduling methods. There are also clear needs for methods that address:

- The link between chemical route development and process design.
- The difficulties in obtaining accurate physical properties data and kinetic information (or conversely methods that do not rely on such data).
- The need for rapid selection and design of technologies (especially unusual ones) so as to fit into the short project times typical of batch processes.

It seems unlikely that without an initiative on the part of industry the problems of design will be resolved. The gap between academia and industry is such that academic research alone is unlikely to provide the necessary tools.

The problems of physical properties representation may in future be addressed through the use of molecular simulation. Newsham and his co-workers have demonstrated the potential of molecular dynamics to help with the prediction of liquid–liquid equilibrium [7] and the selection of solvents for extraction [8]. These methods could substantially reduce the amount of experimentation required during process development by allowing *a priori* predictions to estimate physical properties, solvent selectivities, solvent effects on reaction, etc.

11.5 New skill requirements

There has been much debate as to the core skills required for effective development of batch organic chemicals processes. The key disciplines are chemistry and chemical engineering, with contributions to be made by control, mechanical and civil engineers. Sufficient mathematical, modelling and/or computing skills are required to carry out scheduling calculations and possibly discrete event or dynamic simulations. At undergraduate level, the skills taught in these disciplines do not always match the requirements of the batch process industries. Deficiencies fall into two categories – an individual may lack skills within his own discipline or lack sufficient knowledge of other disciplines to be able to communicate effectively with others. The development chemist will often have a strong background in synthetic chemistry, but may well lack sufficient skills in manipulation of reaction kinetics and chemical thermodynamics to benefit from those techniques when needed. Concepts of mass transfer, heat transfer and coupled processes are also likely to be weak or missing. Chemicals engineers often have a limited background in many aspects of chemistry – a legacy of the widespread concentration on oil and petrochemical processing in undergraduate courses.

It is appropriate to consider whether new forms of technological training would be appropriate to span the gap between chemistry and engineering. Certainly, many companies in the batch sector find it advantageous to train their chemists in the basics of chemical engineering.

The new technologies becoming available to solve problems also call for new skills. Molecular simulation as a means to reduce experimentation, predict properties and understand key phenomena will require greater knowledge of physical chemistry, intermolecular forces and quantum mechanics.

References

1. Yacubowicz, J. and Crocker, P. (1996) Super-stable nanofiltration membranes open the way to new applications. In *Case Studies in Environmental Technology* (Sharratt, P.N. and Sparshott, M., eds). IChemE, Rugby.
2. Bowe, M.J. (1994) Continuous crystallisation using power fluidics. *IChemE NW Branch Papers*, **3**, 7.1.
3. Sharratt, P.N. (1993) Clean chemistry. In *The Expanding World of Chemical Engineering* (Garside, J.G. and Furusaki, S., eds). Gordon and Breach, London.
4. Anon (1993) *Chem. Eng.*, June, 102.
5. Tullet, C. (1996) The electronic measurement of odours and aromas. In *Case Studies in Environmental Technology* (Sharratt, P.N. and Sparshott, M., eds). IChemE, Rugby.
6. Clarke, J.H. (ed.) (1995) *Chemistry of Waster Minimization*. Blackie Academic and Professional, London.
7. Meniai, A.-H. and Newsham, D.M.T. (1992) *Trans. IChemE*, **70A**, 78.
8. Meniai, A.-H. and Newsham, D.M.T. (1996) *Trans. IChemE*, **74A**, 695.

Index

Absorption for solvent recovery 95
Acceptor number (AC) 66
Activity model, control 231
Adsorption for solvent recovery 94, 97
Agitation 107–38
Agitator
  anchor 108
  discharge flow 115
  duties 107
  helical screw 108
  selection 107
  type 108
  see also Turbine
Air stripping of solvents 97
Anchor agitator 108
Angle-bladed turbine 112
Apparent viscosity 117
Aprotic solvent 76
Artificial intelligence (AI) methods in scheduling 33
Autoignition temperature 276
Automation 235
Avoidance of ignition 283
Axial impeller 109–10
Azeotropic distillation 100

Basis of safety 253, 254, 283
Batch filter 154, 177
Batch process
  definition 1
  industries 1
Batch size, in scheduling 25
BATCHCAD simulator 16
BATCHES simulator 51
BATNEEC 289
Bingham plastic 117
Biological treatment of waste solvent 101
Blanketing (inert gas) 285
Blending time 116
Boiling point, solvent 63
Bottleneck cycle 26
BPEO 289
Brush discharge 281

CAD, *see* Computer-aided design
Cake
  drainage 168
  drying 298
  filtration 157
  resistance 155, 161
  squeezing 185
  thickness 183
  washing 166, 298
  see also Filtration
Calorimetry for hazard evaluation 262
Campaign 26, 28
Candle filter 178, 181
Cartridge filter 185
Cation solvation 76
Cement kiln fuels 101
Centrifugal filters 185
Centrifugal filtration 165
Characterization of mixers 147
Chemical reaction hazards 257–70
Chiral intermediates 5
Clarification 169
  laboratory testing 171
Cleaning (waste minimization during) 303
Coalescence 128
  rate 123
Code-driven structure 50
Coils, heating/cooling 132
Commissioning 193
Competitive–consecutive reactions 139
Competitive reactions 140
Computer
  control 235–43
  hardware 238–43
  software 16, 39, 48, 243
  systems architecture 236–7
Computer-aided design 205
Condensation for solvent recovery 95
Cone discharge 282
Construction 193
Containment of an explosion 286
Continuous control loops 220–2
Continuous/discrete simulation 41
Contractors 194
Contracts 194
Control 18, 219–52
  basic 228
  continuous 220–2
  logic 52
  models 225–35
  modules 228
Corona discharge 281

316 INDEX

Crash cooling 269
Cross-flow filtration 174–6
Crystallization 18, 293–4
  waste minimization 295
Curved-bladed turbine 111, 112, 130
Cycle time 25
  filtration 153

Decanting of solvents 96
Decision logic 52
Depth filtration 170
Design
  for ease of construction 207
  plant 21–2, 193–218, 310
  process 13, 312
Dewar calorimetry 262
Dewatering of filter cake 168, 169
Diafiltration 188
Diaphragm filter 184
Dielectric constant 65
DIERS 269
Differential/algebraic equations 43
Differential scanning calorimetry 262
Differential thermal analysis 262
Differentiation of products 2
Dilution for hazard protection 269
Dipole–dipole force 73
Dipole-induced dipole force 73
Dipole moment 65
Disc turbine 111
Discharge flow from agitator 115
Discrete event simulation 33
Discrete simulation 33
  languages 48
Discrete simulators 49
Dismantling 215
Dispersion forces 74
Displacement washing 166, 296–8
Distillation 99
Donor–acceptor interaction 74
Donor number (DN) 65
Draw-down of floating solid 120
Draw-down of gas 131
Droplet break-up 127
Droplet sizes, immiscible liquids 127
Drying 18, 299
DSC, see Differential scanning calorimetry
DTA, see Differential thermal anaylsis
Dump tank 269
Dust cloud flammability 274
Dynamic modelling 16, 41
Dynamics of batch processes 41

Effect chemicals 3
Effluent 13
Electrical hazards, area classification 204
Electron pair acceptor (AC) 74

Electron pair donor (EPD) 65, 74
Electrostatic discharges 280–2
Elimination reactions, solvent effect 78
Energy consumption 15, 290, 300–1
Energy of activation of reaction 79
Engulfment model of mixing 142
Environment 61, 67, 219, 289–307
Enviromental
  impact assessment 204
  protection 289–307
Equilibrium, solvent effect 84
Equipment network in simulation 51
Erection costs 213
Et(30) polarity scale 66
Exotherm detection by DCS/DTA 263
Explosion
  containment 286
  suppression 286
  venting 286
  violence 278
Extractive distillation 101
Extractive reaction 88

Fast track project 195, 198–201
  problems 201
Filter
  aids 160, 174
  cartridges 171–4
  cloths 163–5
  leaf test apparatus 160
  media, laboratory tests 171
  medium 155, 159, 163
  types 177
  wash curve 167
  waste minimization 296–300
Filter cake
  dewatering 168
  washing 166, 298
Filtration 18, 153–92
  cycle time 153
  laboratory tests 159, 171
  rate 158
Fine chemicals 3
Fire and explosion hazards 204, 270–86
Fire point 274
Fire triangle 270
Flames 280
Flammability 271–2
  characteristics of common materials 275
  in non-oxygen atmospheres 279
  limits 272
  of mists 279
Flammable atmospheres, avoidance 284
Flash point 273
Floating solids incorporation 120
Flowcharts 245–6
Food and Drug Administration 6

INDEX

Frank–Kamenetskii model 259
  applied to powders 276
Free radical reaction 84
Friction as an ignition source 280
Fugitive emissions 301

Gantt chart 25
Gas-inducing impellers 131
Gas dispersion in stirred tank 130
Gas evolution, for hazard assessment 263
Gas sparging 130
Generic manufacture 4
GPSS 48

Hazard
  assessment 283
  identification 254
  solvent 70
Hazard and operability study 17, 202, 206
Hazardous area classification 204
Hazards 204, 253–88
HAZOP, see Hazard and operability study
Health and Safety Executive (HSE) 6
Health hazards of solvents 70
Heat loss
  from laboratory equipment 261
  from process vessel 261
Heat of reaction, for common reactions 262
Heat transfer coefficient
  process side 136
  service side 133
Heat transfer in stirred vessels 132
Helical screw agitator 108
Heterogeneous catalysis 89
Heterogeneous reactions 87
Hierarchical process synthesis tools 14
High shear agitator 112
Hildebrand solubility parameter 66
Hot surfaces as an ignition source 280
Hughes–Ingold Rules 78
Hydrogen bonding 74

Ignition
  sources 271, 275, 280
  temperature of powders 276
Impeller selection 108
In-phase operation 27
Incineration of solvents 101
Inerting 285
Inherent safety 267
Integrated project task force 196
Interlocks, safety 234
Intermolecular forces 72–7
Ion–dipole force 73
Ionic solvents 62
ISA SP88 standard 225
Isothermal reaction calorimetry 262

Jacket, heating/cooling 132
Just-dispersed condition, immiscible liquids 126

Knowledge-based systems in design 14

Laboratory safety 255–6
Landfill and solvents 101–2
Layout 205, 208–12
LEL, see Lower explosive limit
Lightning 282
Linear programming in scheduling 33
Liquid–liquid extraction 98
Local area networks 241
Lower explosive limit (LEL) 272

Manual intervention in control 222
Manual operation 19
Mass transfer
  in gas–liquid systems 131
  in liquid–liquid systems 129
Material of construction, impeller 113
Maximum temperature for the synthetic reaction (MTSR) 265
Medicines Control Agency 6
Membrane(s) 170
  filter 186
  filtration 174–6
  filtration, flux decay 187
  materials 174–5
  separation of solvents 96
Mesomixing 141
Metzner–Otto technique 117
Microfiltration 174
Micromixing 141
Microporous membranes 170
Minimum ignition energy 275
Minimum oxygen for combustion (MOC) 274
Miscibility chart 122
Miscibility of solvents 121, 122
Mixed flow turbine 111
Mixing
  and agitation 17–18, 107–38, 292
  and reaction 139–53, 292
  immiscible liquids 121
  time 116
Model-based optimization 33
Modular plant 217
Molten salt solvents 62
Monte-Carlo sampling 41, 44
Multiproduct plant 21, 28, 223
Multipurpose plant 21, 28, 223
Multistream plant 223

NAMUR 226
New processes 312
New technology 308–13

318   INDEX

Non-Newtonian fluid 117, 135, 149
Nucleophilic substitution, solvent effect 78
Nutsche filter 177

Operations 19
Operator
   intervention 222
   training 19
Optimization combined with simulation 54
Out-of-phase operation 27
Overall heat transfer coefficient 134
Oxygen balance, exothermic decomposition 263

Parallel reactions 140
Parameter uncertainty in simulation 41
Particle suspension 119
Patent protection 4, 6
PDMS 205–7
Peeler centrifuge 185
Pervaporation 96
Pesticides 3
Pharmaceuticals 3
Phase
   boundary diagram 125
   continuity 123–4
   equilibrium 64
   inversion 123
   transfer catalysis 88–9
Photochemical ozone creation potential (POCP) 68
Physical model for control 227
Physical properties 16, 64
   solvent 67
Pilot plant safety 255–6
Pipe bridges 208
Pipeless plant 310
Pitched blade turbine 112
Plant
   design 21–2, 193–218, 310
   modifications 8, 257
   relocation 212–17
Plate and frame filter 179
   flow arrangements 182
PLCs, *see* Programmable logic controllers
POCP, *see* Photochemical ozone creation potential
Polarity, solvent 66
Power for agitation 114
   typical values 114
Power number 114
Pre-coat for filtration 174
Pressure–vacuum filter 177
Procedural control 228, 243–51

Process
   chemistry 9
   design 13
   design, novel methods 312
   development 9
   intensification 311
   management 233
   model 226
   operation 18
   synthesis methods 14
   variability in simulation 45
Procurement policy 200
Product information management 232
Product life cycle 5
Programmable logic controllers 242–3
Project(s) 6–7, 193
   co-ordination 195
   organization 195
   strategy 194
   success criteria 194
Propagating brush discharge 281
Protective measures against explosion 268
Pull-type decisions in simulation 44
Push-type decisions in simulation 44
Pyrophoric materials 282

Radial flow turbine 110
Randomized search methods in scheduling 31
Reaction calorimetry
   for data acquisition 16
   for safety assessment 262
Reaction kinetics, for hazard assessment 263
Reaction rate, solvent effect 77
Reaction selectivity
   and waste generation 292
   mixing effects 139–53
   solvent effects 86
Reactions 77–90, 139–53
Reactive tracers 147
Reactor
   emergency venting 269
   intensification 293–4
   waste minimization 291–2
Recessed plate filter 179
Recipe 12, 24, 25
   management 232
   model 229
   network in simulation 51
Regeneration, thermal 15
Regulations 6, 202
Reslurry washing 166
Resource, in scheduling 24
Resource constraints 36
Resource sharing, control issues 224
Reuse of equipment 212–17
Reverse osmosis 174
Reynolds number in stirred vessel 114

Rheology, solvent 63
Rieger and Novak method 117
Road dimensions 208
Robustness of batch plant 8
Route
　development 12, 312
　selection 9–10
Rule-based dispatching methods 31
Rule-based methods in scheduling 31
Runaway reaction 258, 265
Rushton turbine 111

Safety 219, 253–88
　construction 205
　inherent 267
　interlocks 234
　measures, selection 266
　process control 268
　and process design 254–7
Safety and environment 16–17, 20–1
　in design 11
Sampling 19–20
Scale-up
　effect on reaction hazards 261
　of mixing-sensitive reactions 148
Scale of production 3
Scheduling 14, 24–40
　features of the problem 29
　in design 57
　systems 39
　tools 39
Scrubber 95
Search methods in scheduling 31
Second hand plant 213–14
Selective solvation 76
Selectivity
　of reactions 86, 139–53
　solvent effect 86
Semenov model 259
Separation distances 208–10
Sequential function charts 247–9
Simulation 14, 15, 40–60
SLAM 48
Slurry washing 166
Solid–liquid separation 294–300
Solids
　production 292–301
　suspension 119
Solubility 71
Solvation 65, 71
　dynamics 81
Solvent(s) 11, 16, 61–106, 122, 304
　aprotic 76
　design for recoverability 93
　destruction 101
　effect on equilibrium 84
　effect on reaction rate 77, 83
　effect on reaction selectivity 86

environmental effects 61
extraction for solvent recovery 98
management 304
mixtures 76
polarity 66
properties 62
recovery 90–101, 304
recovery from gases 94
recycling 90, 304
shell 71
waste 90, 304
Solvophobic interaction 75
Sources of ignition 280
SP88 control standard 225
Sparks 282
SPC, see Statistical process control
Speciality chemicals 3
State event (in simulation) 42, 43
Static electricity 280–2, 284
Statistical process control 241
Steam distillation 99
Steam stripping of solvents 97
Sterilization using membranes 172
Supercritical solvents 77
Sussmeyer process 100

Table and leaf filter 177
Task force, project 195–6
Thermal
　explosion 259
　runaway 260
　stability 263
Thermodynamic properties, solvent 64
Time event (in simulation) 42
Train firing test 274
Training, operator 19
Training requirements, process
　technologists 2, 313
Transition state model 79
Turbine
　angle-bladed 112
　axial flow 109
　curved blades 111
　disc 111
　high speed 109
　large diameter 108
　mixed flow 111
　Rushton 111
Two-phase systems, agitation 121

UEL, see Upper explosive limit
Ultrafiltration 174
Undifferentiated products 2
Uniform discretization scheduling 35
Unit operations 17
Unit supervision 233
Upper explosive limit (UEL) 272
Utility demand 15

Vacuum distillation 100
van Heuven and Beek's method 126
Vapour distillation 100
Variable chamber press 183
Vendor data 200
Venting, emergency 269
Vertical leaf filter 178, 180
Viscometry 117
Viscosity 108
   apparent for non-Newtonian fluids 117
   of dispersion 125
VOC, see Volatile organic compound
Volatile organic compound (VOC) 67–8, 305

Wall resistance in heat transfer 136
Wash recycling 298

Waste minimization 289–307
   filtration 294–300
   programme 290
Waste treatment 90–103, 304
Waste water 91
Weather, effect on projects 201
Wet air oxidation 101
Work breakdown structure for fast track project 198

Yield pseudoplastic fluid 117
Yield stress fluids, cavern formation 118

Zones for hazardous area classification 204
Zweitering correlation 119